P9-CNF-654

Computational Molecular Biology

Computational Molecular Biology

Sorin Istrail, Pavel Pevzner, and Michael Waterman, editors

Computational molecular biology is a new discipline, bringing together computational, statistical, experimental, and technological methods, which is energizing and dramatically accelerating the discovery of new technologies and tools for molecular biology. The MIT Press Series on Computational Molecular Biology is intended to provide a unique and effective venue for the rapid publication of monographs, textbooks, edited collections, reference works, and lecture notes of the highest quality.

Computational Modeling of Genetic and Biochemical Networks, edited by James Bower and Hamid Bolouri, 2000

Computational Molecular Biology: An Algorithmic Approach, Pavel Pevzner, 2000

Computational Molecular Biology

An Algorithmic Approach

Pavel A. Pevzner

The MIT Press
Cambridge, Massachusetts
London, England

Second Printing, 2001

Printed and bound in the United States of America.

Library of Congress Cataloging-in-Publication Data

Pevzner, Pavel.
Computational molecular biology : an algorithmic approach / Pavel A. Pevzner.
p. cm. — (Computational molecular biology)
Includes bibliographical references and index.
ISBN 0-262-16197-4 (hc. : alk. paper)
1. Molecular biology—Mathematical models. 2. DNA microarrays.
3. Algorithms. I. Title. II. Computational molecular biology series.
QH506 .P47 2000
572.8—dc21 00-032461

To the memory of my father

Contents

Preface

In 1985 I was looking for a job in Moscow, Russia, and I was facing a difficult choice. On the one hand I had an offer from a prestigious Electrical Engineering Institute to do research in applied combinatorics. On the other hand there was Russian Biotechnology Center NIIGENETIKA on the outskirts of Moscow, which was building a group in computational biology. The second job paid half the salary and did not even have a weekly "zakaz," a food package that was the most important job benefit in empty-shelved Moscow at that time. I still don't know what kind of classified research the folks at the Electrical Engineering Institute did as they were not at liberty to tell me before I signed the clearance papers. In contrast, Andrey Mironov at NIIGENETIKA spent a few hours talking about the algorithmic problems in a new futuristic discipline called computational molecular biology, and I made my choice. I never regretted it, although for some time I had to supplement my income at NIIGENETIKA by gathering empty bottles at Moscow railway stations, one of the very few legal ways to make extra money in pre-perestroika Moscow.

Computational biology was new to me, and I spent weekends in Lenin's library in Moscow, the only place I could find computational biology papers. The only book available at that time was Sankoff and Kruskal's classical *Time Warps, String Edits and Biomolecules: The Theory and Practice of Sequence Comparison.* Since Xerox machines were practically nonexistent in Moscow in 1985, I copied this book almost page by page in my notebooks. Half a year later I realized that I had read all or almost all computational biology papers in the world. Well, that was not such a big deal: a large fraction of these papers was written by the "founding fathers" of computational molecular biology, David Sankoff and Michael Waterman, and there were just half a dozen journals I had to scan. For the next seven years I visited the library once a month and read everything published in the area. This situation did not last long. By 1992 I realized that the explosion had begun: for the first time I did not have time to read all published computational biology papers.

Since some journals were not available even in Lenin's library, I sent requests for papers to foreign scientists, and many of them were kind enough to send their preprints. In 1989 I received a heavy package from Michael Waterman with a dozen forthcoming manuscripts. One of them formulated an open problem that I solved, and I sent my solution to Mike without worrying much about proofs. Mike later told me that the letter was written in a very "Russian English" and impossible to understand, but he was surprised that somebody was able to read his own paper through to the point where the open problem was stated. Shortly afterward Mike invited me to work with him at the University of Southern California, and in 1992 I taught my first computational biology course.

This book is based on the *Computational Molecular Biology* course that I taught yearly at the Computer Science Department at Pennsylvania State University (1992–1995) and then at the Mathematics Department at the University of Southern California (1996–1999). It is directed toward computer science and mathematics graduate and upper-level undergraduate students. Parts of the book will also be of interest to molecular biologists interested in bioinformatics. I also hope that the book will be useful for computational biology and bioinformatics professionals.

The rationale of the book is to present algorithmic ideas in computational biology and to show how they are connected to molecular biology and to biotechnology. To achieve this goal, the book has a substantial "computational biology without formulas" component that presents biological motivation and computational ideas in a simple way. This simplified presentation of biology and computing aims to make the book accessible to computer scientists entering this new area and to biologists who do not have sufficient background for more involved computational techniques. For example, the chapter entitled *Computational Gene Hunting* describes many computational issues associated with the search for the cystic fibrosis gene and formulates combinatorial problems motivated by these issues. Every chapter has an introductory section that describes both computational and biological ideas without any formulas. The book concentrates on computational ideas rather than details of the algorithms and makes special efforts to present these ideas in a simple way. Of course, the only way to achieve this goal is to hide some computational and biological details and to be blamed later for "vulgarization" of computational biology. Another feature of the book is that the last section in each chapter briefly describes the important recent developments that are outside the body of the chapter.

Computational biology courses in Computer Science departments often start with a 2- to 3-week "Molecular Biology for Dummies" introduction. My observation is that the interest of computer science students (who usually know nothing about biology) diffuses quickly if they are confronted with an introduction to biology first without any links to computational issues. The same thing happens to biologists if they are presented with algorithms without links to real biological problems. I found it very important to introduce biology and algorithms simultaneously to keep students' interest in place. The chapter entitled *Computational Gene Hunting* serves this goal, although it presents an intentionally simplified view of both biology and algorithms. I have also found that some computational biologists do not have a clear vision of the interconnections between different areas of computational biology. For example, researchers working on gene prediction may have a limited knowledge of, let's say, sequence comparison algorithms. I attempted to illustrate the connections between computational ideas from different areas of computational molecular biology.

The book covers both new and rather old areas of computational biology. For example, the material in the chapter entitled *Computational Proteomics,* and most of material in *Genome Rearrangements, Sequence Comparison* and *DNA Arrays* have never been published in a book before. At the same time the topics such as those in *Restriction Mapping* are rather old-fashioned and describe experimental approaches that are rarely used these days. The reason for including these rather old computational ideas is twofold. First, it shows newcomers the history of ideas in the area and warns them that the hot areas in computational biology come and go very fast. Second, these computational ideas often have second lives in different application domains. For example, almost forgotten techniques for restriction mapping find a new life in the hot area of computational proteomics. There are a number of other examples of this kind (e.g., some ideas related to Sequencing By Hybridization are currently being used in large-scale shotgun assembly), and I feel that it is important to show both old and new computational approaches.

A few words about a trade-off between applied and theoretical components in this book. There is no doubt that biologists in the 21st century will have to know the elements of discrete mathematics and algorithms–at least they should be able to formulate the algorithmic problems motivated by their research. In computational biology, the adequate formulation of biological problems is probably the most difficult component of research, at least as difficult as the solution of the problems. How can we teach students to formulate biological problems in computational terms? Since I don't know, I offer a story instead.

Twenty years ago, after graduating from a university, I placed an ad for "Mathematical consulting" in Moscow. My clients were mainly Cand. Sci. (Russian analog of Ph.D.) trainees in different applied areas who did not have a good mathematical background and who were hoping to get help with their diplomas (or, at least, their mathematical components). I was exposed to a wild collection of topics ranging from "optimization of inventory of airport snow cleaning equipment" to "scheduling of car delivery to dealerships." In all those projects the most difficult part was to figure out what the computational problem was and to formulate it; coming up with the solution was a matter of straightforward application of known techniques.

I will never forget one visitor, a 40-year-old, polite, well-built man. In contrast to others, this one came with a differential equation for me to solve instead of a description of his research area. At first I was happy, but then it turned out that the equation did not make sense. The only way to figure out what to do was to go back to the original applied problem and to derive a new equation. The visitor hesitated to do so, but since it was his only way to a Cand. Sci. degree, he started to reveal some details about his research area. By the end of the day I had figured out that he was interested in landing some objects on a shaky platform. It also became clear to me why he never gave me his phone number: he was an officer doing classified research: the shaking platform was a ship and the landing objects were planes. I trust that revealing this story 20 years later will not hurt his military career.

Nature is even less open about the formulation of biological problems than this officer. Moreover, some biological problems, when formulated adequately, have many bells and whistles that may sometimes overshadow and disguise the computational ideas. Since this is a book about computational ideas rather than technical details, I intentionally used simplified formulations that allow presentation of the ideas in a clear way. It may create an impression that the book is too theoretical, but I don't know any other way to teach computational ideas in biology. In other words, before landing real planes on real ships, students have to learn how to land toy planes on toy ships.

I'd like to emphasize that the book does not intend to uniformly cover all areas of computational biology. Of course, the choice of topics is influenced by my taste and my research interests. Some large areas of computational biology are not covered—most notably, DNA statistics, genetic mapping, molecular evolution, protein structure prediction, and functional genomics. Each of these areas deserves a separate book, and some of them have been written already. For example, Waterman 1995 [357] contains excellent coverage of DNA statistics, Gusfield

1997 [145] includes an encyclopedia of string algorithms, and Salzberg et al. 1998 [296] has some chapters with extensive coverage of protein structure prediction. Durbin et al. 1998 [93] and Baldi and Brunak 1997 [24] are more specialized books that emphasize Hidden Markov Models and machine learning. Baxevanis and Ouellette 1998 [28] is an excellent practical guide in bioinformatics directed more toward applications of algorithms than algorithms themselves.

I'd like to thank several people who taught me different aspects of computational molecular biology. Andrey Mironov taught me that common sense is perhaps the most important ingredient of any applied research. Mike Waterman was a terrific teacher at the time I moved from Moscow to Los Angeles, both in science and life. In particular, he patiently taught me that every paper should pass through at least a dozen iterations before it is ready for publishing. Although this rule delayed the publication of this book by a few years, I religiously teach it to my students. My former students Vineet Bafna and Sridhar Hannenhalli were kind enough to teach me what they know and to join me in difficult long-term projects. I also would like to thank Alexander Karzanov, who taught me combinatorial optimization, including the ideas that were most useful in my computational biology research.

I would like to thank my collaborators and co-authors: Mark Borodovsky, with whom I worked on DNA statistics and who convinced me in 1985 that computational biology had a great future; Earl Hubbell, Rob Lipshutz, Yuri Lysov, Andrey Mirzabekov, and Steve Skiena, my collaborators in DNA array research; Eugene Koonin, with whom I tried to analyze complete genomes even before the first bacterial genome was sequenced; Norm Arnheim, Mikhail Gelfand, Melissa Moore, Mikhail Roytberg, and Sing-Hoi Sze, my collaborators in gene finding; Karl Clauser, Vlado Dancik, Maxim Frank-Kamenetsky, Zufar Mulyukov, and Chris Tang, my collaborators in computational proteomics; and the late Eugene Lawler, Xiaoqiu Huang, Webb Miller, Anatoly Vershik, and Martin Vingron, my collaborators in sequence comparison.

I am also thankful to many colleagues with whom I discussed different aspects of computational molecular biology that directly or indirectly influenced this book: Ruben Abagyan, Nick Alexandrov, Stephen Altschul, Alberto Apostolico, Richard Arratia, Ricardo Baeza-Yates, Gary Benson, Piotr Berman, Charles Cantor, Radomir Crkvenjakov, Kun-Mao Chao, Neal Copeland, Andreas Dress, Radoje Drmanac, Mike Fellows, Jim Fickett, Alexei Finkelstein, Steve Fodor, Alan Frieze, Dmitry Frishman, Israel Gelfand, Raffaele Giancarlo, Larry Goldstein, Andy Grigoriev, Dan Gusfield, David Haussler, Sorin Istrail, Tao Jiang,

Sampath Kannan, Samuel Karlin, Dick Karp, John Kececioglu, Alex Kister, George Komatsoulis, Andrzey Konopka, Jenny Kotlerman, Leonid Kruglyak, Jens Lagergren, Gadi Landau, Eric Lander, Gene Myers, Giri Narasimhan, Ravi Ravi, Mireille Regnier, Gesine Reinert, Isidore Rigoutsos, Mikhail Roytberg, Anatoly Rubinov, Andrey Rzhetsky, Chris Sander, David Sankoff, Alejandro Schaffer, David Searls, Ron Shamir, Andrey Shevchenko, Temple Smith, Mike Steel, Lubert Stryer, Elizabeth Sweedyk, Haixi Tang, Simon Tavar` e, Ed Trifonov, Tandy Warnow, Haim Wolfson, Jim Vath, Shibu Yooseph, and others.

It has been a pleasure to work with Bob Prior and Michael Rutter of the MIT Press. I am grateful to Amy Yeager, who copyedited the book, Mikhail Mayofis who designed the cover, and Oksana Khleborodova, who illustrated the steps of the gene prediction algorithm. I also wish to thank those who supported my research: the Department of Energy, the National Institutes of Health, and the National Science Foundation.

Last but not least, many thanks to Paulina and Arkasha Pevzner, who were kind enough to keep their voices down and to tolerate my absent-mindedness while I was writing this book.

Chapter 1

Computational Gene Hunting

1.1 Introduction

Cystic fibrosis is a fatal disease associated with recurrent respiratory infections and abnormal secretions. The disease is diagnosed in children with a frequency of 1 per 2500. One per 25 Caucasians carries a faulty cystic fibrosis gene, and children who inherit faulty genes from *both* parents become sick.

In the mid-1980s biologists knew nothing about the gene causing cystic fibrosis, and no reliable prenatal diagnostics existed. The best hope for a cure for many genetic diseases rests with finding the defective genes. The search for the cystic fibrosis (CF) gene started in the early 1980s, and in 1985 three groups of scientists simultaneously and independently proved that the CF gene resides on the 7th chromosome. In 1989 the search was narrowed to a short area of the 7th chromosome, and the 1,480-amino-acids-long CF gene was found. This discovery led to efficient medical diagnostics and a promise for potential therapy for cystic fibrosis. Gene hunting for cystic fibrosis was a painstaking undertaking in late 1980s. Since then thousands of medically important genes have been found, and the search for many others is currently underway. Gene hunting involves many computational problems, and we review some of them below.

1.2 Genetic Mapping

Like cartographers mapping the ancient world, biologists over the past three decades have been laboriously charting human DNA. The aim is to position genes and other milestones on the various chromosomes to understand the genome's geography.

When the search for the CF gene started, scientists had no clue about the nature of the gene or its location in the genome. Gene hunting usually starts with *genetic mapping*, which provides an approximate location of the gene on one of the human chromosomes (usually within an area a few million nucleotides long). To understand the computational problems associated with genetic mapping we use an oversimplified model of genetic mapping in uni-chromosomal *robots*. Every robot has n genes (in unknown order) and every gene may be either in state 0 or in state 1, resulting in two *phenotypes* (physical traits): *red* and *brown*. If we assume that $n = 3$ and the robot's three genes define the color of its hair, eyes, and lips, then 000 is *all-red* robot (red hair, red eyes, and red lips), while 111 is *all-brown* robot. Although we can observe the robots' phenotypes (i.e., the color of their hair, eyes, and lips), we don't know the order of genes in their genomes. Fortunately, robots may have children, and this helps us to construct the robots' genetic maps.

A child of robots $m_1 \ldots m_n$ and $f_1 \ldots f_n$ is either a robot $m_1 \ldots m_i f_{i+1} \ldots f_n$ or a robot $f_1 \ldots f_i m_{i+1} \ldots m_n$ for some *recombination position* i, with $0 \leq i \leq n$. Every pair of robots may have $2(n+1)$ different kinds of children (some of them may be identical), with the probability of recombination at position i equal to $\frac{1}{(n+1)}$.

Genetic Mapping Problem Given the phenotypes of a large number of children of all-red and all-brown robots, find the gene order in the robots.

Analysis of the frequencies of different *pairs* of phenotypes allows one to derive the gene order. Compute the probability p that a child of an all-red and an all-brown robot has hair and eyes of different colors. If the hair gene and the eye gene are consecutive in the genome, then the probability of recombination between these genes is $\frac{1}{n+1}$. If the hair gene and the eye gene are not consecutive, then the probability that a child has hair and eyes of different colors is $p = \frac{i}{n+1}$, where i is the *distance* between these genes in the genome. Measuring p in the population of children helps one to estimate the distances between genes, to find gene order, and to reconstruct the genetic map.

In the world of robots a child's chromosome consists of two fragments: one fragment from mother-robot and another one from father-robot. In a more accurate (but still unrealistic) model of recombination, a child's genome is defined as a mosaic of an arbitrary number of fragments of a mother's and a father's genomes, such as $m_1 \ldots m_i f_{i+1} \ldots f_j m_{j+1} \ldots m_k f_{k+1} \ldots$. In this case, the probability of recombination between two genes is proportional to the distance between these genes and, just as before, the farther apart the genes are, the more often a recombination between them occurs. If two genes are very close together, recombination between them will be rare. Therefore, neighboring genes in children of all-red and all-brown robots imply the same phenotype (both red or both brown) more frequently, and thus biologists can infer the order by considering the frequency of phenotypes in pairs. Using such arguments, Sturtevant constructed the first genetic map for six genes in fruit flies in 1913.

Although human genetics is more complicated than robot genetics, the silly robot model captures many computational ideas behind genetic mapping algorithms. One of the complications is that human genes come in pairs (not to mention that they are distributed over 23 chromosomes). In every pair one gene is inherited from the mother and the other from the father. Therefore, the human genome may contain a gene in state 1 (red eye) on one chromosome and a gene in state 0 (brown eye) on the other chromosome from the same pair. If $F_1 \ldots F_n | \mathcal{F}_1 \ldots \mathcal{F}_n$ represents a father genome (every gene is present in two copies F_i and $\mathcal{F}_i$) and $M_1 \ldots M_n | \mathcal{M}_1 \ldots \mathcal{M}_n$ represents a mother genome, then a child genome is represented by $f_1 \ldots f_n | m_1 \ldots m_n$, with f_i equal to either F_i or $\mathcal{F}_i$ and m_i equal to either M_i or $\mathcal{M}_i$. For example, the father $11|00$ and mother $00|00$ may have four different kinds of children: $11|00$ (no recombination), $10|00$ (recombination), $01|00$ (recombination), and $00|00$ (no recombination). The basic ideas behind human and robot genetic mapping are similar: since recombination between close genes is rare, the proportion of recombinants among children gives an indication of the distance between genes along the chromosome.

Another complication is that differences in genotypes do not always lead to differences in phenotypes. For example, humans have a gene called *ABO blood type* which has three states—A, B, and O—in the human population. There exist six possible genotypes for this gene—AA, AB, AO, BB, BO, and OO—but only four phenotypes. In this case the phenotype does not allow one to deduce the genotype unambiguously. From this perspective, eye colors or blood types may not be the best milestones to use to build genetic maps. Biologists proposed using *genetic markers* as a convenient substitute for genes in genetic mapping. To map a new gene it is necessary to have a large number of already mapped markers, ideally evenly spaced along the chromosomes.

Our ability to map the genes in robots is based on the variability of phenotypes in different robots. For example, if all robots had brown eyes, the eye gene would be impossible to map. There are a lot of variations in the human genome that are not directly expressed in phenotypes. For example, if half of all humans had nucleotide A at a certain position in the genome, while the other half had nucleotide T at the same position, it would be a good marker for genetic mapping. Such mutation can occur outside of any gene and may not affect the phenotype at all. Botstein et al., 1980 [44] suggested using such variable positions as genetic markers for mapping. Since sampling letters at a given position of the genome is experimentally infeasible, they suggested a technique called *restriction fragment length polymorphism* (RFLP) to study variability.

Hamilton Smith discovered in 1970 that the *restriction enzyme Hind*II cleaves DNA molecules at every occurrence of a sequence GTGCAC or GTTAAC (restriction sites). In RFLP analysis, human DNA is cut by a restriction enzyme like *Hind*II at every occurrence of the restriction site into about a million restriction fragments, each a few thousand nucleotides long. However, any mutation that affects one of the restriction sites (GTGCAC or GTTAAC for *Hind*II) disables one of the cuts and merges two restriction fragments A and B separated by this site into a single fragment $A + B$. The crux of RFLP analysis is the detection of the change in the length of the restriction fragments.

Gel-electrophoresis separates restriction fragments, and a labeled DNA probe is used to determine the size of the restriction fragment hybridized with this probe. The variability in length of these restriction fragments in different individuals serves as a genetic marker because a mutation of a single nucleotide may destroy (or create) the site for a restriction enzyme and alter the length of the corresponding fragment. For example, if a labeled DNA probe hybridizes to a fragment A and a restriction site separating fragments A and B is destroyed by a mutation, then the probe detects $A + B$ instead of A. Kan and Dozy, 1978 [183] found a new diagnostic for sickle-cell anemia by identifying an RFLP marker located close to the sickle-cell anemia gene.

RFLP analysis transformed genetic mapping into a highly competitive race and the successes were followed in short order by finding genes responsible for Huntington's disease (Gusella et al., 1983 [143]), Duchenne muscular dystrophy (Davies et al., 1983 [81]), and retinoblastoma (Cavenee et al., 1985 [60]). In a landmark publication, Donis-Keller et al., 1987 [88] constructed the first RFLP map of the human genome, positioning one RFLP marker per approximately 10 million nucleotides. In this study, 393 random probes were used to study RFLP in 21 families over 3 generations. Finally, a computational analysis of recombination led to ordering RFLP markers on the chromosomes.

In 1985 the recombination studies narrowed the search for the cystic fibrosis gene to an area of chromosome 7 between markers *met* (a gene involved in cancer) and D7S8 (RFLP marker). The length of the area was approximately 1 million nucleotides, and some time would elapse before the cystic fibrosis gene was found. Physical mapping follows genetic mapping to further narrow the search.

1.3 Physical Mapping

Physical mapping can be understood in terms of the following analogy. Imagine several copies of a book cut by scissors into thousands of pieces. Each copy is cut in an individual way such that a piece from one copy may overlap a piece from another copy. For each piece and each word from a list of key words, we are told whether the piece contains the key word. Given this data, we wish to determine the pattern of overlaps of the pieces.

The process starts with breaking the DNA molecule into small pieces (e.g., with restriction enzymes); in the CF project DNA was broken into pieces roughly 50 Kb long. To study individual pieces, biologists need to obtain each of them in many copies. This is achieved by *cloning* the pieces. Cloning incorporates a fragment of DNA into some self-replicating host. The self-replication process then creates large numbers of copies of the fragment, thus enabling its structure to be investigated. A fragment reproduced in this way is called a *clone*.

As a result, biologists obtain a *clone library* consisting of thousands of clones (each representing a short DNA fragment) from the same DNA molecule. Clones from the library may overlap (this can be achieved by cutting the DNA with distinct enzymes producing overlapping restriction fragments). After a clone library is constructed, biologists want to *order* the clones, i.e., to reconstruct the relative placement of the clones along the DNA molecule. This information is lost in the construction of the clone library, and the reconstruction starts with *fingerprinting* the clones. The idea is to describe each clone using an easily determined fingerprint, which can be thought of as a set of "key words" for the clone. If two clones have substantial overlap, their fingerprints should be similar. If non-overlapping clones are unlikely to have similar fingerprints then fingerprints would allow a biologist to distinguish between overlapping and non-overlapping clones and to reconstruct the order of the clones (physical map). The sizes of the restriction fragments of the clones or the lists of probes hybridizing to a clone provide such fingerprints.

To map the cystic fibrosis gene, biologists used physical mapping techniques called *chromosome walking* and *chromosome jumping*. Recall that the CF gene was linked to RFLP D7S8. The probe corresponding to this RFLP can be used to find a clone containing this RFLP. This clone can be sequenced, and one of its ends can be used to design a new probe located even closer to the CF gene. These probes can be used to find new clones and to *walk* from D7S8 to the CF gene. After multiple iterations, hundreds of kilobases of DNA can be sequenced from a region surrounding the marker gene. If the marker is closely linked to the gene of interest, eventually that gene, too, will be sequenced. In the CF project, a total distance of 249 Kb was cloned in 58 DNA fragments.

Gene walking projects are rather complex and tedious. One obstacle is that not all regions of DNA will be present in the clone library, since some genomic regions tend to be unstable when cloned in bacteria. Collins et al., 1987 [73] developed *chromosome jumping*, which was successfully used to map the area containing the CF gene.

Although conceptually attractive, chromosome walking and jumping are too laborious for mapping entire genomes and are tailored to mapping individual genes. A pre-constructed map covering the entire genome would save significant effort for mapping *any* new genes.

Different fingerprints lead to different mapping problems. In the case of fingerprints based on hybridization with short probes, a probe may hybridize with many clones. For the map assembly problem with n clones and m probes, the hybridization data consists of an $n \times m$ matrix (d_{ij}), where $d_{ij} = 1$ if clone C_i contains probe p_j, and $d_{ij} = 0$ otherwise (Figure 1.1). Note that the data does not indicate how many times a probe occurs on a given clone, nor does it give the order of occurrence of the probes in a clone.

The simplest approximation of physical mapping is the Shortest Covering String Problem. Let S be a string over the alphabet of probes $p_1, \ldots, p_m$. A string S *covers* a clone C if there exists a substring of S containing exactly the same set of probes as C (order and multiplicities of probes in the substring are ignored). A string in Figure 1.1 covers each of nine clones corresponding to the hybridization data.

Shortest Covering String Problem Given hybridization data, find a shortest string in the alphabet of probes that covers all clones.

Before using probes for DNA mapping, biologists constructed restriction maps of clones and used them as fingerprints for clone ordering. The *restriction map* of a clone is an ordered list of restriction fragments. If two clones have restriction maps that share several consecutive fragments, they are likely to overlap. With

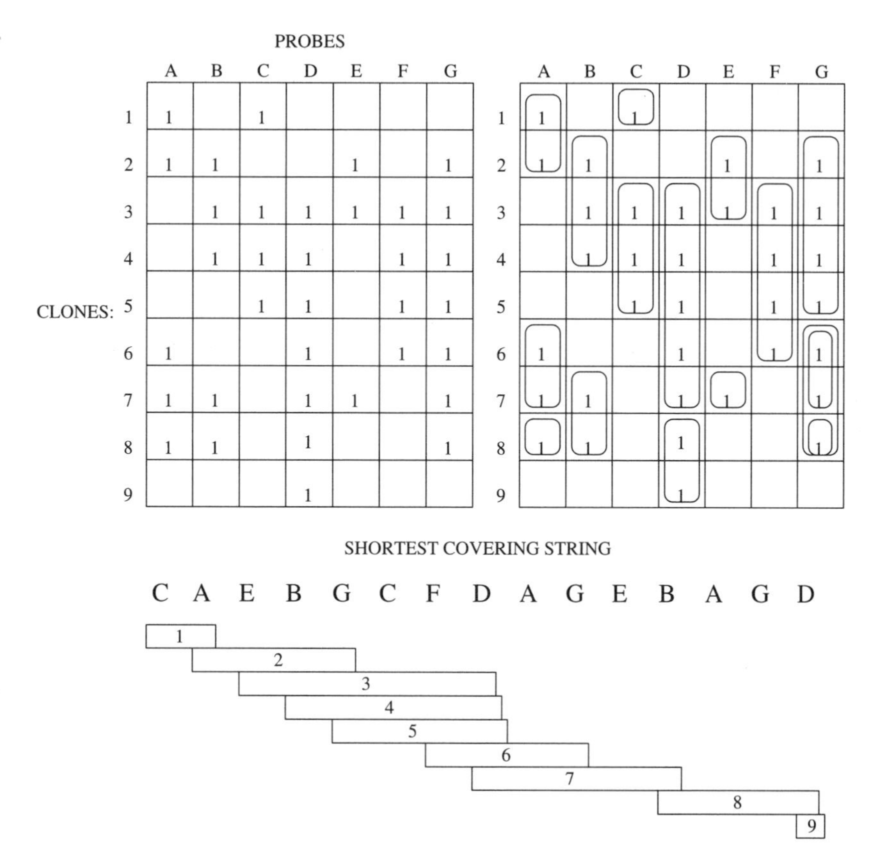

Figure 1.1: Hybridization data and Shortest Covering String.

this strategy, Kohara et al., 1987 [204] assembled a restriction map of the *E. coli* genome with 5 million base pairs.

To build a restriction map of a clone, biologists use different biochemical techniques to derive indirect information about the map and combinatorial methods to reconstruct the map from these data. The problem often might be formulated as recovering positions of points when only some pairwise distances between points are known.

Many mapping techniques lead to the following combinatorial problem. If X is a set of points on a line, then ΔX denotes the multiset of all pairwise distances between points in X: $\Delta X = \{|x_1 - x_2| : x_1, x_2 \in X\}$. In restriction mapping a subset $E \subset \Delta X$, corresponding to the experimental data about fragment lengths, is given, and the problem is to reconstruct X from the knowledge of E alone. In the *Partial Digest Problem (PDP)*, the experiment provides data about *all* pairwise distances between restriction sites and $E = \Delta X$.

Partial Digest Problem Given ΔX, reconstruct X.

The problem is also known as the *turnpike* problem in computer science. Suppose you know the set of all distances between every pair of exits on a highway. Could you reconstruct the "geography" of that highway from these data, i.e., find the distances from the start of the highway to every exit? If you consider instead of highway exits the sites of DNA cleavage by a restriction enzyme, and if you manage to digest DNA in such a way that the fragments formed by *every* two cuts are present in the digestion, then the sizes of the resulting DNA fragments correspond to distances between highway exits.

For this seemingly trivial puzzle no polynomial algorithm is yet known.

1.4 Sequencing

Imagine several copies of a book cut by scissors into 10 million small pieces. Each copy is cut in an individual way so that a piece from one copy may overlap a piece from another copy. Assuming that 1 million pieces are lost and the remaining 9 million are splashed with ink, try to recover the original text. After doing this you'll get a feeling of what a DNA sequencing problem is like. Classical sequencing technology allows a biologist to read short (300- to 500-letter) fragments per experiment (each of these fragments corresponds to one of the 10 million pieces). Computational biologists have to assemble the entire genome from these short fragments, a task not unlike assembling the book from millions of slips of paper. The problem is complicated by unavoidable experimental errors (ink splashes).

The simplest, naive approximation of DNA sequencing corresponds to the following problem:

Shortest Superstring Problem Given a set of strings $s_1, \ldots, s_n$, find the shortest string s such that each s_i appears as a substring of s.

Figure 1.2 presents two superstrings for the set of all eight three-letter strings in a 0-1 alphabet. The first (trivial) superstring is obtained by concatenation of these eight strings, while the second one is a shortest superstring. This superstring is related to the solution of the "Clever Thief and Coding Lock" problem (the minimum number of tests a thief has to conduct to try all possible k-letter passwords).

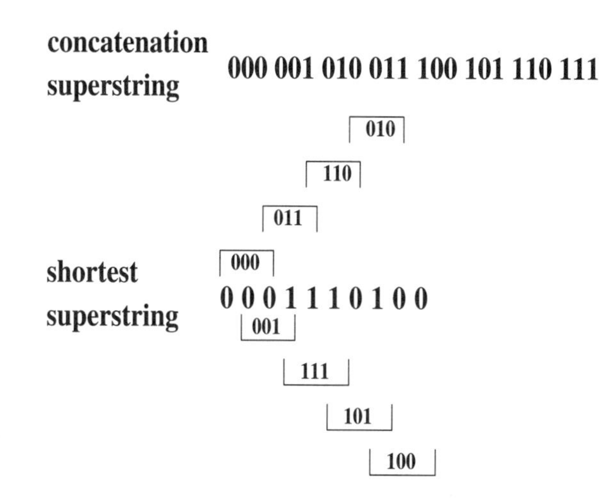

Figure 1.2: Superstrings for the set of eight three-letter strings in a 0-1 alphabet.

Since the Shortest Superstring Problem is known to be NP-hard, a number of heuristics have been proposed. The early DNA sequencing algorithms used a simple *greedy* strategy: repeatedly merge a pair of strings with maximum overlap until only one string remains.

Although conventional DNA sequencing is a fast and efficient procedure now, it was rather time consuming and hard to automate 10 years ago. In 1988 four groups of biologists independently and simultaneously suggested a new approach called Sequencing by Hybridization (SBH). They proposed building a miniature *DNA Chip (Array)* containing thousands of short DNA fragments working like the chip's memory. Each of these short fragments reveals some information about an unknown DNA fragment, and all these pieces of information combined together were supposed to solve the DNA sequencing puzzle. In 1988 almost nobody believed that the idea would work; both biochemical problems (synthesizing thousands of short DNA fragments on the surface of the array) and combinatorial problems (sequence reconstruction by array output) looked too complicated. Now, building DNA arrays with thousands of probes has become an industry.

Given a DNA fragment with an unknown sequence of nucleotides, a DNA array provides l-tuple composition, i.e., information about all substrings of length l contained in this fragment (the positions of these substrings are unknown).

Sequencing by Hybridization Problem Reconstruct a string by its l-tuple composition.

Although DNA arrays were originally invented for DNA sequencing, very few fragments have been sequenced with this technology (Drmanac et al., 1993 [90]). The problem is that the infidelity of hybridization process leads to errors in deriving l-tuple composition. As often happens in biology, DNA arrays first proved successful not for a problem for which they were originally invented, but for different applications in functional genomics and mutation detection.

Although conventional DNA sequencing and SBH are very different approaches, the corresponding computational problems are similar. In fact, SBH is a particular case of the Shortest Superstring Problem when all strings $s_1, \ldots, s_n$ represent the set of all substrings of s of fixed size. However, in contrast to the Shortest Superstring Problem, there exists a simple linear-time algorithm for the SBH Problem.

1.5 Similarity Search

After sequencing, biologists usually have no idea about the function of found genes. Hoping to find a clue to genes' functions, they try to find similarities between newly sequenced genes and previously sequenced genes with known functions. A striking example of a biological discovery made through a similarity search happened in 1984 when scientists used a simple computational technique to compare the newly discovered cancer-causing $v\text{-}sys$ oncogene to all known genes. To their astonishment, the cancer-causing gene matched a normal gene involved in growth and development. Suddenly, it became clear that cancer might be caused by a normal growth gene being switched on at the wrong time (Doolittle et al., 1983 [89], Waterfield et al., 1983 [353]).

In 1879 Lewis Carroll proposed to the readers of *Vanity Fair* the following puzzle: transform one English word into another one by going through a series of intermediate English words where each word differs from the next by only one letter. To transform *head* into *tail* one needs just four such intermediates: $head \to heal \to teal \to tell \to tall \to tail$. Levenshtein, 1966 [219] introduced a notion of *edit distance* between strings as the minimum number of elementary operations needed to transform one string into another where the elementary operations are insertion of a symbol, deletion of a symbol, and substitution of a symbol by another one. Most sequence comparison algorithms are related to computing edit distance with this or a slightly different set of elementary operations.

Since mutation in DNA represents a natural evolutionary process, edit distance is a natural measure of similarity between DNA fragments. Similarity between DNA sequences can be a clue to common evolutionary origin (like similarity between globin genes in humans and chimpanzees) or a clue to common function (like similarity between the ν-sys oncogene and a growth-stimulating hormone).

If the edit operations are limited to insertions and deletions (no substitutions), then the edit distance problem is equivalent to the *longest common subsequence (LCS)* problem. Given two strings $V = v_1 \ldots v_n$ and $W = w_1 \ldots w_m$, a *common subsequence* of V and W of length k is a sequence of indices $1 \leq i_1 < \ldots < i_k \leq n$ and $1 \leq j_1 < \ldots < j_k \leq m$ such that

$$v_{i_t} = w_{j_t} \text{ for } 1 \leq t \leq k$$

Let $LCS(V, W)$ be the length of a *longest common subsequence* (LCS) of V and W. For example, LCS (**A**T**C**T**GAT**, **TGCATA**)=4 (the letters forming the LCS are in bold). Clearly $n + m - 2LCS(V, W)$ is the minimum number of insertions and deletions needed to transform V into W.

Longest Common Subsequence Problem Given two strings, find their longest common subsequence.

When the area around the cystic fibrosis gene was sequenced, biologists compared it with the database of all known genes and found some similarities between a fragment approximately 6500 nucleotides long and so-called *ATP binding proteins* that had already been discovered. These proteins were known to span the cell membrane multiple times and to work as channels for the transport of ions across the membrane. This seemed a plausible function for a CF gene, given the fact that the disease involves abnormal secretions. The similarity also pointed to two conserved ATP binding sites (ATP proteins provide energy for many reactions in the cell) and shed light on the mechanism that is damaged in faulty CF genes. As a result the cystic fibrosis gene was called *cystic fibrosis transmembrane conductance regulator*.

1.6 Gene Prediction

Knowing the approximate gene location does not lead yet to the gene itself. For example, Huntington's disease gene was mapped in 1983 but remained elusive until 1993. In contrast, the CF gene was mapped in 1985 and found in 1989.

In simple life forms, such as bacteria, genes are written in DNA as continuous strings. In humans (and other mammals), the situation is much less straightforward. A human gene, consisting of roughly 2,000 letters, is typically broken into subfragments called *exons*. These exons may be shuffled, seemingly at random, into a section of chromosomal DNA as long as a million letters. A typical human gene can have 10 exons or more. The *BRCA1* gene, linked to breast cancer, has 27 exons.

This situation is comparable to a magazine article that begins on page 1, continues on page 13, then takes up again on pages 43, 51, 53, 74, 80, and 91, with pages of advertising and other articles appearing in between. We don't understand why these jumps occur or what purpose they serve. Ninety-seven percent of the human genome is advertising or so-called "junk" DNA.

The jumps are inconsistent from species to species. An "article" in an insect edition of the genetic magazine will be printed differently from the same article appearing in a worm edition. The pagination will be completely different: the information that appears on a single page in the human edition may be broken up into two in the wheat version, or vice versa. The genes themselves, while related, are quite different. The mouse-edition gene is written in mouse language, the human-edition gene in human language. It's a little like German and English: many words are similar, but many others are not.

Prediction of a new gene in a newly sequenced DNA sequence is a difficult problem. Many methods for deciding what is advertising and what is story depend on statistics. To continue the magazine analogy, it is something like going through back issues of the magazine and finding that human-gene "stories" are less likely to contain phrases like "for sale," telephone numbers, and dollar signs. In contrast, a combinatorial approach to gene prediction uses previously sequenced genes as a template for recognition of newly sequenced genes. Instead of employing statistical properties of exons, this method attempts to solve the combinatorial puzzle: find a set of blocks (candidate exons) in a genomic sequence whose concatenation (splicing) fits one of the known proteins. Figure 1.3 illustrates this puzzle for a "genomic" sequence

'twas brilliant thrilling morning and the slimy hellish lithe doves gyrated and gambled nimbly in the waves

whose different blocks "make up" Lewis Carroll's famous "target protein":

't was brillig, and the slithy toves did gyre and gimble in the wabe

'T WAS BRILLIG, AND THE SLITHY TOVES DID GYRE AND GIMBLE IN THE WABE

T WAS BRILLI | G, AND THE SL | THE DOVES GYRATED AND GAMBLED | IN THE WAVE

T WAS BRILLI | G, AND THE SL | THE DOVES GYRATED | NIMBLY IN THE WAVE

T HRILLING | AND | HEL LISH | DOVES GYRATED AND GAMBLED | IN THE WAVE

T HRILLING | AND | HEL LISH | DOVES GYRATED | NIMBLY IN THE WAVE

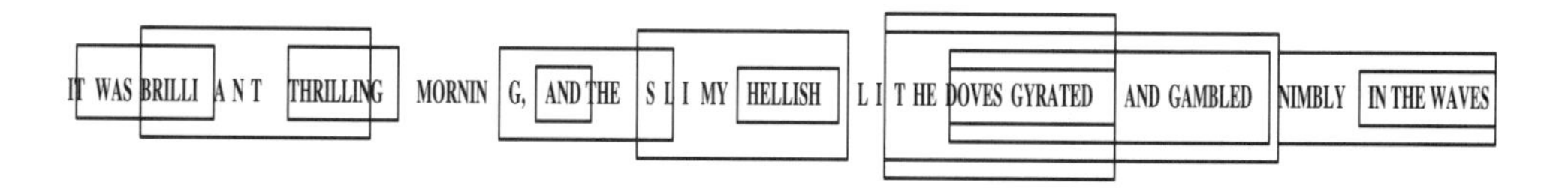

Figure 1.3: Spliced Alignment Problem: block assemblies with the best fit to the Lewis Carroll's "target protein."

This combinatorial puzzle leads to the following

Spliced Alignment Problem Let G be a string called *genomic sequence*, T be a string called *target sequence*, and $\mathcal{B}$ be a set of substrings of G. Given G, T, and $\mathcal{B}$, find a set of non-overlapping strings from $\mathcal{B}$ whose concatenation fits the target sequence the best (i.e., the edit distance between the concatenation of these strings and the target is minimum among all sets of blocks from $\mathcal{B}$).

1.7 Mutation Analysis

One of the challenges in gene hunting is knowing when the gene of interest has been sequenced, given that nothing is known about the structure of that gene. In the cystic fibrosis case, gene predictions and sequence similarity provided some clues for the gene but did not rule out other candidate genes. In particular, three other fragments were suspects. If a suspected gene were really a disease gene, the affected individuals would have mutations in this gene. Every such gene will be subject to re-sequencing in many individuals to check this hypothesis. One mutation (deletion of three nucleotides, causing a deletion of one amino acid) in the CF gene was found to be common in affected individuals. This was a lead, and PCR primers were set up to screen a large number of individuals for this mutation. This mutation was found in 70% of cystic fibrosis patients, thus convincingly proving that it causes cystic fibrosis. Hundreds of diverse mutations comprise the additional 30% of faulty cystic fibrosis genes, making medical diagnostics of cystic fibrosis difficult. Dedicated DNA arrays for cystic fibrosis may be very efficient for screening populations for mutation.

Similarity search, gene recognition, and mutation analysis raise a number of statistical problems. If two sequences are 45% similar, is it likely that they are genuinely related, or is it just a matter of chance? Genes are frequently found in the DNA fragments with a high frequency of CG dinucleotides (*CG-islands*). The cystic fibrosis gene, in particular, is located inside a CG-island. What level of CG-content is an indication of a CG-island and what is just a matter of chance? Examples of corresponding statistical problems are given below:

Expected Length of LCS Problem Find the expected length of the LCS for two random strings of length n.

String Statistics Problem Find the expectation and variance of the number of occurrences of a given string in a random text.

1.8 Comparative Genomics

As we have seen with cystic fibrosis, hunting for human genes may be a slow and laborious undertaking. Frequently, genetic studies of similar genetic disorders in animals can speed up the process.

Waardenburg's syndrome is an inherited genetic disorder resulting in hearing loss and pigmentary dysplasia. Genetic mapping narrowed the search for the Waardenburg's syndrome gene to human chromosome 2, but its exact location remained unknown. There was another clue that directed attention to chromosome 2. For a long time, breeders scrutinized mice for mutants, and one of these, designated *splotch*, had patches of white spots, a disease considered to be similar to Waardenburg's syndrome. Through breeding (which is easier in mice than in humans) the *splotch* gene was mapped to mouse chromosome 2. As gene mapping proceeded it became clear that there are groups of genes that are closely linked to one another in both species. The shuffling of the genome during evolution is not complete; blocks of genetic material remain intact even as multiple chromosomal rearrangements occur. For example, chromosome 2 in humans is built from fragments that are similar to fragments from mouse DNA residing on chromosomes 1, 2, 6, 8, 11, 12, and 17 (Figure 1.4). Therefore, mapping a gene in mice often gives a clue to the location of a related human gene.

Despite some differences in appearance and habits, men and mice are genetically very similar. In a pioneering paper, Nadeau and Taylor, 1984 [248] estimated that surprisingly few genomic rearrangements (178 ± 39) have happened since the divergence of human and mouse 80 million years ago. Mouse and human genomes can be viewed as a collection of about 200 fragments which are shuffled (rearranged) in mice as compared to humans. If a mouse gene is mapped in one of those fragments, then the corresponding human gene will be located in a chromosomal fragment that is linked to this mouse gene. A comparative mouse-human genetic map gives the position of a human gene given the location of a related mouse gene.

Genome rearrangements are a rather common chromosomal abnormality which are associated with such genetic diseases as Down syndrome. Frequently, genome rearrangements are asymptomatic: it is estimated that 0.2% of individuals carry an asymptomatic chromosomal rearrangement.

The analysis of genome rearrangements in molecular biology was pioneered by Dobzhansky and Sturtevant, 1938 [87], who published a milestone paper presenting a rearrangement scenario with 17 inversions for the species of *Drosophila* fruit fly. In the simplest form, rearrangements can be modeled by using a combinatorial problem of finding a shortest series of *reversals* to transform one genome into another. The order of genes in an organism is represented by a permutation $\pi = \pi_1\pi_2\ldots\pi_n$. A *reversal* $\rho(i,j)$ has the effect of reversing the order of genes $\pi_i\pi_{i+1}\ldots\pi_j$ and transforms $\pi = \pi_1\ldots\pi_{i-1}\pi_i\ldots\pi_j\pi_{j+1}\ldots\pi_n$ into

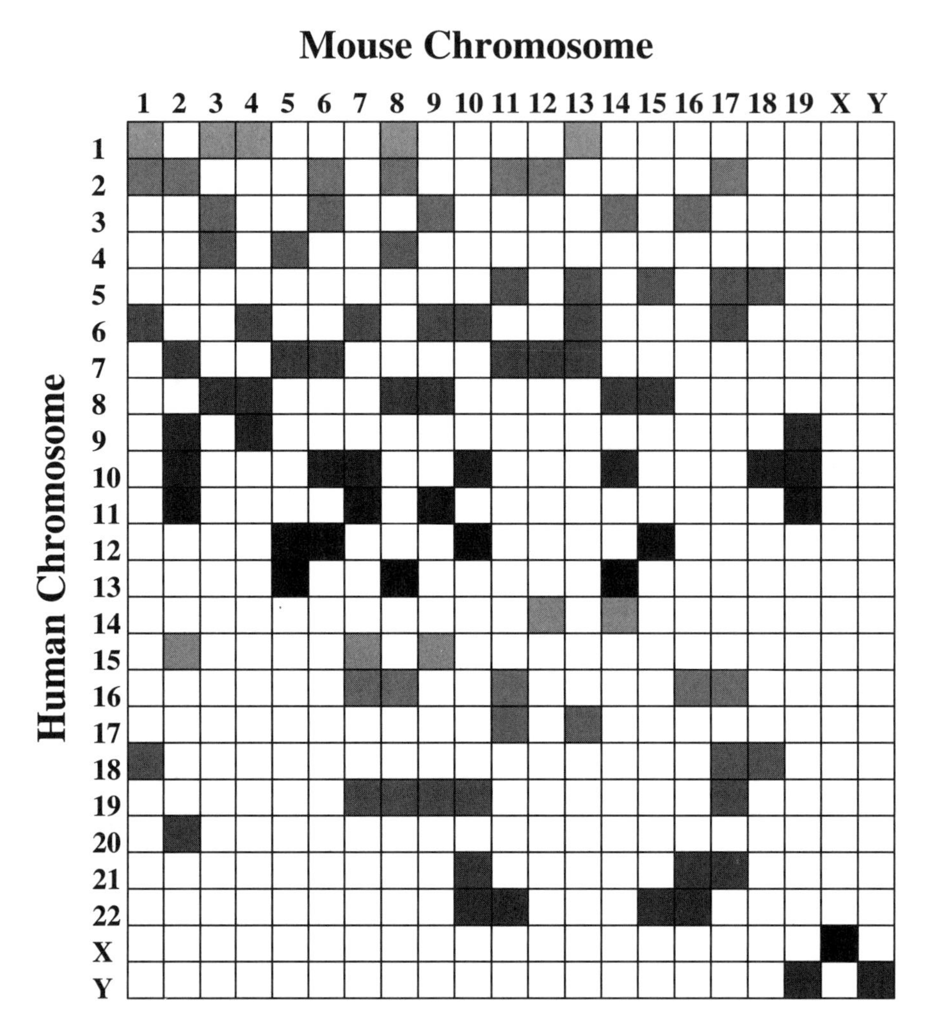

Figure 1.4: Man-mouse comparative physical map.

$\pi \cdot \rho(i,j) = \pi_1 \ldots \pi_{i-1}\pi_j \ldots \pi_i\pi_{j+1} \ldots \pi_n$. Figure 1.5 presents a *rearrangement scenario* describing a transformation of a human X chromosome into a mouse X chromosome.

Reversal Distance Problem Given permutations π and σ, find a series of reversals $\rho_1, \rho_2, \ldots, \rho_t$ such that $\pi \cdot \rho_1 \cdot \rho_2 \cdots \rho_t = \sigma$ and t is minimum.

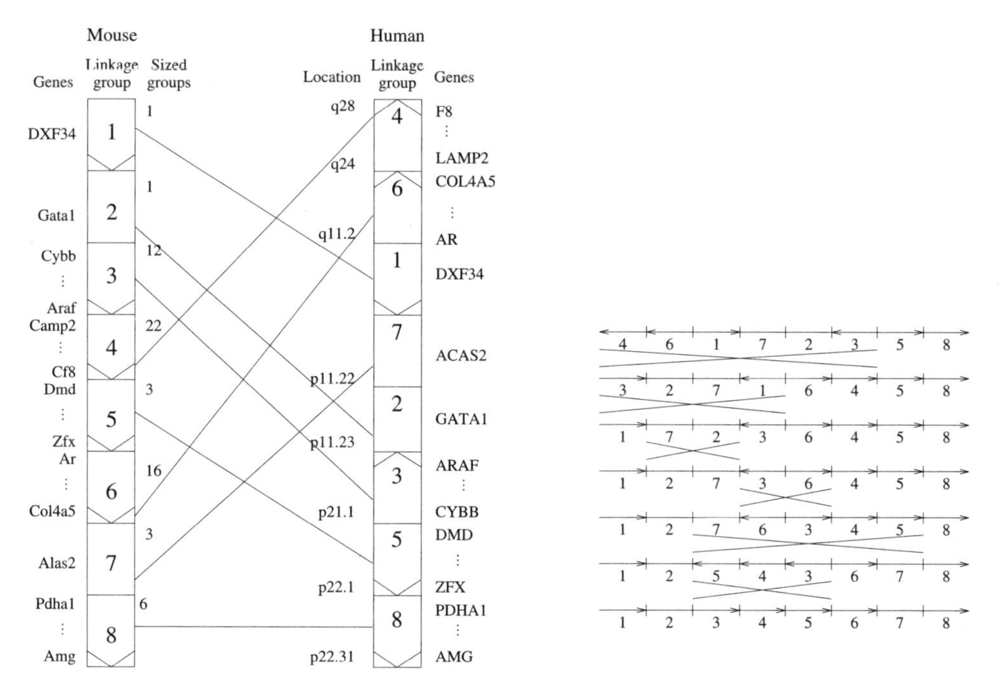

Figure 1.5: "Transformation" of a human X chromosome into a mouse X chromosome.

1.9 Proteomics

In many developing organisms, cells die at particular times as part of a normal process called *programmed cell death.* Death may occur as a result of a failure to acquire survival factors and may be initiated by the expression of certain genes. For example, in a developing nematode, the death of individual cells in the nervous system may be prevented by mutations in several genes whose function is under active investigation. However, the previously described DNA-based approaches are not well suited for finding genes involved in programmed cell death.

The cell death machinery is a complex system that is composed of many genes. While many proteins corresponding to these candidate genes have been identified, their roles and the ways they interact in programmed cell death are poorly understood. The difficulty is that the DNA of these candidate genes is hard to isolate, at least much harder than the corresponding proteins. However, there are no reliable methods for protein sequencing yet, and the sequence of these candidate genes remained unknown until recently.

Recently a new approach to protein sequencing via mass-spectrometry emerged that allowed sequencing of many proteins involved in programmed cell death. In 1996 protein sequencing led to the identification of the *FLICE* protein, which is involved in death-inducing signaling complex (Muzio et al., 1996 [244]). In this case gene hunting started from a protein (rather than DNA) sequencing, and subsequently led to cloning of the *FLICE* gene. The exceptional sensitivity of mass-spectrometry opened up new experimental and computational vistas for protein sequencing and made this technique a method of choice in many areas.

Protein sequencing has long fascinated mass-spectrometrists (Johnson and Biemann, 1989 [182]). However, only now, with the development of mass spectrometry automation systems and *de novo* algorithms, may high-throughout protein sequencing become a reality and even open a door to "proteome sequencing". Currently, most proteins are identified by database search (Eng et al., 1994 [97], Mann and Wilm, 1994 [230]) that relies on the ability to "look the answer up in the back of the book". Although database search is very useful in extensively sequenced genomes, a biologist who attempts to find a *new* gene needs *de novo* rather than database search algorithms.

In a few seconds, a mass spectrometer is capable of breaking a peptide into pieces (*ions*) and measuring their *masses*. The resulting set of masses forms the *spectrum* of a peptide. The *Peptide Sequencing Problem* is to reconstruct the peptide given its spectrum. For an "ideal" fragmentation process and an "ideal" mass-spectrometer, the peptide sequencing problem is simple. In practice, *de novo* peptide sequencing remains an open problem since spectra are difficult to interpret.

In the simplest form, protein sequencing by mass-spectrometry corresponds to the following problem. Let A be the set of amino acids with molecular masses $m(a)$, $a \in A$. A (parent) *peptide* $P = p_1, \ldots, p_n$ is a sequence of amino acids, and the mass of peptide P is $m(P) = \sum m(p_i)$. A *partial peptide* $P' \subset P$ is a substring $p_i \ldots p_j$ of P of mass $\sum_{i \le t \le j} m(p_t)$. *Theoretical spectrum* $E(P)$ of peptide P is a set of masses of its partial peptides. An (experimental) *spectrum* $S = \{s_1, \ldots, s_m\}$ is a set of masses of (fragment) ions. A *match* between spectrum S and peptide P is the number of masses that experimental and theoretical spectra have in common.

Peptide Sequencing Problem Given spectrum S and a parent mass m, find a peptide of mass m with the maximal match to spectrum S.

Chapter 2

Restriction Mapping

2.1 Introduction

Hamilton Smith discovered in 1970 that the *restriction enzyme Hind*II cleaves DNA molecules at every occurrence of a sequence GTGCAC or GTTAAC (Smith and Wilcox, 1970 [319]). Soon afterward Danna et al., 1973 [80] constructed the first restriction map for Simian Virus 40 DNA. Since that time, *restriction maps* (sometimes also called *physical maps*) representing DNA molecules with points of cleavage (*sites*) by restriction enzymes have become fundamental data structures in molecular biology.

To build a restriction map, biologists use different *biochemical* techniques to derive indirect information about the map and *combinatorial* methods to reconstruct the map from these data. Several experimental approaches to restriction mapping exist, each with its own advantages and disadvantages. They lead to different combinatorial problems that frequently may be formulated as recovering positions of points when only some pairwise distances between points are known.

Most restriction mapping problems correspond to the following problem. If X is a set of points on a line, let ΔX denote the multiset of *all* pairwise distances between points in X: $\Delta X = \{|x_1 - x_2| : \ x_1, x_2 \in X\}$. In restriction mapping some subset $E \subset \Delta X$ corresponding to the experimental data about fragment lengths is given, and the problem is to reconstruct X from E.

For the *Partial Digest Problem* (PDP), the experiment provides data about *all* pairwise distances between restriction sites ($E = \Delta X$). In this method DNA is digested in such a way that fragments are formed by every two cuts. No polynomial algorithm for PDP is yet known. The difficulty is that it may not be possible to uniquely reconstruct X from ΔX: two multisets X and Y are *ho-*

mometric if $\Delta X = \Delta Y$. For example, X, $-X$ (reflection of X) and $X + a$ for every number a (translation of X) are homometric. There are less trivial examples of this non-uniqueness; for example, the sets $\{0, 1, 3, 8, 9, 11, 13, 15\}$ and $\{0, 1, 3, 4, 5, 7, 12, 13, 15\}$ are homometric and are not transformed into each other by reflections and translations (*strongly homometric sets*). Rosenblatt and Seymour, 1982 [289] studied strongly homometric sets and gave an elegant pseudo-polynomial algorithm for PDP based on factorization of polynomials. Later Skiena et al., 1990 [314] proposed a simple *backtracking* algorithm which performs very well in practice but in some cases may require exponential time.

The backtracking algorithm easily solves the PDP problem for all inputs of practical size. However, PDP has never been the favorite mapping method in biological laboratories because it is difficult to digest DNA in such a way that the cuts between *every* two sites are formed.

Double Digest is a much simpler experimental mapping technique than Partial Digest. In this approach, a biologist maps the positions of the sites of two restriction enzymes by complete digestion of DNA in such a way that only fragments between *consecutive* sites are formed. One way to construct such a map is to measure the fragment lengths (not the order) from a complete digestion of the DNA by each of the two enzymes singly, and then by the two enzymes applied together. The problem of determining the positions of the cuts from fragment length data is known as the *Double Digest Problem* or *DDP*.

For an arbitrary set X of n elements, let δX be the set of $n - 1$ distances between *consecutive* elements of X. In the Double Digest Problem, a multiset $X \subset [0, t]$ is partitioned into two subsets $X = A \bigcup B$ with $0 \in A, B$ and $t \in A, B$, and the experiment provides three sets of length: $\delta A, \delta B$, and δX (A and B correspond to the single digests while X corresponds to the double digest). The Double Digest Problem is to reconstruct A and B from these data.

The first attempts to solve the Double Digest Problem (Stefik, 1978 [329]) were far from successful. The reason for this is that the number of potential maps and computational complexity of DDP grow very rapidly with the number of sites. The problem is complicated by experimental errors, and all DDP algorithms encounter computational difficulties even for small maps with fewer than 10 sites for each restriction enzyme.

Goldstein and Waterman, 1987 [130] proved that DDP is NP-complete and showed that the number of solutions to DDP increases exponentially as the number of sites increases. Of course NP-completeness and exponential growth of the number of solutions are the bottlenecks for DDP algorithms. Nevertheless, Schmitt and Waterman, 1991 [309] noticed that even though the number of solutions grows very quickly as the number of sites grows, most of the solutions are very similar (could be transformed into each other by simple transformations). Since mapping algorithms generate a lot of "very similar maps," it would seem reasonable to partition the entire set of physical maps into equivalence classes and to generate only one basic map in every equivalence class. Subsequently, all solutions could be generated from the basic maps using simple transformations. If the number of equivalence classes were significantly smaller than the number of physical maps, then this approach would allow reduction of computational time for the DDP algorithm.

Schmitt and Waterman, 1991 [309] took the first step in this direction and introduced an equivalence relation on physical maps. All maps of the same equivalence class are transformed into one another by means of *cassette transformations*. Nevertheless, the problem of the constructive generation of all equivalence classes for DDP remained open and an algorithm for a transformation of equivalent maps was also unknown. Pevzner, 1995 [267] proved a characterization theorem for equivalent transformations of physical maps and described how to generate all solutions of a DDP problem. This result is based on the relationships between DDP solutions and alternating Eulerian cycles in edge-colored graphs.

As we have seen, the combinatorial algorithms for PDP are very fast in practice, but the experimental PDP data are hard to obtain. In contrast, the experiments for DDP are very simple but the combinatorial algorithms are too slow. This is the reason why restriction mapping is not a very popular experimental technique today.

2.2 Double Digest Problem

Figure 2.1 shows "DNA" cut by restriction enzymes A and B. When Danna et al., 1973 [80] constructed the first physical map there was no experimental technique to directly find the positions of cuts. However, they were able to measure the sizes (but not the order!) of the restriction fragments using the experimental technique known as *gel-electrophoresis*. Through gel-electrophoresis experiments with two restriction enzymes A and B (Figure 2.1), a biologist obtains information about the sizes of restriction fragments 2, 3, 4 for A and 1, 3, 5 for B, but there are many orderings (maps) corresponding to these sizes (Figure 2.2 shows two of them). To find out which of the maps shown in Figure 2.2 is the correct one, biologists use *Double Digest* $A + B$—cleavage of DNA by *both* enzymes, A and B. Two maps presented in Figure 2.2 produce the same single digests A and B but different double digests $A + B$ (1, 1, 2, 2, 3 and 1, 1, 1, 2, 4). The double digest that fits

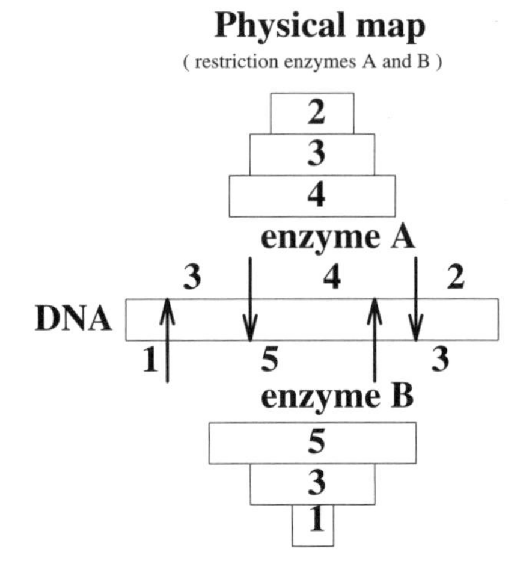

Figure 2.1: Physical map of two restriction enzymes. Gel-electrophoresis provides information about the sizes (but not the order) of restriction fragments.

experimental data corresponds to the correct map. The Double Digest Problem is to find a physical map, given three "stacks" of fragments: A, B, and $A+B$ (Figure 2.3).

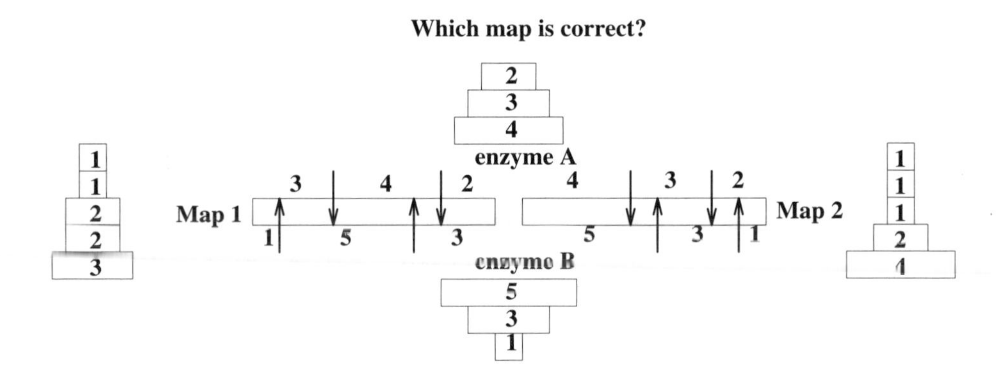

Figure 2.2: Data on A and B do not allow a biologist to find a true map. $A+B$ data help to find the correct map.

2.3 Multiple Solutions of the Double Digest Problem

Figure 2.3 presents two solutions of the Double Digest Problem. Although they look very different, they can be transformed one into another by a simple operation called *cassette exchange* (Figure 2.4). Another example of multiple solutions is given in Figure 2.5. Although these solutions cannot be transformed into one another by cassette exchanges, they can be transformed one into another through a different operation called *cassette reflection* (Figure 2.6). A surprising result is that these two simple operations, in some sense, are sufficient to enable a transformation between any two "similar" solutions of the Double Digest Problem.

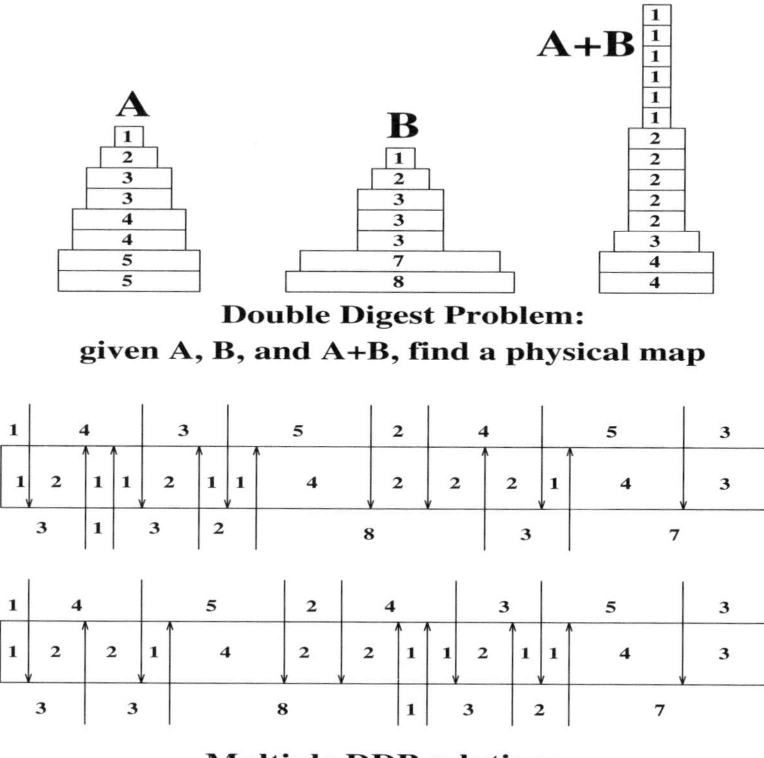

Figure 2.3: The Double Digest Problem may have multiple solutions.

A physical map is represented by the ordered sequence of fragments of single digests $A_1, \ldots, A_n$ and $B_1, \ldots, B_m$ and double digest $C_1, \ldots, C_l$ (Figure 2.4).

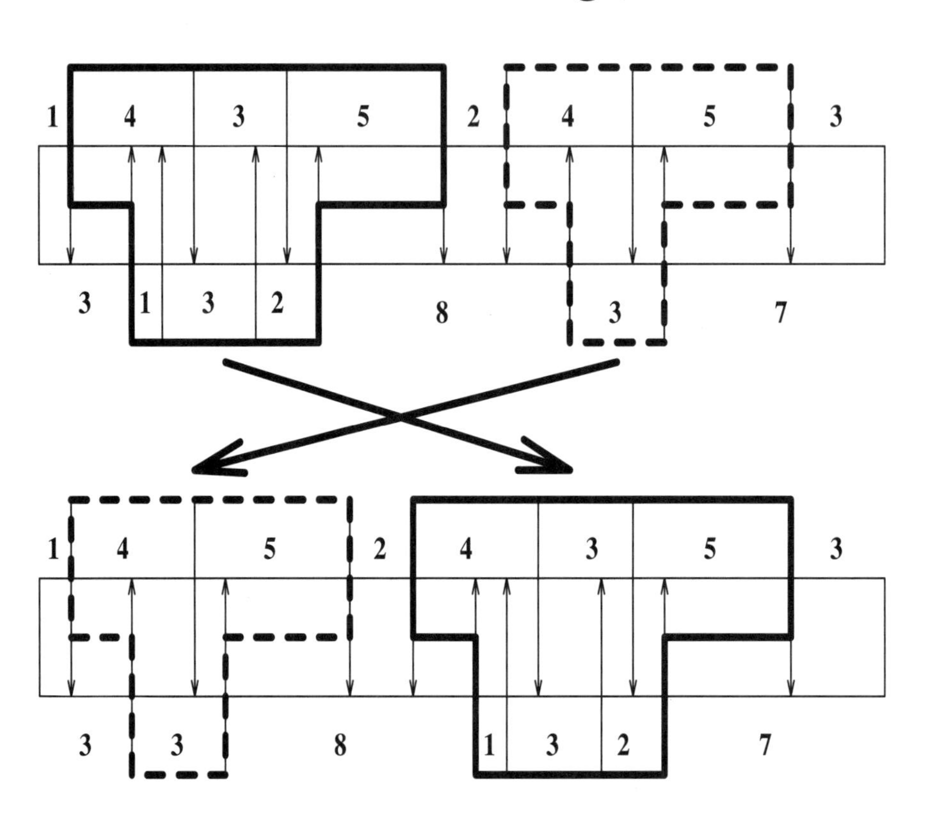

Figure 2.4: Cassette exchange. The upper map is defined by the ordered sequences of fragment sizes for restriction enzyme A ($\{1,4,3,5,2,4,5,3\}$), restriction enzyme B ($\{3,1,3,2,8,3,7\}$), and restriction enzymes $A+B=C$ ($\{1,2,1,1,2,1,1,4,2,2,2,1,4,3\}$). The interval $I=[3,7]$ defines the set of double digest fragments $I_C=\{C_3,C_4,C_5,C_6,C_7\}$ of length 1, 1, 2, 1, 1. I_C defines a cassette (I_A,I_B) where $I_A=\{A_2,A_3,A_4\}=\{4,3,5\}$ and $I_B=\{B_2,B_3,B_4\}=\{1,3,2\}$. The left overlap of (I_A,I_B) equals $m_A-m_B=1-3=-2$. The right overlap of (I_A,I_B) equals $13-9=4$.

For an interval $I=[i,j]$ with $1\le i\le j\le l$, define $I_C=\{C_k: \quad i\le k\le j\}$ as the set of fragments between C_i and C_j. The *cassette* defined by I_C is the pair of sets of fragments (I_A,I_B), where I_A and I_B are the sets of all fragments of A and B respectively that contain a fragment from I_C (Figure 2.4). Let m_A and m_B

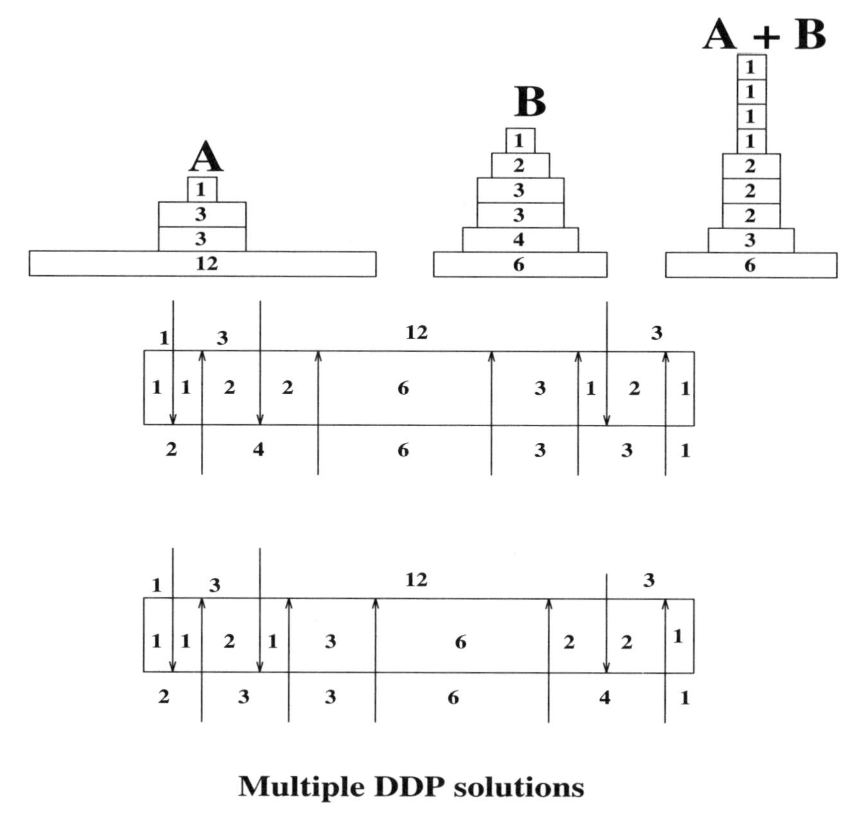

Figure 2.5: Multiple DDP solutions that cannot be transformed into one another by cassette exchange.

be the starting positions of the leftmost fragments of I_A and I_B respectively. The *left overlap* of (I_A, I_B) is the distance $m_A - m_B$. The *right overlap* of (I_A, I_B) is defined similarly, by substituting the words "ending" and "rightmost" for the words "starting" and "leftmost" in the definition above.

Suppose two cassettes within the solution to DDP have the same left overlaps and the same right overlaps. If these cassettes do not intersect (have no common fragments), then they can be *exchanged* as in Figure 2.4, and one obtains a new solution of DDP. Also, if the left and right overlaps of a cassette (I_A, I_B) have the same size but different signs, then the cassette may be *reflected* as shown in Figure 2.6, and one obtains a new solution of DDP.

Cassette reflection

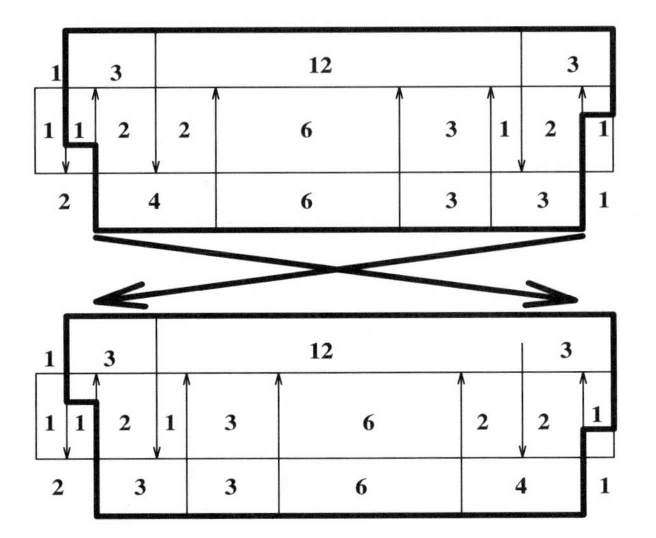

Figure 2.6: Cassette reflection. The left and the right overlaps have the same size but different signs.

Schmitt and Waterman, 1991 [309] raised the question of how to transform one map into another by cassette transformations. The following section introduces a graph-theoretic technique to analyze the combinatorics of cassette transformations and to answer the Schmitt-Waterman question.

2.4 Alternating Cycles in Colored Graphs

Consider an *undirected graph* $G(V, E)$ with the *edge* set E edge-colored in l *colors*. A sequence of vertices $P = x_1 x_2 \dots x_m$ is called a *path* in G if $(x_i, x_{i+1}) \in E$ for $1 \le i \le m - 1$. A path P is called a *cycle* if $x_1 = x_m$. Paths and cycles can be vertex self-intersecting. We denote $P^- = x_m x_{m-1} \dots x_1$.

A path (cycle) in G is called *alternating* if the colors of every two consecutive edges (x_i, x_{i+1}) and (x_{i+1}, x_{i+2}) of this path (cycle) are distinct (if P is a cycle we consider (x_{m-1}, x_m) and (x_1, x_2) to be consecutive edges). A path (cycle) P in G is called *Eulerian* if every $e \in E$ is traversed by P exactly once. Let $d_c(v)$ be the number of c-colored edges of E incident to v and $d(v) = \sum_{c=1}^{l} d_c(v)$ be the *degree* of vertex v in the graph G. A vertex v in the graph G is called *balanced* if $\max_c d_c(v) \leq d(v)/2$. A *balanced graph* is a graph whose every vertex is balanced.

Theorem 2.1 *(Kotzig, 1968 [206]) Let G be a colored connected graph with even degrees of vertices. Then there is an alternating Eulerian cycle in G if and only if G is balanced.*

Proof To construct an alternating Eulerian cycle in G, partition $d(v)$ edges incident to vertex v into $d(v)/2$ pairs such that two edges in the same pair have different colors (it can be done for every balanced vertex). Starting from an arbitrary edge in G, form a trail C_1 using at every step an edge paired with the last edge of the trail. The process stops when an edge paired with the last edge of the trail has already been used in the trail. Since every vertex in G has an even degree, every such trail starting from vertex v ends at v. With some luck the trail will be Eulerian, but if not, it must contain a node w that still has a number of untraversed edges. Since the graph of untraversed edges is balanced, we can start from w and form another trail C_2 from untraversed edges using the same rule. We can now combine cycles C_1 and C_2 as follows: insert the trail C_2 into the trail C_1 at the point where w is reached. This needs to be done with caution to preserve the alternation of colors at vertex w. One can see that if inserting the trail C_2 in direct order destroys the alternation of colors, then inserting it in reverse order preserves the alternation of colors. Repeating this will eventually yield an alternating Eulerian cycle. ■

We will use the following corollary from the Kotzig theorem:

Lemma 2.1 *Let G be a bicolored connected graph. Then there is an alternating Eulerian cycle in G if and only if* $d_1(v) = d_2(v)$ *for every vertex in G.*

2.5 Transformations of Alternating Eulerian Cycles

In this section we introduce *order transformations* of alternating paths and demonstrate that every two alternating Eulerian cycles in a bicolored graph can be transformed into each other by means of order transformations. This result implies the characterization of Schmitt-Waterman cassette transformations.

Let $F = \ldots x \ldots y \ldots x \ldots y \ldots$ be an alternating path in a bicolored graph G. Vertices x and y partitions F into five subpaths $F = F_1 F_2 F_3 F_4 F_5$ (Figure 2.7).

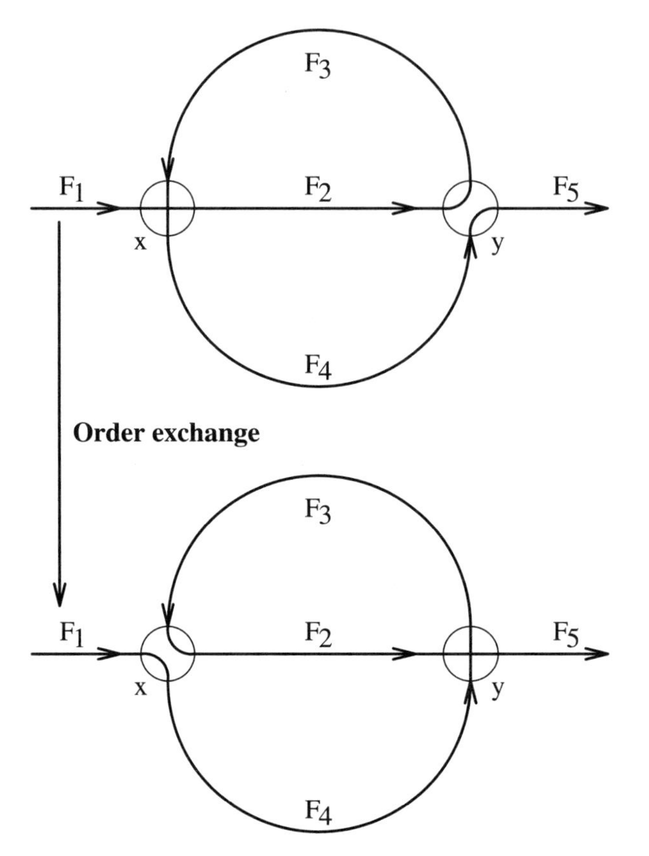

Figure 2.7: Order exchange.

The transformation $F = F_1F_2F_3F_4F_5 \longrightarrow F^* = F_1F_4F_3F_2F_5$ is called an *order exchange* if F^* is an alternating path.

Let $F = \ldots x \ldots x \ldots$ be an alternating path in a bicolored graph G. Vertex x partition F into three subpaths $F = F_1F_2F_3$ (Figure 2.8). The transformation $F = F_1F_2F_3 \longrightarrow F^* = F_1F_2^-F_3$ is called an *order reflection* if F^* is an alternating path. Obviously, the order reflection $F \longrightarrow F^*$ in a bicolored graph exists if and only if F_2 is an odd cycle.

Theorem 2.2 *Every two alternating Eulerian cycles in a bicolored graph G can be transformed into each other by a series of order transformations (exchanges and reflections).*

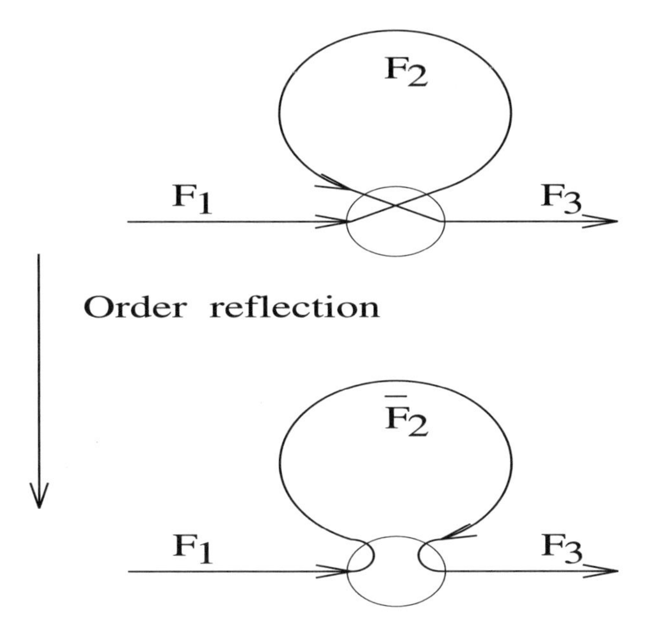

Figure 2.8: Order reflection.

Proof Let X and Y be two alternating Eulerian cycles in G. Consider the set of alternating Eulerian cycles $\mathcal{C}$ obtained from X by all possible series of order transformations. Let $X^* = x_1 \dots x_m$ be a cycle in $\mathcal{C}$ having the longest common prefix with $Y = y_1 \dots y_m$, i.e., $x_1 \dots x_l = y_1 \dots y_l$ for $l \leq m$. If $l = m$, the theorem holds: otherwise let $v = x_l = y_l$ (i.e., $e_1 = (v, x_{l+1})$ and $e_2 = (v, y_{l+1})$ are the first different edges in X^* and Y, respectively (Figure 2.9)).

Since X^* and Y are alternating paths, the edges e_1 and e_2 have the same color. Since X^* is Eulerian path, X^* contains the edge e_2. Clearly, e_2 succeeds e_1 in X^*. There are two cases (Figure 2.9) depending on the direction of the edge e_2 in the path X^* (toward or from vertex v):

Case 1. Edge $e_2 = (y_{l+1}, v)$ in the path X^* is directed toward v. In this case $X^* = x_1 \dots v x_{l+1} \dots y_{l+1}\ v \dots x_m$. Since the colors of the edges e_1 and e_2 coincide, the transformation $X^* = F_1 F_2 F_3 \longrightarrow F_1 F_2^- F_3 = X^{**}$ is an order reflection (Figure 2.10). Therefore $X^{**} \in \mathcal{C}$ and at least $(l+1)$ initial vertices in X^{**} and Y coincide, a contradiction to the choice of X^*.

Case 2. Edge $e_2 = (v, y_{l+1})$ in the path X^* is directed from v. In this case, vertex v partitions the path X^* into three parts, prefix X_1 ending at v, cycle X_2,

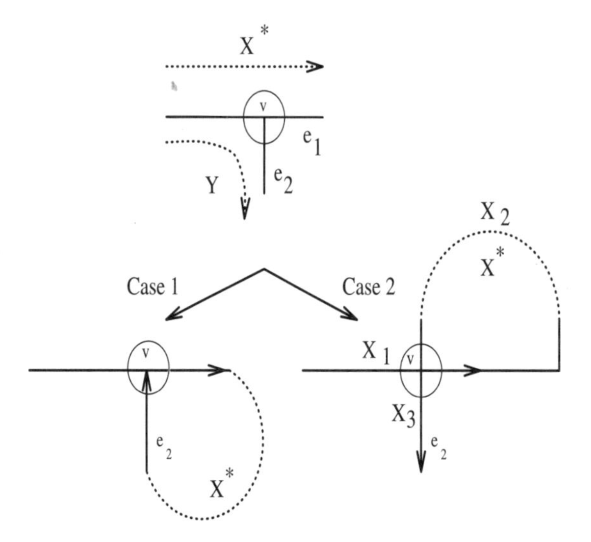

Figure 2.9: Two cases in theorem 2.

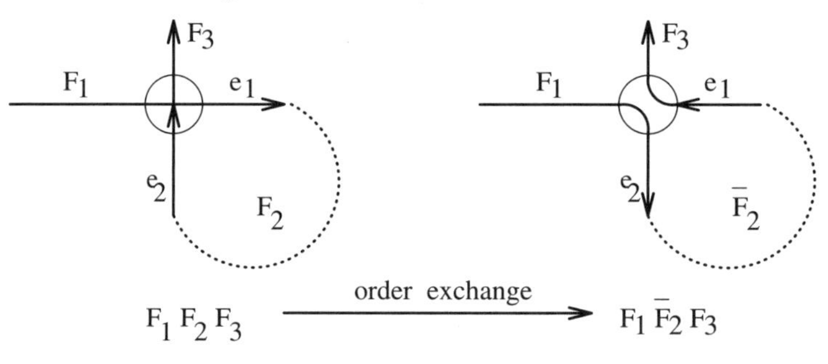

Figure 2.10: Case 1: Order exchange.

and suffix X_3 starting at v. It is easy to see that X_2 and X_3 have a vertex $x_j = x_k$ $(l < j < k < m)$ in common (otherwise, Y would not be an Eulerian cycle). Therefore, the cycle X^* can now be rewritten as $X^* = F_1F_2F_3F_4F_5$ (Figure 2.11).

Consider the edges (x_k, x_{k+1}) and (x_{j-1}, x_j) that are shown by thick lines in Figure 2.11. If the colors of these edges are different, then $X^{**} = F_1F_4F_3F_2F_5$ is the alternating cycle obtained from X^* by means of the order exchange shown in Figure 2.11 (top). At least $(l + 1)$ initial vertices of X^{**} and Y coincide, a contradiction to the choice of X^*.

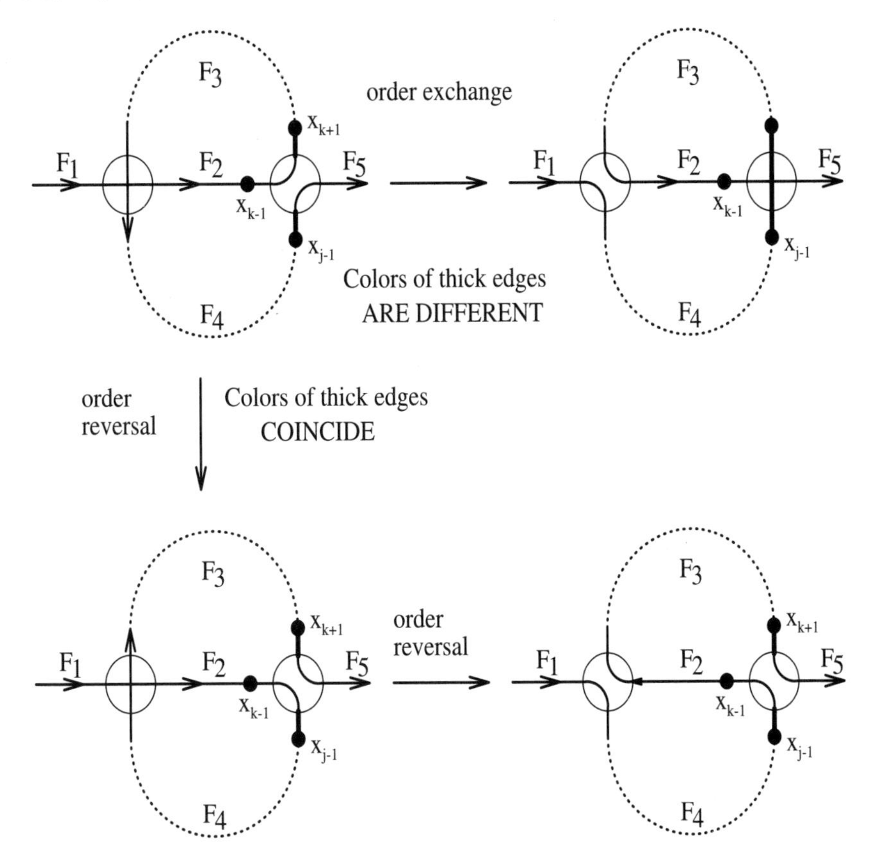

Figure 2.11: Case 2: Depending on the colors of the thick edges, there exists either an order exchange or two order reflections transforming X^* into a cycle with a longer common prefix with Y.

If the colors of the edges (x_k, x_{k+1}) and (x_{j-1}, x_j) coincide (Figure 2.11, bottom), then $X^{**} = F_1 F_4 F_2^- F_3^- F_5$ is obtained from X^* by means of two order reflections g and h:

$$F_1 F_2 F_3 F_4 F_5 \xrightarrow{g} F_1 F_2 (F_3 F_4)^- F_5 = F_1 F_2 F_4^- F_3^- F_5 \xrightarrow{h}$$

$$F_1 (F_2 F_4^-)^- F_3^- F_5 = F_1 F_4^{--} F_2^- F_3^- F_5 = F_1 F_4 F_2^- F_3^- F_5$$

At least $(l+1)$ initial vertices of X^{**} and Y coincide, a contradiction to the choice of X^*. ■

2.6 Physical Maps and Alternating Eulerian Cycles

This section introduces *fork graphs* of physical maps and demonstrates that every physical map corresponds to an alternating Eulerian path in the fork graph.

Consider a physical map given by (ordered) fragments of single digests A and B and double digest $C = A+B$: $\{A_1, \ldots, A_n\}$, $\{B_1, \ldots, B_m\}$, and $\{C_1, \ldots, C_l\}$. Below, for the sake of simplicity, we assume that A and B do not cut DNA at the same positions, i.e., $l = n + m - 1$. A *fork* of fragment A_i is the set of double digest fragments C_j contained in A_i:

$$F(A_i) = \{C_j : \quad C_j \subset A_i\}$$

(a fork of B_i is defined analogously). For example, $F(A_3)$ consists of two fragments C_5 and C_6 of sizes 4 and 1 (Figure 2.12). Obviously every two forks $F(A_i)$ and $F(B_j)$ have at most one common fragment. A fork containing at least two fragments is called a *multifork.*

Leftmost and rightmost fragments of multiforks are called *border fragments.* Obviously, C_1 and C_l are border fragments.

Lemma 2.2 *Every border fragment, excluding C_1 and C_l, belongs to exactly two multiforks $F(A_i)$ and $F(B_j)$. Border fragments C_1 and C_l belong to exactly one multifork.*

Lemma 2.2 motivates the construction of the *fork graph* with vertex set of lengths of border fragments (two border fragments of the same length correspond to the same vertex). The edge set of the fork graph corresponds to all multiforks (every multifork is represented by an edge connecting the vertices corresponding to the length of its border fragments). Color edges corresponding to multiforks of A with color A and edges corresponding to multiforks of B with color B (Figure 2.12).

All vertices of G are balanced, except perhaps vertices $|C_1|$ and $|C_l|$ which are semi-balanced, i.e., $|d_A(|C_1|) - d_B(|C_1|)| = |d_A(|C_l|) - d_B(|C_l|)| = 1$. The graph G may be transformed into a balanced graph by adding an edge or two edges. Therefore G contains an alternating Eulerian path.

Every physical map (A, B) defines an alternating Eulerian path in its fork graph. Cassette transformations of a physical map do not change the set of forks of this map. The question arises whether two maps with the same set of forks can be transformed into each other by cassette transformations. Fig 2.12 presents two

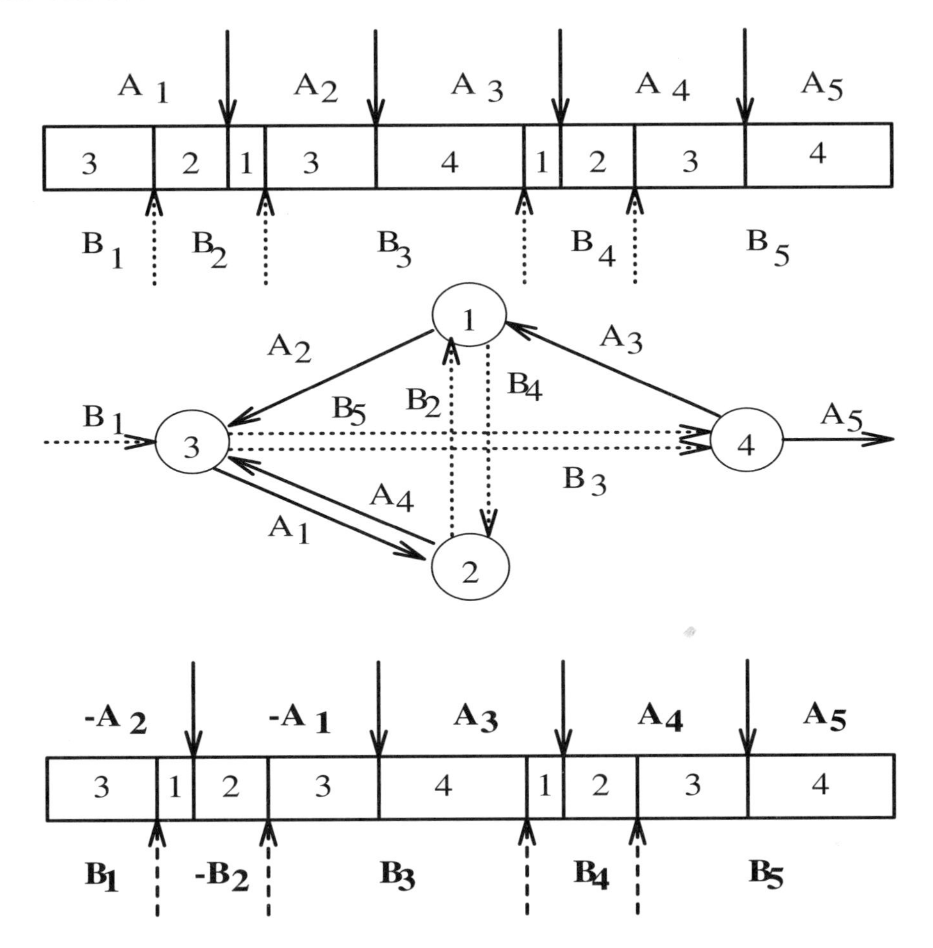

Figure 2.12: Fork graph of a physical map with added extra edges B_1 and A_5. Solid (dotted) edges correspond to multiforks of A (B). Arrows on the edges of this (undirected) graph follow the path $B_1A_1B_2A_2B_3A_3B_4A_4B_5A_5$, corresponding to the map at the top. A map at the bottom $B_1 - A_2 - B_2 - A_1B_3A_3B_4A_4B_5A_5$ is obtained by changing the direction of edges in the triangle A_1, B_2, A_2 (cassette reflection).

maps with the same set of forks that correspond to two alternating Eulerian cycles in the fork graph. It is easy to see that cassette transformations of the physical maps correspond to order transformations in the fork graph. Therefore every alternating Eulerian path in the fork graph of (A, B) corresponds to a map obtained from (A, B) by cassette transformations (Theorem 2.2).

2.7 Partial Digest Problem

The Partial Digest Problem is to reconstruct the positions of n restriction sites from the set of the $\binom{n}{2}$ distances between all pairs of these sites. If ΔX is the (multi)set of distances between all pairs of points of X, then the PDP problem is to reconstruct X given ΔX. Rosenblatt and Seymour, 1982 [289] gave a pseudo-polynomial algorithm for this problem using factoring of polynomials. Skiena et al., 1990 [314] described the following simple backtracking algorithm, which was further modified by Skiena and Sundaram, 1994 [315] for the case of data with errors.

First find the longest distance in ΔX, which decides the two outermost points of X, and then delete this distance from ΔX. Then repeatedly position the longest remaining distance of ΔX. Since for each step the longest distance in ΔX must be realized from one of the outermost points, there are only two possible positions (left or right) to put the point. At each step, for each of the two positions, check whether all the distances from the position to the points already selected are in ΔX. If they are, delete all those distances before going to next step. Backtrack if they are not for both of the two positions. A solution has been found when ΔX is empty.

For example, suppose $\Delta X = \{2, 2, 3, 3, 4, 5, 6, 7, 8, 10\}$. Since ΔX includes all the pairwise distances, then $|\Delta X| = \binom{n}{2}$, where n is the number of points in the solution. First set $L = \Delta X$ and $x_1 = 0$. Since 10 is the largest distance in L, it is clear that $x_5 = 10$. Removing distance $x_5 - x_1 = 10$ from L, we obtain

$$X = \{0, 10\} \qquad L = \{2, 2, 3, 3, 4, 5, 6, 7, 8\}.$$

The largest remaining distance is 8. Now we have two choices: either $x_4 = 8$ or $x_2 = 2$. Since those two cases are mirror images of each other, without loss of generality, we can assume $x_2 = 2$. After removal of distances $x_5 - x_2 = 8$ and $x_2 - x_1 = 2$ from L, we obtain

$$X = \{0, 2, 10\} \qquad L = \{2, 3, 3, 4, 5, 6, 7\}.$$

Since 7 is the largest remaining distance, we have either $x_4 = 7$ or $x_3 = 3$. If $x_3 = 3$, distance $x_3 - x_2 = 1$ must be in L, but it is not, so we can only set $x_4 = 7$. After removing distances $x_5 - x_4 = 3$, $x_4 - x_2 = 5$, and $x_4 - x_1 = 7$ from L, we obtain

$$X = \{0, 2, 7, 10\} \qquad L = \{2, 3, 4, 6\}.$$

Now 6 is the largest remaining distance. Once again we have two choices: either $x_3 = 4$ or $x_3 = 6$. If $x_3 = 6$, the distance $x_4 - x_3 = 1$ must be in L, but it is not. So that leaves us only the choice $x_3 = 4$ and provides a solution $\{0, 2, 4, 7, 10\}$ of the Partial Digest Problem.

The pseudo-code for the described algorithm is given below. Here the function **Delete_Max**(L) returns the maximum value of L and removes it from list L, and two global variables X and $width$ are used. $\Delta(X, Y)$ is the (multi)set of all distances between a point of X and a point of Y.

```
set X
int width
Partial_Digest(List L)
        width = Delete_Max(L)
        X = {0, width}
        Place(L)

Place(List L)
        if L = ∅ then
                output solution X
                exit
        y = Delete_Max(L)
        if Δ({y}, X) ⊂ L then
                X = X ∪ {y}
                Place(L \ Δ({y}, X))          /* place a point at right position */
                X = X \ {y}                    /* backtracking */
        if Δ({width − y}, X) ⊂ L then
                X = X ∪ {width − y}
                Place(L \ Δ({width − y}, X)) /* place a point at left position */
                X = X \ {width − y}                 /* backtracking */
```

This algorithm runs in $O(n^2 \log n)$ expected time if L arises from real points in general positions, because in this case at each step, one of the two choices will be pruned with probability 1. However, the running time of the algorithm may be exponential in the worst case (Zhang, 1994 [377]).

2.8 Homometric Sets

It is not always possible to uniquely reconstruct a set X from ΔX. Sets A and B are *homometric* if $\Delta A = \Delta B$. Let U and V be two multisets. It is easy to verify that the multisets $U + V = \{u + v : u \in U, v \in V\}$ and $U - V = \{u - v : u \in U, v \in V\}$ are homometric. The example presented in Figure 2.13 arises from this construction for $U = \{6, 7, 9\}$ and $V = \{-6, 2, 6\}$.

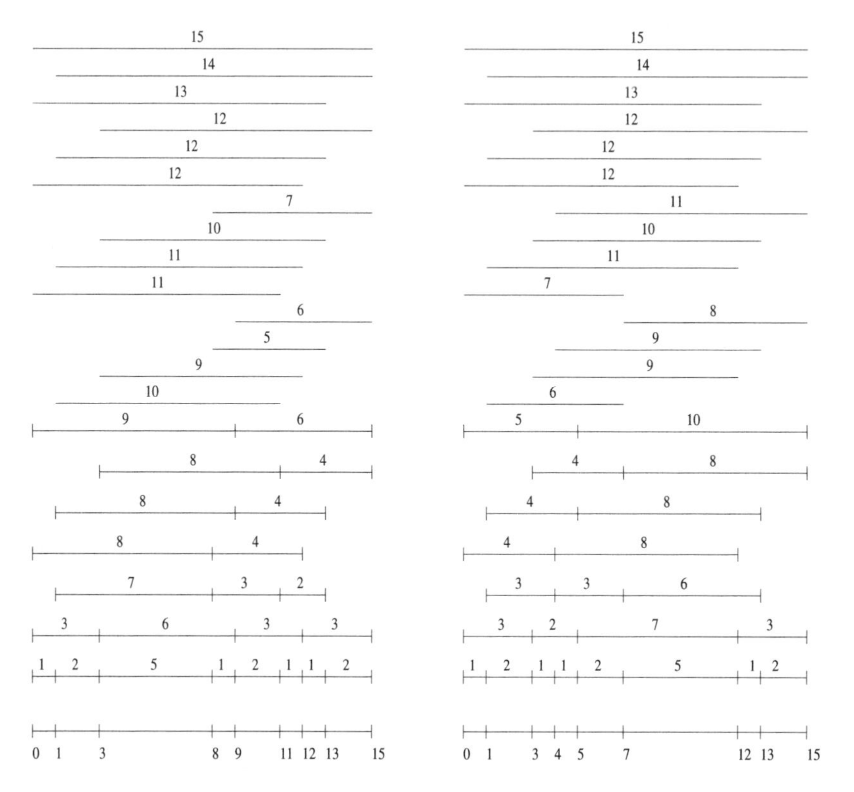

Figure 2.13: Homometric sets $U + V = \{0, 1, 3, 8, 9, 11, 12, 13, 15\}$ and $U - V = \{0, 1, 3, 4, 5, 7, 12, 13, 15\}$.

It is natural to ask if every pair of homometric sets represents an instance of this construction. The answer is negative: the homometric sets $\{0, 1, 2, 5, 7, 9, 12\}$ and $\{0, 1, 5, 7, 8, 10, 12\}$ provide a counterexample. Nevertheless this conjecture is true if we define U and V as "multisets with possibly negative multiplicities."

Given a multiset of integers $A = \{a_i\}$, let $A(x) = \sum_i x^{a_i}$ be a *generating function* for A. It is easy to see that the generating function for ΔA is $\Delta A(x) = A(x)A(x^{-1})$. Let $A(x)$ and $B(x)$ be generating functions for multisets A and B such that $A(x) = U(x)V(x)$ and $B(x) = U(x)V(x^{-1})$. Then $A(x)A(x^{-1}) = B(x)B(x^{-1}) = U(x)V(x)U(x^{-1})V(x^{-1})$, implying that A and B are homometric.

Theorem 2.3 *(Rosenblatt and Seymour, 1982 [289]) Two sets A and B are homometric if and only if there exist generating functions $U(x)$ and $V(x)$ and an integer β such that $A(x) = U(x)V(x)$ and $B(x) = \pm x^{\beta}U(x)V(x^{-1})$.*

Proof Let A and B be homometric sets. Let $P(x)$ be the greatest common divisor of $A(x)$ and $B(x)$ and let $A(x) = P(x)Q_A(x)$ and $B(x) = P(x)Q_B(x)$, where $Q_A(x)$ and $Q_B(x)$ are relatively prime. Let $V(x)$ be the greatest common divisor of $Q_A(x)$ and $Q_B(x^{-1})$ and let $Q_A(x) = V(x)S_A(x)$ and $Q_B(x^{-1}) = V(x)S_B(x)$, where $S_A(x)$ and $S_B(x)$ are relatively prime. Clearly $S_A(x)$ and $S_A(x^{-1})$ are relatively prime to *both* $S_B(x)$ and $S_B(x^{-1})$.

Since A and B are homometric,

$$P(x)V(x)S_A(x)P(x^{-1})V(x^{-1})S_A(x^{-1}) = P(x)V(x^{-1})S_B(x^{-1})P(x^{-1})V(x)S_B(x)$$

implying that $S_A(x)S_A(x^{-1}) = S_B(x)S_B(x^{-1})$. Since $S_A(x)$ and $S_A(x^{-1})$ are relatively prime to both $S_B(x)$ and $S_B(x^{-1})$, $S_A(x) = \pm x^a$ and $S_B(x) = \pm x^b$. Therefore, $A(x) = \pm x^a P(x)V(x)$ and $B(x) = \pm x^b P(x)V(x^{-1})$. Substitution $U(x) = \pm x^a P(x)$ proves the theorem. ■

It can happen that some of the coefficients in the decomposition given in theorem 2.3 are negative, corresponding to multisets with negative multiplicities. For example, if $A = \{0, 1, 2, 5, 7, 9, 12\}$ and $B = \{0, 1, 5, 7, 8, 10, 12\}$, $U(x) = (1 + x + x^2 + x^3 + x^4 + x^5 + x^7)$ and $V(x) = x^{-5}(1 - x^3 + x^5)$.

We say that a set A is *reconstructible* if whenever B is homometric to A, we have $B = A + \{v\}$ or $B = -A + \{v\}$ for a number v. A set A is called *symmetric* if $-A = A + v$ for some number v. A polynomial $A(x)$ is *symmetric* if the corresponding set is symmetric, i.e., $A(x^{-1}) = x^v A(x)$. Theorem 2.3 implies the following:

Theorem 2.4 *A set A is reconstructible if and only if $A(x)$ has at most one prime factor (counting multiplicities) that is not symmetric.*

Rosenblatt and Seymour, 1982 [289] gave the following pseudo-polynomial algorithm for the Partial Digest Problem with n points. Given a set of $\binom{n}{2}$ distances $\Delta A = \{d_i\}$, we form the generating function $\Delta A(x) = n + \sum_i (x^{d_i} + x^{-d_i})$. We factor this polynomial into irreducibles over the ring of polynomials with integer coefficients using a factoring algorithm with runtime polynomial in $\max_i d_i$. The solution $A(x)$ of the PDP problem must have a form $\Delta A(x) = A(x)A(x^{-1})$. Therefore, we try all 2^F possible subsets S of the F irreducible nonreciprocal factors of $\Delta A(x)$ as putative factors of $A(x)$. As a result, we find a set of all the possible sets $A(x)$. Finally, we eliminate the sets $A(x)$ with negative coefficients and sort to remove possible redundant copies.

2.9 Some Other Problems and Approaches

2.9.1 Optical mapping

Schwartz et al., 1993 [311] developed the *optical mapping* technique for construction of restriction maps. In optical mapping, single copies of DNA molecules are stretched and attached to a glass support under a microscope. When restriction enzymes are activated, they cleave the DNA molecules at their restriction sites. The molecules remain attached to the surface, but the elasticity of the stretched DNA pulls back the molecule ends at the cleaved sites. These can be identified under the microscope as tiny gaps in the fluorescent line of the molecule. Thus a "photograph" of the DNA molecule with gaps at the positions of cleavage sites gives a snapshot of the restriction map.

Optical mapping bypasses the problem of reconstructing the order of restriction fragments, but raises new computational challenges. The problem is that not all sites are cleaved in each molecule (false negative) and that some may incorrectly appear to be cut (false positive). In addition, inaccuracies in measuring the length of fragments, difficulties in analyzing proximal restriction sites, and the unknown orientation of each molecule (left to right or vice versa) make the reconstruction difficult. In practice, data from many molecules is gathered to build a consensus restriction map.

The problem of unknown orientation was formalized as *Binary Flip-Cut* (BFC) Problem by Muthukrishnan and Parida, 1997 [243]. In the BFC problem, a set of n binary 0-1 strings is given (each string represents a snapshot of a DNA molecule with 1s corresponding to restriction sites). The problem is to assign a *flip* or *no-flip* state to each string so that the number of *consensus* sites is minimized. A site is called a consensus under the assignment of flips if at least cn 1s are present at that site if the molecules are flipped accordingly, for some small constant parameter c.

Handling real optical mapping data is considerably harder than the BFC problem. Efficient algorithms for the optical mapping problem were developed by Anantharaman et al., 1997 [8], Karp and Shamir, 1998 [190], and Lee et al., 1998 [218].

2.9.2 Probed Partial Digest mapping

Another technique used to derive a physical map leads to the *Probed Partial Digest Problem (PPDP)*. In this method DNA is partially digested with a restriction enzyme, thus generating a collection of DNA fragments between any two cutting sites. Afterward a labeled probe, which attaches to the DNA between two cutting sites, is hybridized to the partially digested DNA, and the sizes of fragments to which the probe hybridizes are measured. The problem is to reconstruct the positions of the sites from the multiset of measured lengths.

In the PPDP problem, multiset $X \subset [-s, t]$ is partitioned into two subsets $X = A \bigcup B$ with $A \subset [-s, 0]$ and $B \subset [0, t]$ corresponding to the restriction sites to the left and to the right of the probe. The PPDP experiment provides the multiset $E = \{b - a : a \in A, b \in B\}$. The problem is to find X given E. Newberg and Naor, 1993 [252] showed that the number of PPDP solutions can grow quickly, at least more quickly than $n^{1.72}$.

Chapter 3

Map Assembly

3.1 Introduction

The map assembly problem can be understood in terms of the following analogy. Imagine several copies of a book cut by scissors into thousands of pieces. Each copy is cut in an individual way such that a piece from one copy may overlap a piece from another copy. For each piece and each word from a list of key words, we are told whether the piece contains the key word. Given this data, we wish to determine the pattern of overlaps of the pieces.

Double Digest and Partial Digest techniques allow a biologist to construct restriction (physical) maps of small DNA molecules, such as viral, chloroplast, or mitochondrial DNA. However, these methods do not work (experimentally or computationally) for large DNA molecules. Although the first restriction map of a viral genome was constructed in 1973, it took more than a decade to construct the first physical maps of a bacterial genome by assembling restriction maps of small fragments. To study a large DNA molecule, biologists break it into smaller pieces, map or fingerprint each piece, and then assemble the pieces to determine the map of the entire molecule. This mapping strategy was originally developed by Olson et al., 1986 [257] for yeast and by Coulson et al., 1986 [76] for nematode. However, the first large-scale physical map was constructed by Kohara et al., 1987 [204] for *E. Coli* bacteria.

Mapping usually starts with breaking a DNA molecule into small pieces using restriction enzymes. To study individual pieces, biologists obtain many identical copies of each piece by *cloning* them. Cloning incorporates a fragment of DNA into a *cloning vector*, a small, artificially constructed DNA molecule that originates from a virus or other organism. Cloning vectors with DNA inserts are introduced into a bacterial self-replicating host. The self-replication process then creates an enormous number of copies of the fragment, thus enabling its structure to be investigated. A fragment reproduced in this way is called a *clone*.

As a result, biologists obtain a *clone library* consisting of thousands of clones (each representing a short DNA fragment) from the same DNA molecule. Clones from the clone library may overlap (overlapping can be achieved by using a few restriction enzymes). After the clone library is constructed biologists want to *order* the clones, i.e., to reconstruct the relative placement of the clones along the DNA molecule. This information is lost in the construction of the clone library, and the process of reconstruction starts with *fingerprinting* the clones. The idea is to describe each clone using an easily determined fingerprint, which can be thought of as a set of "key words" present in a clone. If two clones have substantial overlap, their fingerprints should be similar. If non-overlapping clones are unlikely to have similar fingerprints then fingerprints would allow a biologist to distinguish between overlapping and non-overlapping clones and to reconstruct the order of the clones. The following fingerprints have been used in many mapping projects.

- *Restriction maps.* The restriction map of a clone provides an ordered list of restriction fragments. If two clones have restrictions maps that share several consecutive fragments, they are likely to overlap. With this strategy Kohara et al., 1987 [204] constructed a physical map of the *E. coli* genome.

- *Restriction fragment sizes.* Restriction fragment sizes are obtained by cutting a clone with a restriction enzyme and measuring the sizes of the resulting fragments. This is simpler than constructing a restriction map. Although an unordered list of fragment sizes contains less information than an ordered list, it still provides an adequate fingerprint. This type of fingerprint was used by Olson et al., 1986 [257] in the yeast mapping project.

- *Hybridization data.* In this approach a clone is exposed to a number of probes, and it is determined which of these probes hybridize to the clone. Probes may be short random sequences or practically any previously identified piece of DNA. One particularly useful type of probe is the *Sequence Tag Site* (STS). STSs are extracted from the DNA strand itself, often from the endpoints of clones. Each STS is sufficiently long that it is unlikely to occur a second time on the DNA strand; thus, it identifies a unique site along the DNA strand. Using STS mapping, Chumakov et al., 1992 [68] and Foote et al., 1992 [111] constructed the first physical map of the human genome.

The STS technique leads to *mapping with unique probes.* If the probes are short random sequences, they may hybridize with DNA at many positions, thus leading to *mapping with non-unique probes.* For the map assembly problem with n clones and m probes, the experimental data is an $n \times m$ matrix $D = (d_{ij})$, where $d_{ij} = 1$ if clone C_i contains probe p_j, and $d_{ij} = 0$ otherwise (Figure 1.1). Note that the data does not indicate how many times a probe occurs on a given clone, nor does it give the order of the probes along the clones. A string S *covers* a clone

C if there exists a substring of S containing exactly the same set of probes as C (the order and multiplicities of probes in the substring are ignored). The string in Figure 1.1 covers each of nine clones corresponding to the hybridization matrix D. The Shortest Covering String Problem is to find a shortest string in the alphabet of probes that covers all clones.

The Shortest Covering String Problem is NP-complete. However, if the order of clones is fixed, it can be solved in polynomial time. Alizadeh et al., 1995 [3] suggested a local improvement strategy for the Shortest Covering String Problem that is based on finding optimal interleaving for a fixed clone order.

Given a set of intervals on the line, one can form the *interval graph* of the set by associating a vertex of the graph with each interval and joining two vertices by an edge if the corresponding intervals overlap (Figure 3.2). In the case of unique probes, every error-free hybridization matrix defines an interval graph on the vertex set of clones in which clones i and j are joined by an edge if they have a probe in common. The study of interval graphs was initiated by Benzer, who obtained data on the overlaps between pairs of fragments of bacteriophage T4 DNA. He was successful in arranging the overlap data in a way that implied the linear nature of the gene. Benzer's problem can be formulated as follows: Given information about whether or not two fragments of a genome overlap, is the data consistent with the hypothesis that the genes are arranged in linear order? This is equivalent to the question of whether the overlap graph is an interval graph.

Interval graphs are closely related to matrices with the consecutive ones property. A $(0, 1)$ matrix has the *consecutive ones* property if its columns can be permuted in such a way that 1s in each row occur in consecutive positions. In the case of unique probes, every error-free hybridization matrix has the consecutive ones property (the required permutation of columns corresponds to ordering probes from left to right). Given an arbitrary matrix, we are interested in an algorithm to test whether it has the consecutive ones property. Characterization of interval graphs and matrices with the consecutive ones property was given by Gilmore and Hoffman, 1964 [128] and Fulkerson and Gross, 1965 [114]. Booth and Leuker, 1976 [40] developed a data structure called a *PQ-tree* that leads to a linear-time algorithm for recognition of the consecutive ones property. Given an error-free hybridization matrix, the Booth-Leuker algorithm constructs a compact representation of all correct orderings of probes in linear time. However, their approach does not tolerate experimental errors. Alizadeh et al., 1995 [3] devised an alternative simple procedure for probe ordering. In the presence of experimental errors the procedure may fail, but in most cases it remains a good heuristic for the construction of probe ordering.

The most common type of hybridization error is a *false negative*, where the incidence between a probe and a clone occurs but is not observed. In addition, the hybridization data are subject to false positives and errors due to *clone abnormalities*. Different cloning technologies suffer from different clone abnormalities.

Early clone libraries were based on bacteriophage λ vectors and accommodated up to 25 Kb of DNA. *Cosmids* represent another vector that combines DNA sequences from *plasmids* and a region of λ genome. With cosmids, the largest size of a DNA insert is 45 Kb, and it would take 70,000 cosmids to cover the human genome. To reduce this number, *Yeast Artificial Chromosomes (YAC)* were developed to clone longer DNA fragments (up to 1,000 Kb). Although YACs have been used in many mapping projects, there are a number of clone abnormalities associated with them. The most common abnormality is *chimerism*. A chimeric clone consists of two distinct segments of DNA joined together by an error in a cloning process. It is estimated that 10–60% of clones in YAC libraries are chimeric. Chimeras may arise by *co-ligation* of different DNA fragments or by *recombination* of two DNA molecules. Another problem with YACs is that many clones are unstable and tend to delete internal regions. BAC (*Bacterial Artificial Chromosome*) cloning systems based on the *E. coli* genome significantly reduce the chimerism problem as compared to YACs.

A true ordering of probes corresponds to a permutation π of columns of the hybridization matrix D, which produces a matrix D_π. Each row of D_π corresponding to a normal clone contains one block of 1s, and each row corresponding to a chimeric clone contains two blocks of 1s. Define a *gap* as a block of zeroes in a row, flanked by ones. The number of gaps in D_π is equal to the number of chimeric clones. A false negative error typically splits a block of ones into two parts, thus creating a gap. A false positive error typically splits a gap into two gaps. Thus the number of gaps in D_π tends to be approximately equal to the number of chimeric clones plus the number of hybridization errors. This suggests a heuristic principle that a permutation of the columns that minimizes the number of gaps will correspond to a good probe ordering (Alizadeh et al., 1995 [3], Greenberg and Istrail, 1995 [137]). Minimizing the number of gaps can be cast as a *Traveling Salesman Problem* called the *Hamming Distance TSP* in which the cities are the columns of D together with an additional column of all zeroes, and the distance between two cities is the Hamming distance between the corresponding columns, i.e., the number of positions in which the two columns differ (Figure 3.1).

3.2 Mapping with Non-Unique Probes

Physical mapping using hybridization fingerprints with short probes was suggested by Poustka et al., 1986 [279]. The advantage of this approach is that probe generation is cheap and straightforward. However, the error rate of hybridization experiments with short probes is very high, and as a result, there were very few successful mapping projects with non-unique probes (Hoheisel et al., 1993 [165]). Hybridization errors and a lack of good algorithms for map construction were the major obstacles to using this method in large-scale mapping projects. Recently, Alizadeh

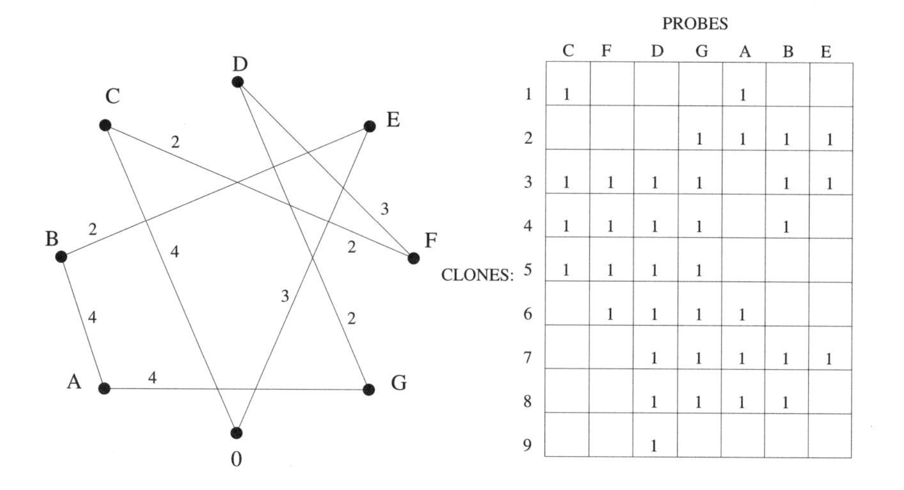

CLONES	C	F	D	G	A	B	E
1	1				1		
2				1	1	1	1
3	1	1	1	1		1	1
4	1	1	1	1		1	
5	1	1	1	1			
6		1	1	1	1		
7			1	1	1	1	1
8			1	1	1	1	
9			1				

Figure 3.1: Shortest cycle for the Hamming Distance TSP corresponding to the hybridization matrix in Figure 1.1 with a different order of columns. The shown minimum cycle defines the ordering of clones with the minimum number of gaps. Clones 1, 3, and 4 in this ordering are chimeric.

et al., 1995 [3] and Mayraz and Shamir, 1999 [234] designed algorithms that work well in the presence of hybridization errors.

A *placement* is an assignment of an interval on the line $[0, N]$ to each clone (the line $[0, N]$ corresponds to the entire DNA molecule). An *interleaving* is a specification of the linear ordering of the $2n$ endpoints of these n intervals. An interleaving may be viewed as an equivalence class of placements with a common topological structure. Given matrix D, the map assembly problem is to determine the most likely interleaving (Alizadeh et al., 1995 [3]):

Alizadeh et al., 1995 [3] gave a precise meaning to the notion "most likely" for the Lander and Waterman, 1988 [214] stochastic model and showed that it corresponds to the shortest covering string. In this model, clones of the same length are thrown at independent random positions along the DNA, and the probes along the DNA are positioned according to mutually independent Poisson processes. Alizadeh et al., 1995 [3] devised a maximum likelihood function for this model. The simplest approximation of this function is to find an interleaving that minimizes the number of occurrences of probes needed to explain the hybridization data, the Shortest Covering String Problem.

In the following discussion, we assume that no clone properly contains another. A string S *covers a permutation of clones* π if it covers all clones in the order given by π. A string $ABACBACDBCE$ covers the permutation $(3, 2, 4, 1)$ of clones

C_1, C_2, C_3, C_4 that hybridize with the following probes A, B, C, D, E:

$$C_1 - \{B, C, E\},\ C_2 - \{A, B, C, D\},\ C_3 - \{A, B, C\},\ C_4 - \{B, C, D\}.$$

Let $c(\pi)$ be the length of the shortest string covering π. Figure 1.1 presents a shortest covering string for the permutation $\pi = (1, 2, \dots, 9)$ of length $c(\pi) = 15$. Alizadeh et al., 1995 [3] devised a polynomial algorithm for finding $c(\pi)$ and a local improvement algorithm for approximating $\min_\pi c(\pi)$.

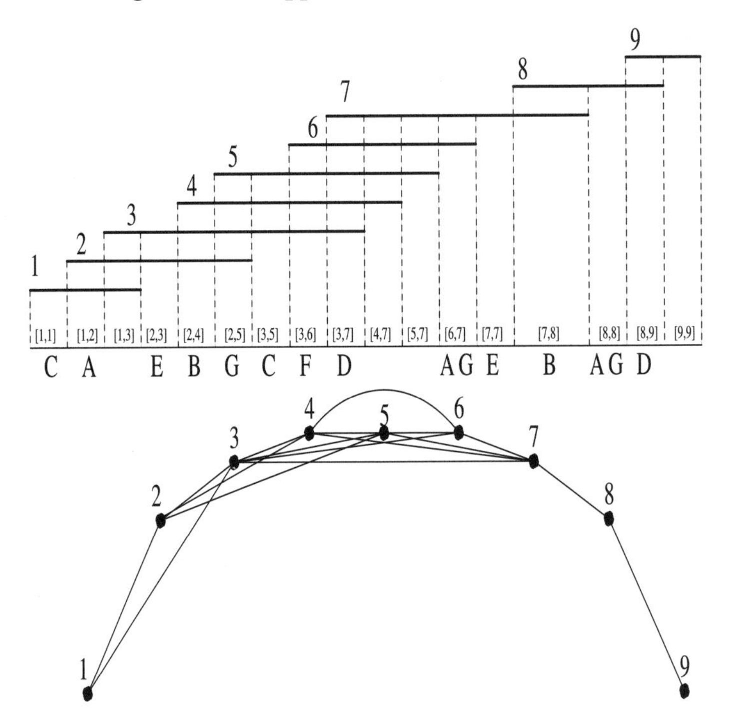

Figure 3.2: Atomic intervals.

The endpoints of clones partition the line into a number of *atomic intervals*, where an atomic interval is a maximal interval that contains no clone endpoints (Figure 3.2). Every atomic interval is contained in clones $i, \dots, j$, where i (j) represents the leftmost (rightmost) clones containing the atomic interval. We denote such atomic interval as $[i, j]$. Intervals $[i, j]$ and $[i', j']$ are *conflicting* if $i < i' < j' < j$. Note that every set of atomic intervals is *conflict-free*, i.e., contains no conflicting intervals (once again, we assume that no clone properly contains another).

A set of intervals $\mathcal{I} = ([i, j])$ is *consistent* if there exists an interleaving of clones with respect to which every interval in $\mathcal{I}$ is atomic.

Lemma 3.1 *A set of intervals is consistent if and only if it is conflict-free.*

Proof Induction on the number of clones. ■

For the data presented in Figure 1.1, the clones containing a probe A are organized into *two* runs: $[1, 2]$ and $[6, 8]$. Every probe A in a covering string defines an atomic interval $[i, j]$ in the corresponding interleaving of clones and therefore generates *one* run of length $j - i + 1$ in the A-column of the hybridization matrix. Therefore A appears at least twice in every covering string. It gives a lower bound of $2 + 2 + 2 + 1 + 2 + 1 + 1 = 11$ for the length of the shortest covering string. However, there is no covering string of length 11 since some of the runs in the hybridization matrix are conflicting (for example, a run $[3, 5]$ for C conflicts with a run $[2, 8]$ for G). Note that every covering string of length t defines a consistent set of t runs (intervals). On the other hand, every consistent set of t runs defines a covering string of length t. This observation and lemma 3.1 imply that the Shortest Covering String Problem can be reformulated as follows: sub-partition a set of runs for a hybridization matrix into a *conflict-free set with a minimum number of runs*. For example, a way to avoid a conflict between runs $[3, 5]$ and $[2, 8]$ is to sub-partition the run $[2, 8]$ into $[2, 5]$ and $[6, 8]$. This sub-partitioning still leaves the interval $[7, 7]$ for E in conflict with intervals $[6, 8]$ for A, $[3, 9]$ for D, and $[6, 8]$ for G. After sub-partitioning of these three intervals, we end up with the conflict-free set of $11 + 4 = 15$ runs (right matrix in Figure 1.1). These 15 runs generate the shortest covering string:

$$\underbrace{[1,1]}_{C}\underbrace{[1,2]}_{A}\underbrace{[2,3]}_{E}\underbrace{[2,4]}_{B}\underbrace{[2,5]}_{G}\underbrace{[3,5]}_{C}\underbrace{[3,6]}_{F}\underbrace{[3,7]}_{D}\underbrace{[6,7]}_{A}\underbrace{[6,7]}_{G}\underbrace{[7,7]}_{E}\underbrace{[7,8]}_{B}\underbrace{[8,8]}_{A}\underbrace{[8,8]}_{G}\underbrace{[8,9]}_{D}$$

Below we describe a greedy approach to sub-partitioning that produces the shortest covering string. Let $[i, j]$ and $[i', j']$ be two conflicting runs with $i < i' \leq j' < j$ and minimal j' among all conflicting runs. Clearly, $[i, j]$ has to be cut "before" j' in every sub-partitioning (i.e., in every subpartitioning of $[i, j]$ there exists an interval $[i, t]$ with $t \leq j'$). This observation suggests a sub-partitioning strategy based on cutting every interval "as late as possible" to avoid the conflicts. More precisely, for a given run $[i, j]$, let $\mathcal{I}$ be a set of intervals contained in $[i, j]$. Let t be a maximal number of mutually non-overlapping intervals from $\mathcal{I}$ such that $[i_1, j_1]$ is an interval with minimal j_1 in $\mathcal{I}$, $[i_2, j_2]$ is an interval with minimal j_2 among intervals with $i_2 > j_1$ in $\mathcal{I}$, ..., and $[i_t, j_t]$ is an interval with minimal j_t among intervals with $i_t > j_{t-1}$ in $\mathcal{I}$. At first glance, it looks like the partition of the run $[i, j]$ into $t+1$ runs $[i, j_1], [j_1+1, j_2], [j_2+1, j_3], \ldots, [j_{t-1}+1, j_t], [j_t+1, j]$ leads to the solution of the Shortest Covering String Problem. Unfortunately, that is not true, as an example in Figure 3.3 (proposed by Tao Jiang) illustrates. However, a simple modification of the last interval among these $t + 1$ intervals leads to the solution of the problem (see Figure 3.3 for an example). Note that the shortest covering string in this case has a clone with a double occurrence of the probe D.

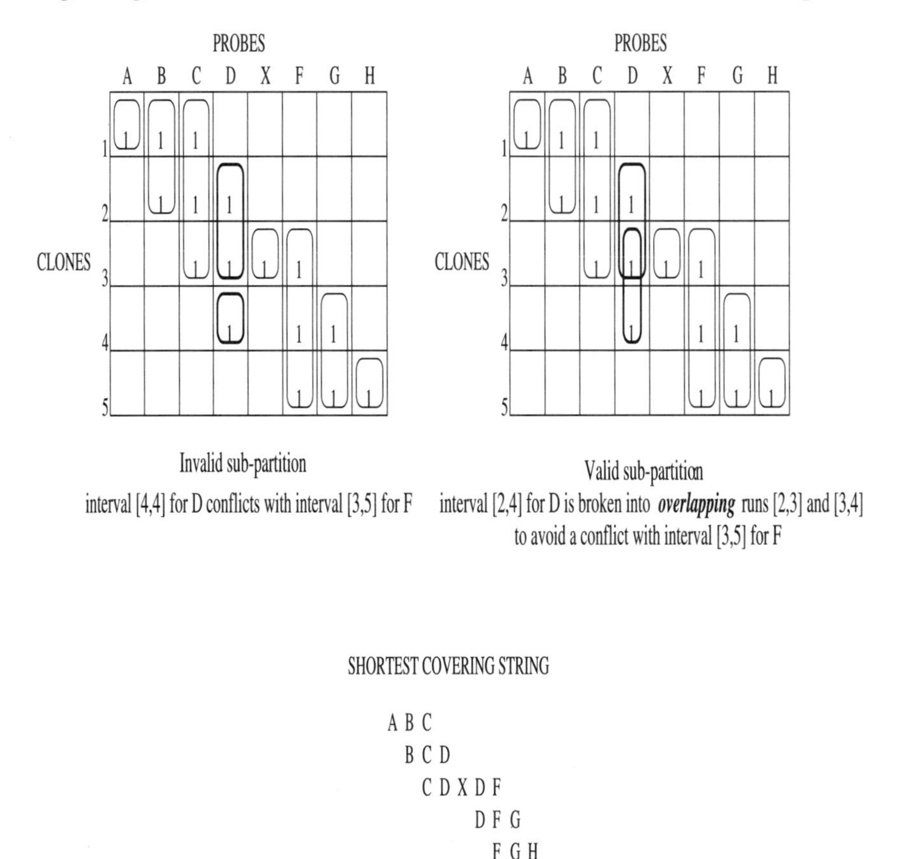

Figure 3.3: Shortest covering string may contain clones with probe repeats (such as CDXDF).

3.3 Mapping with Unique Probes

Let $\{1, \ldots, m\}$ be the set of probes, and let C_i be the set of clones incident with probe i. A clone X *contains* a clone Y if the set of probes incident with X (strictly) contains the set of probes incident with Y. In the following, we assume that the set of clones satisfies the following conditions:

- *Non-inclusion.* There is no clone containing another clone.

- *Connectedness.* For every partition of the set of probes into two non-empty sets A and B, there exist probes $i \in A$ and $j \in B$ such that $C_i \cap C_j$ is not empty.

- *Distinguishability.* $C_i \neq C_j$ for $i \neq j$.

There is no essential loss of generality in these assumptions, since any set of clones may be filtered by eliminating the clones containing inside other clones. The following lemma reformulates the consecutive ones property:

Lemma 3.2 *Let* $(1, \ldots, m)$ *be the correct ordering of the probes and* $1 \leq i < j < k \leq m$. *Then, in the error-free case,* $|C_i \cap C_j| \geq |C_i \cap C_k|$ *and* $|C_k \cap C_j| \geq |C_i \cap C_k|$.

Given a probe i, how can we find adjacent probes $i-1$ and $i+1$ in the correct ordering of probes? Lemma 3.2 suggests that these probes are among the probes that maximize $|C_i \cap C_k|$—i.e., either $|C_i \cap C_{i-1}| = \max_{k \neq i} |C_i \cap C_k|$ or $|C_i \cap C_{i+1}| = \max_{k \neq i} |C_i \cap C_k|$. If $\max_{k \neq i} |C_i \cap C_k|$ is achieved for only probe k, then either $k = i-1$ or $k = i+1$. If $\max_{k \neq i} |C_i \cap C_k|$ is achieved for a few probes, then it is easy to see that one of them with minimal $|C_k|$ corresponds to $i-1$ or $i+1$. These observations lead to an efficient algorithm for ordering the probes that starts from an arbitrary probe i and attempts to find an adjacent probe ($i-1$ or $i+1$) at the next step. At the $(k+1)$-th step, the algorithm attempts to extend an already found block of k consecutive probes to the right or to the left using lemma 3.2.

For each probe i, define the partial ordering $\succ_i$ of the set of probes $j \succ_i k$ if either

$$|C_i \cap C_j| > |C_i \cap C_k|$$

or

$$|C_i \cap C_j| = |C_i \cap C_k| \neq \emptyset \text{ and } |C_j| < |C_k|.$$

Clearly, if $j \succ_i k$, then probe k does not lie between probe i and probe j in the true ordering. Moreover, a maximal element in $\succ_i$ is either $i-1$ or $i+1$.

Let $N(i)$ be the set of probes that occur together with probe i on at least one clone.

Lemma 3.3 *The true ordering of probes is uniquely determined up to reversal by the requirement that the following properties hold for each* i:

- *The probes in* $N(i)$ *occur consecutively in the ordering as a block* $B(i)$.

- *Starting at probe* i *and moving to the right or to the left, the probes in* $B(i)$ *form a decreasing chain in the partial order* $\succ_i$.

The lemma motivates the following algorithm, which finds the true ordering of probes (Alizadeh et al., 1995 [4]). Throughout the algorithm the variable $\pi = \pi_{first} \ldots \pi_{last}$ denotes a sequence of consecutive probes in the true ordering. At the beginning, π is initialized to the sequence consisting of a single probe. Each step adjoins one element to π as follows:

- If every element in $N(\pi_{last})$ lies in π, replace the sequence π by its reversal.
- Choose a probe $k \notin \pi$ that is greatest in the ordering $\succ_{\pi_{last}}$ among probes in $N(\pi_{last})$. If $\pi_{last} \succ_k \pi_{first}$, append k to the end of π; otherwise, append k to the beginning of π.

The algorithm stops when π contains all probes. Lemma 3.2 implies that after every step, the probes in π form a consecutive block in the true ordering. Since a new element is adjoined to π at every step, the algorithm gives the true ordering π after m steps. Alizadeh et al., 1995 [4] report that a simple modification of this algorithm performs well even in the presence of hybridization errors and chimeric clones.

3.4 Interval Graphs

Golumbic, 1980 [132] is an excellent introduction to interval graphs, and our presentation follows that book.

A *triangulated graph* is a graph in which every simple cycle of length larger than 3 has a *chord*. The "house" graph in Figure 3.4 is not triangulated because it contains a chordless 4-cycle.

Lemma 3.4 *Every interval graph is triangulated.*

Not every triangulated graph is an interval graph; for example, the star graph in Figure 3.4 is not an interval graph (prove it!). An undirected graph has a *transitive orientation property* if each edge can be assigned a direction in such a way that the resulting directed graph $G(V, E)$ satisfies the following condition for every triple of vertices a, b, c: $(a, b) \in E$ and $(b, c) \in E$ imply $(a, c) \in E$. An undirected graph that is transitively orientable is called a *comparability graph* (Figure 3.4). The complement of the star graph in Figure 3.4 is not a comparability graph.

Lemma 3.5 *The complement of an interval graph is a comparability graph.*

Proof For non-overlapping intervals $[i, j]$ and $[i', j']$ with $j \leq i'$, direct the corresponding edge in the complement of the interval graph as $[i, j] \rightarrow [i', j']$. ■

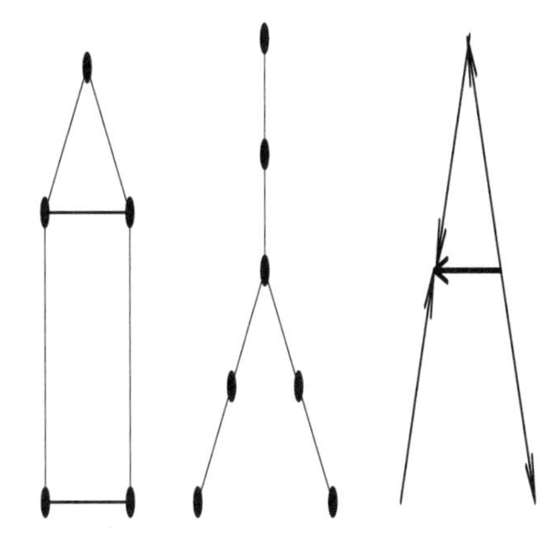

Figure 3.4: (i) The house graph is not an interval graph because it is not triangulated. (ii) The star graph is not an interval graph because its complement is not a comparability graph. (iii) Transitive orientation of a graph "A."

A *complete* graph is a graph in which every pair of vertices forms an edge. A clique of a graph G is a complete subgraph of G. A clique is maximal if it is not contained inside any other clique. The graph in Figure 3.2 has five maximal cliques: $\{1, 2, 3\}$, $\{2, 3, 4, 5\}$, $\{3, 4, 5, 6, 7\}$, $\{7, 8\}$ and $\{8, 9\}$. The following theorem establishes the characterization of interval graphs:

Theorem 3.1 *(Gilmore and Hoffman, 1964 [128]) Let G be an undirected graph. The following statements are equivalent.*
(i) G is an interval graph.
(ii) G is a triangulated graph and its complement is a comparability graph.
(iii) The maximal cliques of G can be linearly ordered such that for every vertex x of G, the maximal cliques containing x occur consecutively.

Proof (i) $\rightarrow$ (ii) follows from lemmas 3.4 and 3.5.

(ii) $\rightarrow$ (iii). Let $G(V, E)$ be a triangulated graph and let F be a transitive orientation of the complement $\overline{G}(V, \overline{E})$. Let A_1 and A_2 be maximal cliques of G. Clearly, there exists an edge in F with one endpoint in A_1 and another endpoint in A_2 (otherwise $A_1 \cup A_2$ would form a clique of G). It is easy to see that all edges of $\overline{E}$ connecting A_1 with A_2 have the same orientation. (Hint: if edges (a_1, a_2) and (a_1', a_2') connect A_1 and A_2 as in Figure 3.5(left), then at least one of the edges (a_1, a_2') and (a_1', a_2) belongs to $\overline{E}$. Which way is it oriented?) Order the maximal cliques according to the direction of edges in F: $A_1 < A_2$ if and only if there exists an edge of F connecting A_1 and A_2 and oriented toward A_2. We claim that this ordering is transitive and therefore defines a linear order of cliques.

Suppose that $A_1 < A_2$ and $A_2 < A_3$. Then there exist edges (a_1, a_2') and (a_2'', a_3) in F with $a_1 \in A_1$, $a_2', a_2'' \in A_2$, and $a_3 \in A_3$ (Figure 3.5(right)). If either $(a_2', a_3) \notin E$ or $(a_1, a_2'') \notin E$, then $(a_1, a_3) \in F$ and $A_1 < A_3$. Therefore, assume that the edges (a_1, a_2''), (a_2'', a_2'), and (a_2', a_3) are all in E. Since G contains no chordless 4-cycle, $(a_1, a_3) \notin E$, and the transitivity of F implies $(a_1, a_3) \in F$. Thus $A_1 < A_3$, proving that the cliques are arranged in linear order.

Let $A_1, \ldots, A_m$ be the linear ordering of maximal cliques. Suppose there exist cliques $A_i < A_j < A_k$ with $x \in A_i$, $x \notin A_j$ and $x \in A_k$. Since $x \notin A_j$, there is a vertex $y \in A_j$ such that $(x, y) \notin E$. But $A_i < A_j$ implies $(x, y) \in F$, whereas $A_j < A_k$ implies $(y, x) \in F$, a contradiction.

(iii) $\rightarrow$ (i). For each vertex x, let $I(x)$ denote the set of all maximal cliques of G that contain x. The sets $I(x)$, for $x \in V$, form the intervals of the interval graph G. ∎

Theorem 3.1 reduces the problem of recognition of an interval graph to the problems of recognition of triangulated and comparability graphs (Fulkerson and Gross, 1965 [114], Pnueli et al., 1971 [277]).

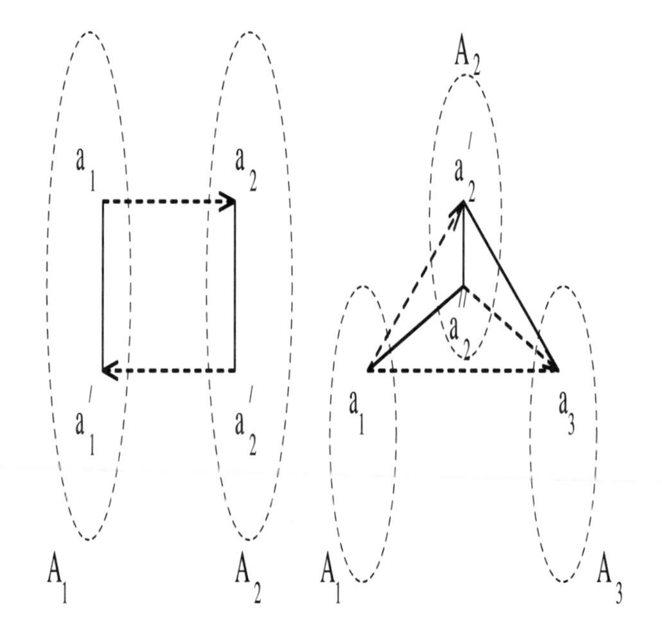

Figure 3.5: Proof of theorem 3.1.

3.5 Mapping with Restriction Fragment Fingerprints

The simplest case of mapping with restriction fragment fingerprints is *Single Complete Digest* (SCD) mapping (Olson et al., 1986 [257]). In this case the fingerprint of a clone is a *multiset* of the sizes of its restriction fragments in a digest by a restriction enzyme. An SCD map (Figure 3.6) is a placement of clones and restriction fragments consistent with the given SCD data (Gillett et al., 1995 [127]).

SCD Mapping Problem Find a most compact map (i.e., a map with the minimum number of restriction fragments) that is consistent with SCD data.

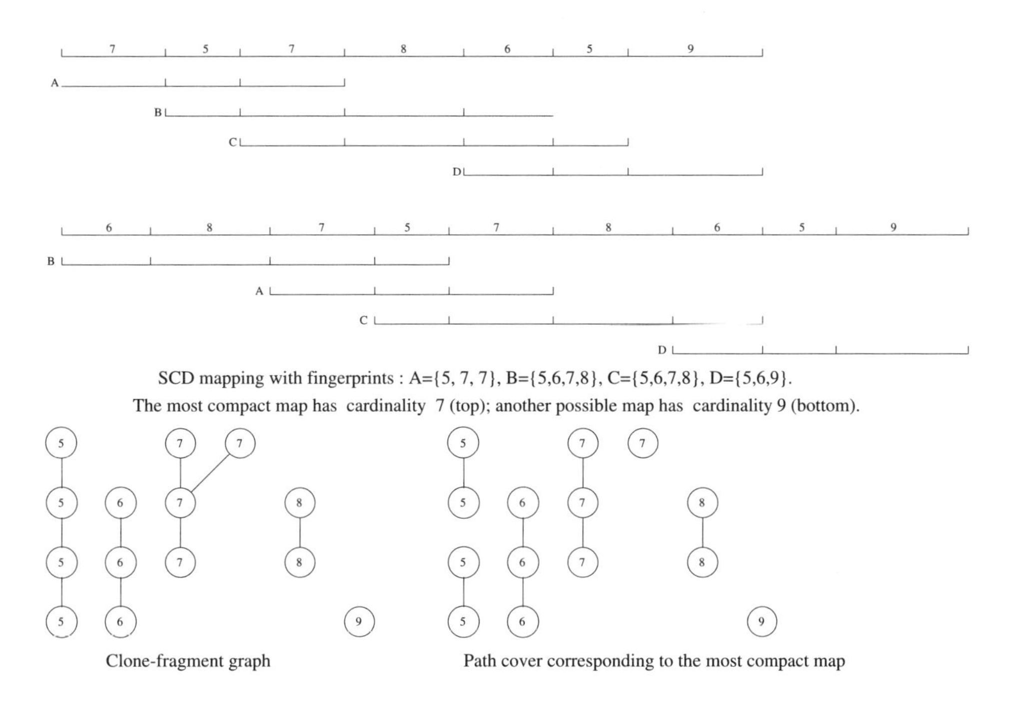

Figure 3.6: SCD mapping.

The problem of finding the most compact map is NP-hard. In practice the fingerprints of clones often provide strong statistical evidence that allows one to estimate the ordering of clones. Jiang and Karp, 1998 [179] studied the problem of finding a most compact map for clones with a given ordering.

Assume that no clone contains another clone, and assume that every clone starts and ends with the restriction enzyme site. Jiang and Karp, 1998 [179] formulated SCD mapping with known clone ordering as a *constrained path cover* problem on a special multistage graph. Let $\mathcal{S} = \{S_1, \ldots, S_n\}$ be an instance of SCD mapping, where each S_i is a multiset representing the fingerprint of the i-th clone in the clone ordering by the left endpoints. A *labeled multistage* graph G (called a *clone-fragment* graph) consists of n stages, with the i-th stage containing $|S_i|$ vertices. At stage i, G has a vertex for each element x of S_i (including duplicates), with label x. Two vertices are connected if they are at adjacent stages and have identical labels (Figure 3.6).

Intuitively, a path in G specifies a set of restriction fragments on consecutive clones that can be placed at the same location in a map. We denote a path running through stages $i, \ldots, j$ as simply $[i, j]$

A *path cover* is a collection of paths such that every vertex is contained in exactly one path. Clearly, any map for $\mathcal{S}$ corresponds to a path cover of G of the same cardinality (Figure 3.6 presents a path cover of cardinality 7). But the converse is not true; for example, the clone-fragment graph in Figure 3.6 has a path cover of cardinality 6 that does not correspond to any map. The reason is that some paths are in conflict (compare with lemma 3.1). Paths $[i, j]$ and $[i', j']$ are *conflicting* if $i < i' < j' < j$. A path cover is conflict-free if it has no conflicting paths.

A path cover of G is *consistent* if it corresponds to a map of $\mathcal{S}$ with the same cardinality. Similarly to lemma 3.1:

Lemma 3.6 *A path cover is consistent if and only if it is conflict-free.*

Hence, constructing a most compact map of $\mathcal{S}$ is equivalent to finding a smallest conflict-free path cover of G. Although the problem of finding the smallest conflict-free path cover of G is similar to the Shortest Covering String Problem, their computational complexities are very different. Jiang and Karp, 1998 [179] described a 2-approximation algorithm for SCD mapping with a given ordering of clones.

3.6 Some Other Problems and Approaches

3.6.1 Lander-Waterman statistics

When a physical mapping project starts, biologists have to decide how many clones they need to construct a map. A minimum requirement is that nearly all of the genome should be covered by clones. In one of the first mapping projects, Olson et al., 1986 [257] constructed a library containing $n = 4946$ clones. Each clone carried a DNA fragment of an average length of $L = 15,000$ nucleotides. This clone library represented a DNA molecule of length $G = 2 \cdot 10^7$; i.e., each nucleotide was represented in $\frac{NL}{G} \approx 4$ clones on average. The number $c = \frac{NL}{G}$ is called the *coverage* of a clone library. A typical library provides a coverage in the range of 5–10. When this project started, it was not clear what percentage of the yeast genome would be covered by 4946 clones.

Lander and Waterman, 1988 [214] studied a probabilistic model in which clones of the same length were thrown at independent random positions along the DNA. Define a gap as an interval that is not covered by any clones, and a contig as a maximal interval without gaps. Lander and Waterman, 1988 [214] demonstrated that the expected number of gaps is $\approx ne^{-c}$ and that the expected fraction of DNA not covered by any clone is $\approx e^{-c}$. The Olson et al., 1986 [257] mapping project resulted in 1422 contigs, which comes close to the Lander-Waterman estimate of 1457 contigs. Arratia et al., 1991 [11] further developed the Lander-Waterman statistics for the case of hybridization fingerprints.

3.6.2 Screening clone libraries

A naive approach to obtaining a hybridization matrix for n clones and m probes requires $n \times m$ hybridization experiments. The aim of *pooling* is to reduce the number of experiments that are needed to obtain the hybridization matrix. If the number of 1s in a hybridization matrix is small (as is the case for mapping with unique probes), then most experiments return negative results. It is clear that pooling clones in groups and testing probes against these pools may save experimental efforts.

Assume for simplicity that $\sqrt{n}$ is an integer, and view n clones as elements of a $\sqrt{n} \times \sqrt{n}$ array. Pool together clones corresponding to every row and every column of this array. The resulting set consists of $2\sqrt{n}$ pools, each pool containing $\sqrt{n}$ clones. This reduces experimental efforts from $n \times m$ to $2\sqrt{n} \times m$ hybridization but makes the computational problem of map assembly more difficult (Evans and Lewis, 1989 [98] and Barillot et al., 1991 [25]). Chumakov et al., 1992 [68] used this pooling strategy for the construction of the first human physical map. Computational analysis of pooling strategies is related to the following problem:

Group Testing Problem Find the distinguished members of a set of objects $\mathcal{L}$ by asking the minimum number of queries of the form "Does the set $Q \subset \mathcal{L}$ contain a distinguished object?"

Asking the query corresponds to testing the pool (set of clones) with a probe. For mapping applications, it is most cost-effective to ask queries in parallel (non-adaptive group testing). Bruno et al., 1995 [51] advocated a "random k-set design" pooling strategy that has advantages over the row-column pooling design. In the "random k-set design," each clone occurs in k pools, and all choices of the k pools are equally likely. See Knill et al., 1998 [200] for analysis of non-adaptive group testing in the presence of errors.

3.6.3 Radiation hybrid mapping

Radiation hybrid (RH) mapping (Cox et al., 1990 [77]) is an experimental strategy that uses random pooling that occurs in nature. RH mapping involves exposing human cells to radiation, which breaks each chromosome into random fragments. These fragments are then "rescued" by fusing the human cells with hamster cells that incorporate a random subset of the human DNA fragments into their chromosomes. One can think about human fragments as clones and about hamster cells as pools of these clones. The resulting hybrid cell can be grown into a cell line containing a pool of the fragment from the human genome. Figure 3.7 presents an RH mapping experiment with three hybrid cell lines and four probes (markers) that results in a 4×3 *hybrid screening matrix.* The Radiation Hybrid Mapping Problem is to reconstruct marker order from the hybrid screening matrix (Slonim et al., 1997 [317]).

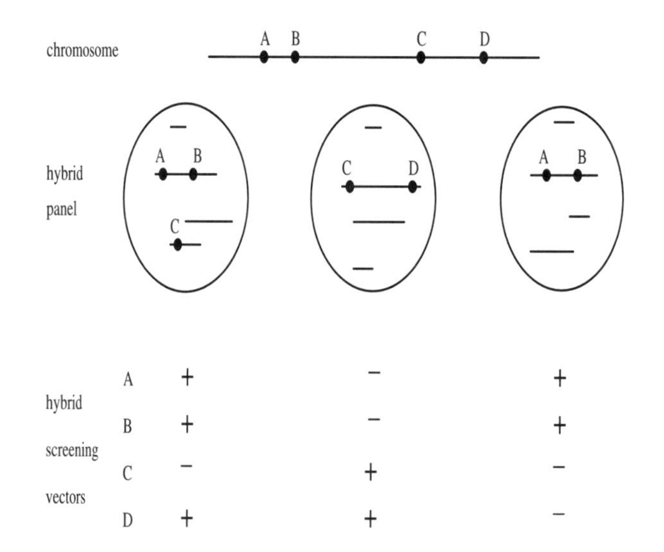

Figure 3.7: Hybrid screening matrix.

If the radiation-induced breaks occur uniformly at random across the chromosome, then breaks between closely located markers (like A and B) are rare. This implies that they are co-retained, i.e., that the hybrid cell contains either both of them or neither of them. This observation allows one to elucidate closely located markers and to use arguments similar to genetic mapping for probe ordering (Boehnke et al., 1991 [39], Lange et al., 1995 [215]). Slonim et al., 1997 [317] used the Hidden Markov Models approach for RH mapping and built the framework of an RH map by examining the triplets of markers. The most likely order of triples of markers can be estimated from the hybrid screening matrix. The example in Figure 3.7 may result in a subset of triples ABC, ABD, ACD and BCD, and the problem is to find a string (ABCD) that contains these triples as subsequences. The problem is complicated by the presence of incorrectly ordered triples and the unknown orientation of triples (the marker order (A-B-C or C-B-A) for triple ABC is not known).

Chapter 4

Sequencing

4.1 Introduction

At a time when the Human Genome Project is nearing completion, few people remember that before DNA sequencing even started, scientists routinely sequenced proteins. Frederick Sanger was awarded his first Nobel Prize for determining the amino acid sequence of insulin, the protein used to treat diabetes. Sequencing insulin in the late 1940s looked more challenging than sequencing an entire bacterial genome looks today. The computational aspects of protein sequencing at that time are very similar to the computational aspects of modern DNA sequencing. The difference is mainly in the length of the sequenced fragments. In the late 1940s biologists learned how to chop one amino acid at a time from the end of a protein and read this amino acid afterward. However, it worked only for a few amino acids from the end, since after 4–5 choppings the results were hard to interpret. To get around this problem Sanger digested insulin with proteases and sequenced each of the resulting fragments. He then used these overlapping fragments to reconstruct the entire sequence, exactly like in the DNA sequencing "break - read the fragments - assemble" method today:

$$\begin{array}{llllllll} Gly & Ile & Val & Glu & & & & \\ & Ile & Val & Glu & Gln & & & \\ & & & & Gln & Cys & Cys & Ala \\ Gly & Ile & Val & Glu & Gln & Cys & Cys & Ala \end{array}$$

Edman degradation that chopped and sequenced one terminal amino acid at a time became the dominant protein sequencing method for the next 20 years, and by the late 1960s protein sequencing machines were on the market.

Sanger's protein sequencing method influenced the work on RNA sequencing. The first RNA was sequenced in 1965 with the same "break - read the fragments -

assemble" approach. It took Holley and collaborators at Cornell University seven years to determine the sequence of 77 nucleotides in tRNA. For many years afterward DNA sequencing was done by first transcribing DNA to RNA and then sequencing RNA.

DNA sequencing methods were invented independently and simultaneously in Cambridge, England and Cambridge, Massachusetts. In 1974 Russian scientist Andrey Mirzabekov was visiting Walter Gilbert's lab at Harvard University and found a way to selectively break DNA at A and G. Later on, Maxam and Gilbert found a method to break DNA at every C and T. After measuring the lengths of the resulting fragments in four separate reactions, they were able to sequence DNA.

Sanger's method takes advantage of how cells make copies of DNA. Cells copy DNA letter by letter, adding one base at a time; Sanger realized that he could make a ladder of DNA fragments of different sizes if he "starved" the reaction of one of the four bases needed to make DNA. The cell would copy the DNA until the point where it ran out of one of the bases. For a sequence ACGTAAGCTA, starving at T would produce the fragments ACG and ACGTAAGC. By running four starvation experiments for A, T, G, and C and separating the resulting DNA fragments by length, one can read DNA. Later Sanger found chemicals that were inserted in place of A, T, G, or C but caused a growing DNA chain to end, preventing further growth. As a result, by 1977 two independent DNA sequencing techniques were developed (Sanger et al., 1977 [297] and Maxam and Gilbert, 1977 [233]) that culminated in sequencing of a 5,386-nucleotide virus and a Nobel Prize in 1980. Since then the amount of DNA sequencing data has been increasing exponentially, and in 1989 the Human Genome Project was launched. It is aimed at determining the approximately 100,000 human genes that comprise the entire 3 billion nucleotides of the human genome. Genetic texts are likely to become the main research tools of biologists over the next decades.

Similar to protein sequencing 50 years ago, modern biologists are able to sequence short (300- to 500-nucleotide) DNA fragments which have to be assembled into continuous genomes. The conventional shotgun sequencing starts with a sample of a large number of copies of a long DNA fragment (i.e., 50 Kb long). The sample is sonicated, randomly partitioning each fragment into *inserts*, and the inserts that are too small or too large are removed from further consideration. A sample of the inserts is then cloned into a vector with subsequent infection of a bacterial host with a single vector. After the bacterium reproduces, it creates a bacterial colony containing millions of copies of the vector and its associated insert. As a result, the cloning process results in the production of a sample of a given insert that is sequenced, typically by the Sanger et al., 1977 [297] method. Usually, only the first 300 to 500 nucleotides of the insert can be interpreted from this experiment. To assemble these fragments, the biologists have to solve a tricky puzzle, not unlike trying to assemble the text of a book from many slips of paper.

4.2 Overlap, Layout, and Consensus

After short DNA fragments (*reads*) are sequenced, we want to assemble them together and reconstruct the entire DNA sequence of the clone (*fragment assembly problem*). The Shortest Superstring Problem is an overly simplified abstraction that does not capture the real fragment assembly problem, since it assumes perfect data and may collapse DNA repeats. The human genome contains many repeats; for example, a 300 bp *Alu* sequence is repeated (with 5–15% variation) about a million times in the human genome. Fortunately, different copies of these repeats mutated differently in the course of evolution, and as a result, they are not exact repeats. This observation gives one a chance to assemble the sequence even in the presence of repeats.

Another complication is the unknown orientation of substrings in the fragment assembly problem. The DNA is double-stranded, and which of the two strands is actually represented by a subclone depends on the arbitrary way the insert is oriented in the vector. Thus it is unknown whether one should use a substring or its Watson-Crick complement in the reconstruction.

The earlier sequencing algorithms followed the greedy strategy and merged the strings together (starting with the strings with the strongest overlaps) until only one string remained. Most fragment assembly algorithms include the following three steps (Peltola et al., 1984 [262], Kececioglu and Myers, 1995 [195]):

- *Overlap.* Finding potentially overlapping fragments.
- *Layout.* Finding the order of fragments.
- *Consensus.* Deriving the DNA sequence from the layout.

The overlap problem is to find the best match between the suffix of one sequence and the prefix of another. If there were no sequencing errors, then we would simply find the longest suffix of one string that exactly matches the prefix of another string. However, sequencing errors force us to use a variation of the dynamic programming algorithm for sequence alignment. Since errors are small (1–3%), the common practice is to use filtration methods and to filter out pairs of fragments that do not share a significantly long common substring.

Constructing the layout is the hardest step in DNA sequence assembly. One can view a DNA sequence of a fragment as an extensive fingerprint of this fragment and use the computational ideas from map assembly. Many fragment assembly algorithms select a pair of fragments with the best overlap at every step. The score of overlap is either the similarity score or a more involved probabilistic score as in the popular Phrap program (Green, 1994 [136]). The selected pair of fragments with the best overlap score is checked for consistency, and if this check is accepted, the two fragments are merged. At the later stages of the algorithm the collections of fragments (*contigs*)–rather than individual fragments–are merged. The difficulty with the layout step is deciding whether two fragments with a good overlap really overlap (i.e., their differences are caused by sequencing errors) or represent a repeat in a genome (i.e., their differences are caused by mutations).

The simplest way to build the consensus is to report the most frequent character in the substring layout that is (implicitly) constructed after the layout step is completed. More sophisticated algorithms optimally align substrings in small windows along the layout. Alternatively, Phrap (Green, 1994 [136]) builds the consensus sequence as a mosaic of the best (in terms of some probabilistic scores) segments from the layout.

4.3 Double-Barreled Shotgun Sequencing

The shotgun sequencing approach described earlier was sequencing cosmid-size DNA fragments on the order of 50 Kb in the early 1990s. As a result, the Human Genome Project originally pursued the "clone-by-clone" strategy, which involved physical mapping, picking a minimal tiling set of clones that cover the genome, and shotgun-sequencing each of the clones in the tiling set.

DNA sequencing moved toward the entire 1800 Kb *H. Influenzae* bacterium genome in mid-1990s. Inspired by this breakthrough, Weber and Myers, 1997 [367] proposed using the shotgun approach to sequence the entire human genome. A year later, a company named Celera Genomics was formed with a goal of completing shotgun sequencing of the human genome by the year 2001.

However, fragment assembly becomes very difficult in large-scale sequencing projects, such as sequencing the entire fruit fly genome. In this case, the standard fragment assembly algorithms tend to collapse repeats that are located in different parts of the genome. Increasing the length of the read would solve the problem, but the sequencing technology has not significantly improved the read length yet. To get around this problem, biologists suggested a virtual increase in the length of the read by a factor of two by obtaining a pair of reads separated by a fixed-size gap. In this method, inserts of approximately the same length are selected, and *both* ends of the inserts are sequenced. This produces a pair of reads (called *mates*) in opposite orientation at a known approximate distance from each other.

Repeats represent a major challenge for whole-genome shotgun sequencing. Repeats occur at several scales. For example, in the human T-cell receptor locus, there are five closely located repeats of the trypsinogen gene, which is 4 Kb long and varies 3–5% between copies. These repeats are difficult to assemble since reads with unique portions outside the repeat cannot span them. The human genome contains an estimated 1 million Alu repeats (300 bp) and 200,000 $LINE$ repeats (1000 bp), not to mention that an estimated 25% of human genes are present in at least two copies.

The computational advantage of double-barreled DNA sequencing is that it is unlikely that both reads of the insert will lie in a large-scale DNA repeat (Roach et al., 1995 [286]). Thus the read in a unique portion of DNA determines which copy of a repeat its mate is in.

Double-barreled shotgun sequencing can be further empowered by using STS maps for fragment assembly. An STS is a unique 300-bp DNA fragment, and the available STS maps order tens of thousands of STSs along human chromosomes (Hudson et al., 1995 [173]). Since the approximate distance between consecutive STSs is known, the positions of STSs can serve as checkpoints in fragment assembly (Weber and Myers, 1997 [367], Myers, 1999 [247]).

The ambitious projects of genomic double-barreled shotgun sequencing raise a challenge of "finishing" sequencing in the areas that remain uncovered after the shotgun stage is completed. Any reasonable amount of shotgun sequencing will leave insufficiently sequenced areas. These areas will include both sequenced areas with low coverage and gaps of unknown length. Finishing is done by *walking* experiments that use primers from the already sequenced contigs to extend a sequenced region by one read length. Optimization of DNA sequencing requires a trade-off between the amount of shotgun sequencing and walking experiments. Another problem is where to walk in order to meet the minimum coverage criteria.

4.4 Some Other Problems and Approaches

4.4.1 Shortest Superstring Problem

Blum et al., 1994 [37] devised an algorithm that finds a superstring that is no more than three times the optimal length. Later Breslauer et al., 1997 [48] described an algorithm with a 2.596 approximation ratio. The question about the approximation ratio of a simple greedy algorithm that merges a pair of strings with maximum overlap remains open. No one has produced an example showing that this algorithm produces a superstring more than twice as long as an optimal one. Thus it is conjectured that the greedy algorithm is a 2-approximation.

4.4.2 Finishing phase of DNA sequencing

The minimum requirement for production of accurate DNA sequences is to have at least three clones covering every DNA position and to use majority rule in consensus construction. However, every genomic sequencing project is likely to result in DNA fragments with low clone coverage. Once the locations of these regions have been established, there is a need for a further finishing phase that is usually done by genome walking. The set of sequenced DNA reads defines a position coverage that is the number of reads covering position x in DNA. Given a requirement of k-fold redundant coverage, the goal of the finishing phase is to find a minimum set of walking-sequenced reads increasing the coverage to k for every position (Czabarka et al., 2000 [78]).

Chapter 5

DNA Arrays

5.1 Introduction

When the Human Genome Project started, DNA sequencing was a routine but time-consuming and hard-to-automate procedure. In 1988 four groups of biologists independently and simultaneously suggested a completely new sequencing technique called *Sequencing by Hybridization (SBH)*. SBH involves building a miniature *DNA array* (also known as DNA chips) containing thousands of short DNA fragments attached to a surface. Each of these short fragments reveals some information about an unknown DNA fragment, and all these pieces of information combined together are supposed to sequence DNA fragments. In 1988 almost nobody believed that the idea would work; both biochemical problems (synthesizing thousands of short DNA fragments on the array) and combinatorial problems (sequence reconstruction by array output) looked too complicated. Shortly after the first paper describing DNA arrays was published, *Science* magazine wrote that given the amount of work involved in synthesizing a DNA array, *"it would simply be substituting one horrendous task for another."* It was not a good prognosis: a major breakthrough in DNA array technology was made by Fodor et al., 1991 [110]. Their approach to array manufacturing is based upon light-directed polymer synthesis, which has many similarities to computer chip manufacturing. Using this technique, building an array with all 4^l probes of length l requires just $4 \cdot l$ separate reactions. With this method, in 1994, a California-based biotechnology company, Affymetrix, built the first 64-Kb DNA array. Shortly afterward building 1-Mb arrays became a routine, and the idea of DNA arrays was transformed from an intellectual game into one of the most promising new biotechnologies, one that revolutionized medical diagnostics and functional genomics.

Every probe p in a DNA array queries a target (unknown) DNA fragment by answering the question of whether p hybridizes with this fragment. Given an unknown DNA fragment, an array provides information about all strings of length l

contained in this fragment (l-tuple composition of the fragment) but does not provide information about the positions of these strings. Combinatorial algorithms are then used to reconstruct the sequence of the fragment from its l-tuple composition.

The SBH problem can be cast as a Hamiltonian path problem, i.e., the problem of finding a path in a graph that visits every vertex exactly once. The vertices of the graph correspond to l-tuples and the edges correspond to pairs of overlapping l-tuples. However, this reduction does not lead to an efficient SBH algorithm, since efficient algorithms for the Hamiltonian path problem are unknown. The SBH problem was actually solved long ago— centuries before the study of molecular biology even existed—by ... Leonhard Euler, the great 18th-century mathematician. Of course, he didn't know he was solving the SBH problem; he was just trying to solve – the "Seven Bridges of Konigsberg" puzzle. Konigsberg was located on a few islands connected by seven bridges (Figure 5.1), and Euler got interested in the problem of finding a path that traveled over each bridge exactly once. The solution of this problem heralded the birth of graph theory and, two centuries later, resulted in the solution of many combinatorial problems; SBH is one of them.

Although DNA arrays were originally proposed as an alternative to conventional gel-based DNA sequencing, *de novo* sequencing with DNA arrays is still a difficult problem. The primary obstacles in applications of DNA arrays for DNA sequencing are inaccuracies in interpreting hybridization data: distinguishing perfect matches from highly stable mismatches. This is a particularly difficult problem for short (8- to 10-nucleotide) probes used in *de novo* sequencing.

As a result, DNA arrays found more applications in *re-sequencing* and *mutation detection* (which can be done with longer probes) than in *de novo* sequencing. In this case the problem is to find the differences between the (known) wild type gene and a mutated (useful) gene. Relatively long (20-nucleotide) probes can be designed to reliably detect mutations and to bypass the still unsolved problem of distinguishing perfect matches from highly stable mismatches. These probes are usually variations of probes hybridizing with known DNA fragment. For example, each 20-tuple in DNA may correspond to four probes: the wild type, and three *middle-mutations* with a central position replaced by one of the alternative nucleotides. Lipshutz et al., 1995 [226] described such *tiling arrays* for detecting mutations in the HIV virus. Although no threshold of hybridization signal can distinguish between perfect and imperfect matches, the distinction between these signals is achieved if we compare the hybridization intensities of a probe with the hybridization intensities of its middle-mutations.

Tiling arrays can be used to explore the genetic diversity of entire populations. Analysis of mutations in human mitochondrial DNA has greatly influenced studies of human evolution and genetic diseases. These studies involve re-sequencing human mitochondrial DNA in many individuals to find the mutations. Because of the cost of conventional sequencing, most of the studies were limited to short hypervariable regions totaling ≈ 600 base pairs. Chee et al., 1996 [65] designed a tiling array for the entire human mitochondrial genome (16,569 base pairs) and were able to successfully detect three disease-causing mutations in a mtDNA sample from a patient with Leber's hereditary optic neuropathy. Prefabricated mtDNA arrays allow us to re-sequence DNA in many individuals and provide an efficient and fast technology for molecular evolution studies (Hacia et al., 1999 [148]). Other applications of DNA arrays include functional genomics (monitoring gene expression) and genetic mapping (Golub et al., 1999 [131] , Wang et al., 1998 [349]).

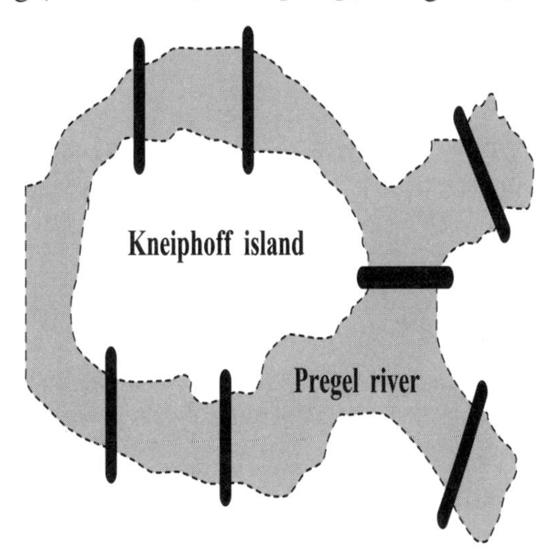

Figure 5.1: Bridges of Konigsberg.

5.2 Sequencing by Hybridization

DNA Arrays, or *DNA Chips*, were proposed simultaneously and independently by Bains and Smith, 1988 [23], Drmanac et al., 1989 [91], Lysov et al, 1988 [228], and Southern, 1988 [325]. The inventors of DNA arrays suggested using them for DNA sequencing, and the original name for this technology was *DNA Sequencing by Hybridization (SBH)*. SBH relies on the hybridization of an (unknown) DNA fragment with a large array of short probes. Given a short (8- to 30-nucleotide) synthetic fragment of DNA, called a *probe*, and a single-stranded target DNA fragment, the target will bind (*hybridize*) to the probe if there is a substring of the target that is the Watson-Crick *complement* of the probe (A is complementary to T and G is complementary to C). For example, a probe $ACCGTGGA$ will hybridize with a target CCC*TGGCACCT*A since it is complementary to the substring $TGGCACCT$ of the target. In this manner, probes can be used to test the unknown target DNA and determine its l-tuple composition. The simplest DNA array, $C(l)$, contains all probes of length l and works as follows (Figure 5.2):

- Attach all possible probes of length l (l=8 in the first SBH papers) to the surface, each probe at a distinct and known location. This set of probes is called the *DNA array*.

- Apply a solution containing a fluorescently labeled DNA fragment to the array.

- The DNA fragment hybridizes with those probes that are complementary to substrings of length l of the fragment.

- Detect probes hybridizing with the DNA fragment (using a spectroscopic detector) and obtain l-tuple composition of the DNA fragment.

- Apply a combinatorial algorithm to reconstruct the sequence of the DNA fragment from the l-tuple composition.

The "all 8-tuples" DNA array $C(8)$ requires synthesizing $4^8 = 65,536$ probes. It did look like a horrendous task in 1988 when DNA arrays were first proposed.

5.3 SBH and the Shortest Superstring Problem

SBH provides information about l-tuples present in DNA, but does not provide information about the positions of these l-tuples. Suppose we are given all substrings of length l of an unknown string (l-tuple composition or *spectrum* of a DNA fragment). How do we reconstruct the target DNA fragment from this data?

SBH may be considered a special case of the *Shortest Superstring Problem*. A *superstring* for a given set of strings $s_1, \ldots, s_m$ is a string that contains each s_i as a substring. Given a set of strings, finding the shortest superstring is NP-complete (Gallant et al., 1980 [117]).

Define $overlap(s_i, s_j)$ as the length of a maximal prefix of s_j that matches a suffix of s_i. The Shortest Superstring Problem can be cast as a Traveling Salesman Problem in a complete directed graph with m vertices corresponding to strings s_i and edges of length $-overlap(s_i, s_j)$. SBH corresponds to the special case when all substrings $s_1, \ldots, s_m$ have the same length. l-tuples p and q *overlap* if the last $l-1$ letters of p coincide with the first $l-1$ letters of q, i.e., $overlap(p, q) = l-1$. Given the spectrum S of a DNA fragment, construct the directed graph H with vertex set S and edge set $E = \{(p, q) : p \text{ and } q \text{ overlap}\}$. The graph H is formed by the lightest edges (of length $-(l-1)$) of the previously defined complete directed graph. There is a one-to-one correspondence between paths that visit each vertex of H at least once and DNA fragments with the spectrum S. The

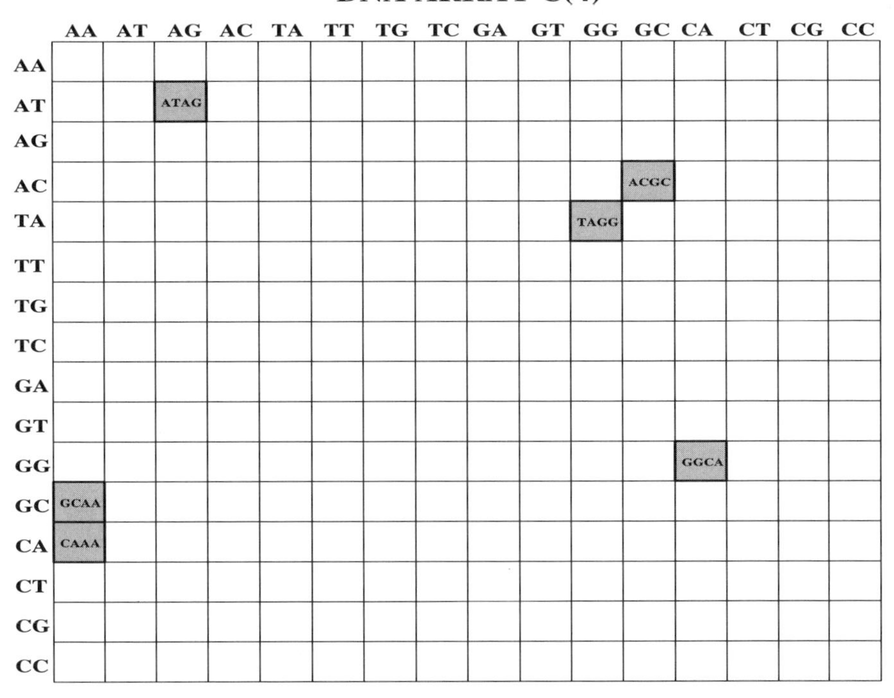

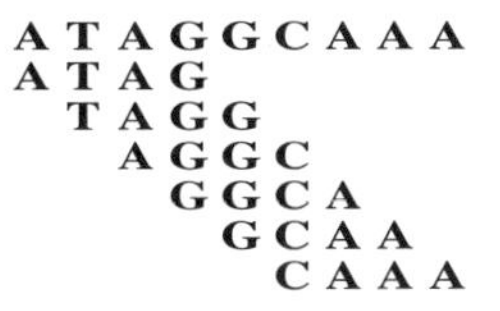

Figure 5.2: Hybridization of TATCCGTTT with DNA array $C(4)$.

spectrum presented in Figure 5.3 yields a *path-graph* H. In this case, the sequence reconstruction ATGCAGGTCC corresponds to the only path

$$ATG \to TGC \to GCA \to CAG \to AGG \to GGT \to GTC \to TCC$$

visiting all vertices of H. A path in a graph visiting every vertex exactly once is called a *Hamiltonian* path.

The spectrum shown in Figure 5.4 yields a more complicated graph with two Hamiltonian paths corresponding to two possible reconstructions.

Sequence reconstruction (Hamiltonian path approach)

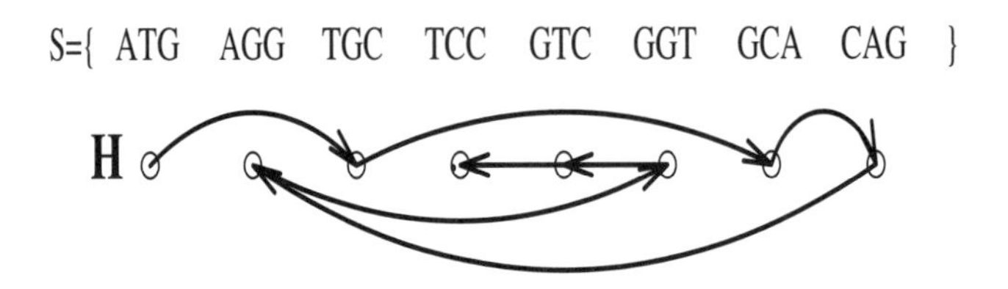

Path visiting ALL VERTICES corresponds to sequence reconstruction ATGCAGGTCC

Figure 5.3: SBH and the Hamiltonian path problem.

For larger DNA fragments, the overlap graphs become rather complicated and hard to analyze. The Hamiltonian path problem is NP-complete, and no efficient algorithm for this problem is known. As a result, the described approach is not practical for long DNA fragments.

5.4 SBH and the Eulerian Path Problem

As we have seen, the reduction of the SBH problem to the Hamiltonian path problem does not lead to efficient algorithms. Fortunately, a different reduction to the *Eulerian path* problem leads to a simple linear-time algorithm for sequence reconstruction. The idea of this approach is to construct a graph whose edges (rather than vertices, as in the previous construction) correspond to l-tuples and to find a path in this graph visiting every edge exactly once. In this approach a graph G on the set of all $(l-1)$-tuples is constructed (Pevzner, 1989 [264]). An $(l-1)$-tuple v is joined by a directed edge with an $(l-1)$-tuple w if the spectrum contains an l-tuple for which the first $l-1$ nucleotides coincide with v and the last $l-1$ nucleotides coincide with w (Figure 5.5). Each probe from the spectrum corresponds to a directed edge in G but not to a *vertex* as in H (compare Figures 5.3 and 5.5). Therefore, finding a DNA fragment containing all probes from the spectrum corresponds to finding a path visiting all *edges* of G, *Eulerian* path. Finding Eulerian paths is a well-known and simple problem.

A directed graph G is *Eulerian* if it contains a cycle that traverses every directed edge of G exactly once. A vertex v in a graph is *balanced* if the number of edges entering v equals the number of edges leaving v: $indegree(v) = outdegree(v)$. The following theorem gives a characterization of Eulerian graphs:

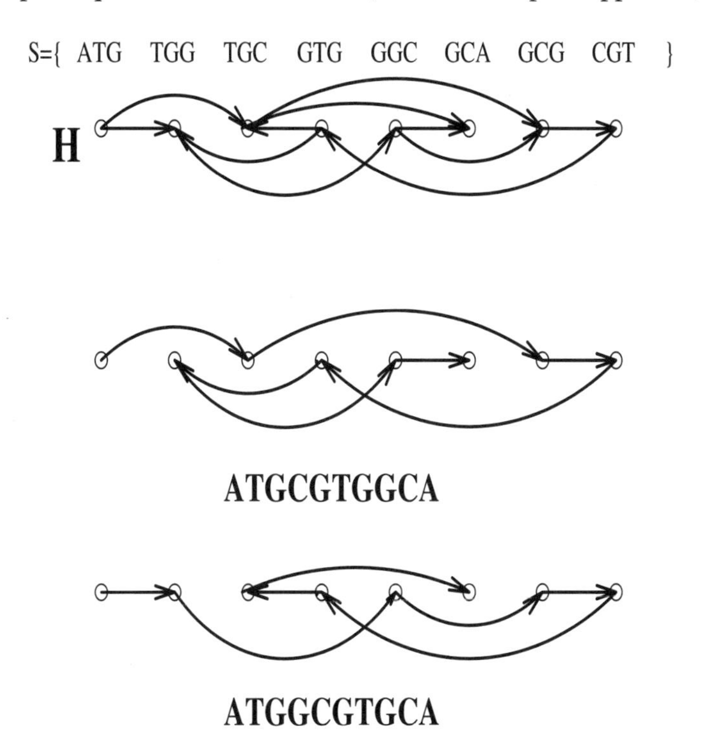

Figure 5.4: Spectrum S yields two possible reconstructions corresponding to distinct Hamiltonian paths.

Theorem 5.1 *A connected graph is Eulerian if and only if each of its vertices is balanced.*

The SBH problem is equivalent to finding an *Eulerian path* in a graph. A vertex v in a graph is called *semi-balanced* if $|indegree(v) - outdegree(v)| = 1$. The Eulerian path problem can be reduced to the Eulerian cycle problem by adding an edge between two semi-balanced vertices.

Theorem 5.2 *A connected graph has an Eulerian path if and only if it contains at most two semi-balanced vertices and all other vertices are balanced.*

To construct an Eulerian cycle, start from an arbitrary edge in G and form a "random" trail by extending the already existing trail with arbitrary new edges.

This procedure ends when all edges incident to a vertex in G are used in the trail. Since every vertex in G is balanced, every such trail starting at vertex v will end at v. With some luck the trail will be Eulerian, but this need not be so. If the trail is not Eulerian, it must contain a vertex w that still has a number of untraversed edges. Note that all vertices in the graph of untraversed edges are balanced and, therefore, there exists a random trail starting and ending at w and containing only untraversed edges. One can now enlarge the random trail as follows: insert a random trail of untraversed edges from w at some point in the random trail from v where w is reached. Repeating this will eventually yield an Eulerian cycle. This algorithm can be implemented in linear time (Fleischner, 1990 [108]).

Sequence reconstruction (Eulerian path approach)

S={ATG, TGG, TGC, GTG, GGC, GCA, GCG , CGT}

Vertices correspond to (l-1)-tuples.

Edges correspond to l-tuples from the spectrum

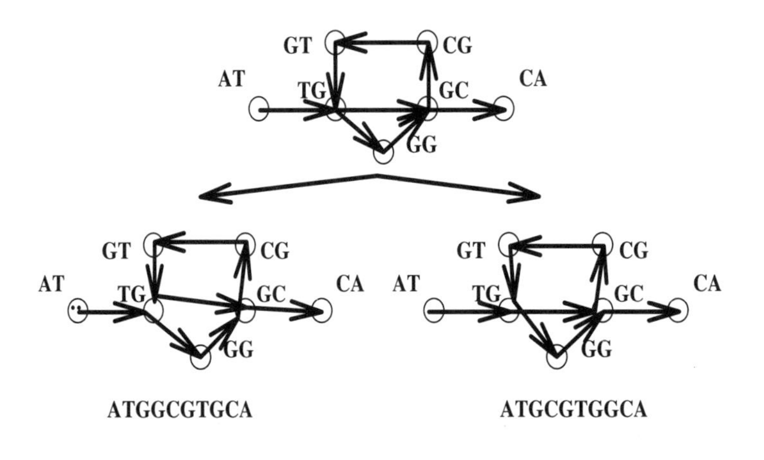

Paths visiting ALL EDGES correspond to sequence reconstructions

Figure 5.5: SBH and the Eulerian path problem.

The number of different sequence reconstructions in SBH is bounded by the number of Eulerian cycles (paths). The formula for the number of Eulerian cycles in a directed graph is known as the *BEST* theorem (Fleischner, 1990 [108]). Let G be a directed Eulerian graph with adjacency matrix $A = (a_{ij})$, where $a_{ij} = 1$ if there is an edge from vertex i into vertex j in G, and $a_{ij} = 0$ otherwise (Figure 5.6). Define a matrix M by replacing the i-th diagonal entry of $-A$ by $indegree(i)$ for all i. An i-cofactor of a matrix M is the determinant $det(M_i)$ of the matrix M_i, obtained from M by deleting its i-th row and i-th column. All cofactors of the matrix M have the same value, denoted $c(G)$.

Theorem 5.3 *The number of Eulerian cycles in an Eulerian graph $G(V, E)$ is*

$$c(G) \cdot \prod_{v \in V} (degree(v) - 1)!$$

There exists an easy way to check whether for a given spectrum there exists a unique sequence reconstruction. Decompose an Eulerian graph G into *simple* cycles $C_1, \ldots, C_t$, i.e., cycles without self-intersections. Each edge of G is used in exactly one cycle (vertices of G may be used in many cycles). For these cycles, define the intersection graph G_I on t vertices $C_1, \ldots, C_t$ where C_i and C_j are connected by k edges if and only if they have k vertices in common.

Theorem 5.4 *Graph G has only one Eulerian cycle if and only if the intersection graph G_I is a tree.*

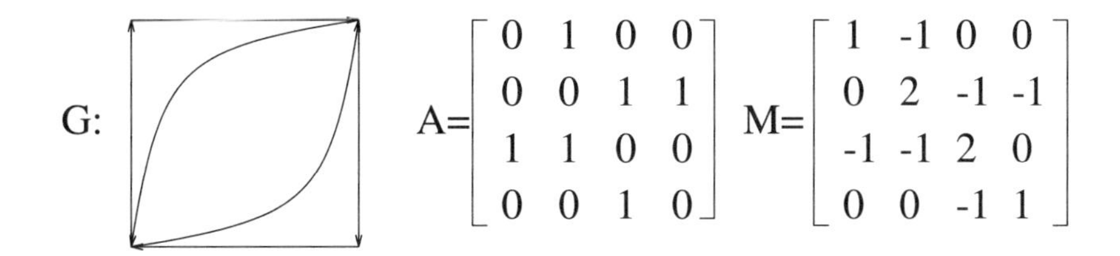

Figure 5.6: Each cofactor of the matrix M is 2. The number of Eulerian cycles in graph G is $2 \cdot 0! \cdot 1! \cdot 1! \cdot 0! = 2$.

The Eulerian path approach works in the case of error-free SBH experiments. However, even in this case, multiple Eulerian paths may exist, leading to multiple reconstructions. For real experiments with DNA arrays, the errors in the spectra make reconstruction even more difficult. In addition, repeats of length l complicate the analysis since it is hard to determine the multiplicity of l-tuples from hybridization intensities.

Figure 5.7 presents the same spectrum as in Figure 5.5 with two trinucleotides missing (false negative) and two possible reconstructions (only one of them is correct). The situation becomes even more complicated in the case of non-specific hybridization, when the spectrum contains l-tuples absent in a target DNA fragment (*false positive*). Several biochemical approaches to the elimination of non-specific hybridization in SBH experiments attempt to better discriminate between perfect and imperfect matches. However, DNA array hybridization data are still much more ambiguous than computer scientists and biologists would like them to be.

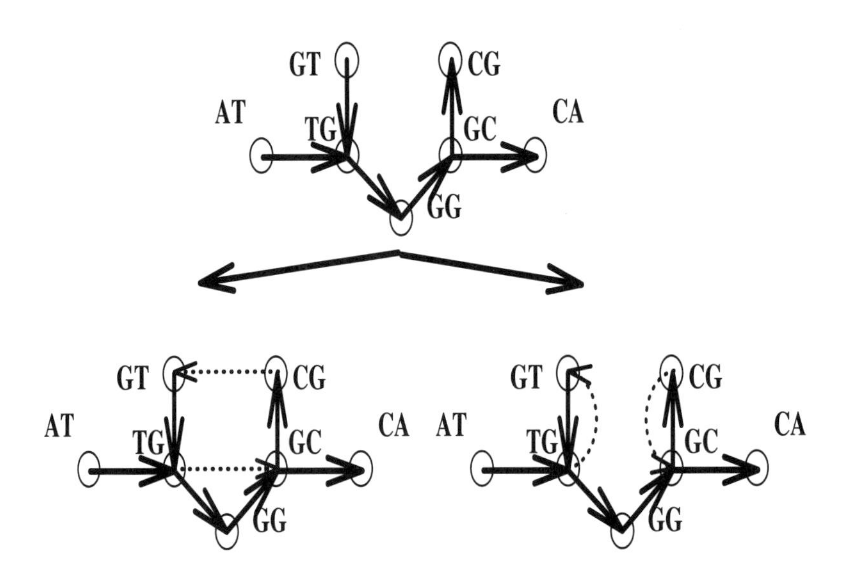

Figure 5.7: Sequence reconstruction in the case of missing l-tuples (false negative).

5.5 Probability of Unique Sequence Reconstruction

What is probability that a DNA fragment of length n can be uniquely reconstructed by a DNA array $C(l)$? Or, in other words, how big must l be to uniquely reconstruct a random sequence of length n from its l-tuple spectrum? For the sake of simplicity, assume that the letters of the DNA fragment are independent and identically distributed with probability $p = \frac{1}{4}$ for each of $A, T, G, andC$.

A crude heuristic is based on the observation that l-tuple repeats often lead to non-unique reconstructions. There are about $\binom{n}{2}p^l$ potential repeats of length

l corresponding to pairs of positions in the DNA fragment of length n. Solving $\binom{n}{2}p^l = 1$ yields a rough estimate $l = \log_{\frac{1}{p}}\binom{n}{2}$ for the minimal probe length needed to reliably reconstruct an n-letter sequence from its l-tuple spectrum. The maximal length of DNA fragment that can be reconstructed with a $C(l)$ array can be roughly estimated as $\sqrt{2 \cdot 4^l}$.

A more careful analysis reveals that l-tuple repeats are not the major cause of non-unique reconstructions (Dyer et al., 1994 [94]). The most likely cause is an interleaved pair of repeated $(l-1)$-tuples (repeats of AGTC and TTGG in Figure 5.9 interleave). Therefore, repeats of length $l-1$ should be considered. Another observation is that repeats may form clumps, and something like "maximal repeats" of length $l-1$ should be considered. The expected number of such repeats is about $\lambda \approx \binom{n}{2}(1-p)p^{l-1}$. A Poisson approximation takes the form $P\{k \text{ repeats}\} \approx e^{-\lambda}\frac{\lambda^k}{k!}$. Arratia et al., 1996 [12] showed that when there are k repeats, the probability of having no interleaved pair is $\approx \frac{k!2^k C_k}{(2k)!}$, where $C_k = \frac{1}{k+1}\binom{2k}{k}$ is the k-th Catalan number. Averaging over k reveals that the probability of unique reconstruction for an n-letter sequence from its l-tuple spectrum is approximately

$$\sum_{k \geq 0} e^{-\lambda} \frac{k!2^k C_k}{(2k)!} \frac{\lambda^k}{k!} = e^{-\lambda} \sum_{k \geq 0} \frac{(2\lambda)^k}{k!(k+1)!}$$

Arratia et al., 1996 [12] transformed these heuristic arguments into accurate estimates for resolving power of DNA arrays.

5.6 String Rearrangements

Figure 5.8 presents two DNA sequences with the same SBH spectrum. The graph G corresponding to the spectrum in Figure 5.8 contains a branching vertex TG. We don't know which 3-tuple (TGC or TGG) follows ATG in the original sequence and cannot distinguish between the correct and the incorrect reconstructions. An additional biochemical experiment (for example, hybridization of a target DNA fragment with ATGC) would find the correct reconstruction (the sequence at the top of Figure 5.8 contains ATGC, while the sequence at the bottom does not).

Additional biochemical experiments to resolve branchings in the reconstruction process were first proposed by Southern, 1988 [325] (using a longer probe for each branching vertex) and Khrapko et al., 1989 [197] (continuous stacking hybridization). *Continuous stacking hybridization* assumes an additional hybridization of short probes that continuously extends duplexes formed by the target DNA fragment and the probes from the array. In this approach, additional hybridization with an m-tuple on the array $C(l)$ provides information about some $(l+m)$-tuples

S={ATG, TGG, TGC, GTG, GGC, GCA, GCG, CGT}

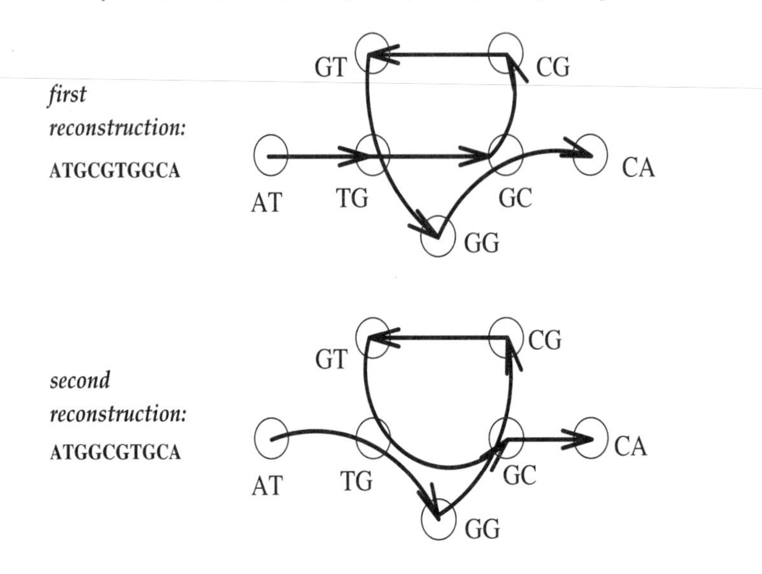

An additional experiment with ATGC reveals the correct reconstruction:

if ATGC hybridizes with a target -	**ATGCGTGGCA**
if ATGC does not hybridize with a target -	**ATGGCGTGCA**

Figure 5.8: Additional biochemical experiments resolve ambiguities in sequence reconstruction.

contained in the target DNA sequence. Computer simulations suggest that continuous stacking hybridization with only three additional experiments provides an unambiguous reconstruction of a 1000-bp fragment in 97% of cases.

To analyze additional biochemical experiments, one needs a *characterization* of all DNA sequences with the given spectrum. In the very first studies of DNA arrays, biologists described *string rearrangements* that do not change the spectrum of the strings (Drmanac et., 1989 [91]). However, the problem of characterizing *all* these rearrangements remained unsolved. Ukkonen, 1992 [340] conjectured that every two strings with the same l-tuple spectrum can be transformed into each other by simple transformations called transpositions and rotations (Figure 5.9). In the following, by "a string written in l-tuple notation," we mean a string written as a sequence of its l-tuples. For example, the 2-tuple notation for the string ATGGGC is AT TG GG GG GC. Ukkonen, 1992 [340] conjectured that any two strings with the same l-tuple composition can be transformed into each other by

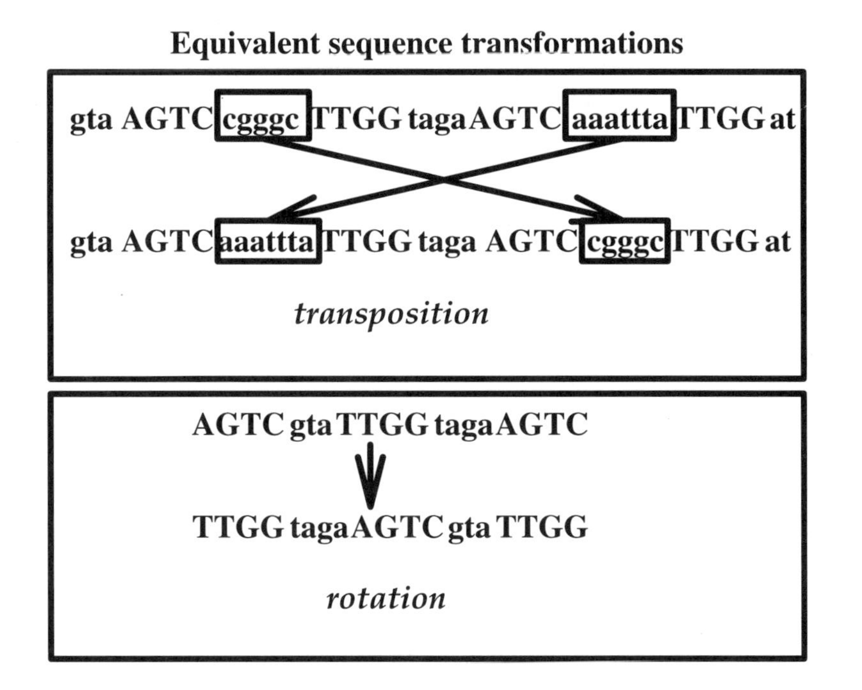

Figure 5.9: Ukkonen's transformations ($l = 5$).

simple operations called transpositions and rotations.

If a string

$$s = \ldots x \underline{\ldots} z \ldots x \underbrace{\ldots} z \ldots$$

(written in $(l-1)$-tuple notation) contains interleaving pairs of $(l-1)$-tuples x and z, then the string $\ldots x \underbrace{\ldots} z \ldots x \underline{\ldots} z \ldots$ where $\underline{\ldots}$ and $\underbrace{\ldots}$ change places is called a *transposition* of s. If $s = \ldots x \underline{\ldots} x \underbrace{\ldots} x \ldots$, where x is an $(l-1)$-tuple, we also call $\ldots x \underbrace{\ldots} x \underline{\ldots} x \ldots$ a transposition of s. If a string

$$s = x \underline{\ldots} z \underbrace{\ldots} x$$

(written in $(l-1)$-tuple notation) starts and ends with the same $(l-1)$-tuple x, then the string $z \underbrace{\ldots} x \underline{\ldots} z$ is called a *rotation* of s.

Clearly, transpositions and rotations do not change l-tuple composition.

Theorem 5.5 *(Pevzner, 1995 [267]) Every two strings with the same l-tuple composition can be transformed into each other by transpositions and rotations.*

Proof Strings with a given l-tuple composition correspond to Eulerian paths in the directed graph G (Figure 5.10). Graph G either is Eulerian or contains an Eulerian path. Notice that only if G is Eulerian does there exist a rotation of the corresponding string. In this case the rotations correspond simply to a choice of the initial vertex of an Eulerian cycle.

Substitute each directed edge $a = (v, w)$ in G with two (undirected) edges, (v, a) colored white and (a, w) colored black (Figure 5.10). Obviously each alternating path in the new graph G^* is a directed path in G and vice versa. According to theorem 2.2, order exchanges and reflections generate all Eulerian paths in G^* and, therefore, all strings with a given l-tuple composition. Notice that Ukkonen's transpositions correspond to order exchanges in G^*. On the other hand, every cycle in G^* is even; therefore there are no order reflections in G^*. This proves Ukkonen's conjecture that transpositions and rotations generate all strings with a given l-tuple composition. ■

5.7 2-optimal Eulerian Cycles

A *2-path* (v_1, v_2, v_3) in a directed graph is a path consisting of two directed edges (v_1, v_2) and (v_2, v_3). Every Eulerian cycle in a graph $G(V, E)$ defines a set of $|E|$ 2-paths corresponding to every pair of consecutive edges in this cycle. A set of $|E|$ 2-paths is *valid* if every edge of E appears in this set twice: once as the beginning of a 2-path and once as the end of a 2-path. Every valid set of 2-paths defines a decomposition of E into cycles. A valid set of 2-paths defining an Eulerian cycle is called an *Euler set* of 2-paths.

Let G be a directed Eulerian graph with a *weight* $w(v_1v_2v_3)$ assigned to every 2-path $(v_1v_2v_3)$ in G. The *weight of an Eulerian cycle* $C = v_1 \ldots v_m$ is the sum of the weights of its 2-paths $w(C) = \sum_{i=1}^{m-1} w(v_iv_{i+1}v_{i+2})$ (we assume that $v_m = v_1$ and $v_{m+1} = v_2$). *The 2-optimal Eulerian cycle problem* is to find an Eulerian cycle of maximal weight (Gusfield et al., 1998 [147]).

Let

$$C = \ldots \bar{v}x\underline{\bar{w}\ldots\bar{\bar{v}}}y\bar{\bar{w}}\ldots\hat{v}x\underbrace{\hat{w}\ldots\hat{\hat{v}}}y\hat{\hat{w}}\ldots$$

be an Eulerian cycle in G traversing vertices x and y in an interleaving order. An order exchange transforming C into

$$C' = \ldots \bar{v}x\underbrace{\hat{w}\ldots\hat{\hat{v}}}y\bar{\bar{w}}\ldots\hat{v}x\underline{\bar{w}\ldots\bar{\bar{v}}}y\hat{\hat{w}}\ldots$$

is called an *Euler switch* of C at x, y. Every two Eulerian cycles in a directed graph can be transformed into each other by means of a sequence of Euler switches (Pevzner, 1995 [267]).

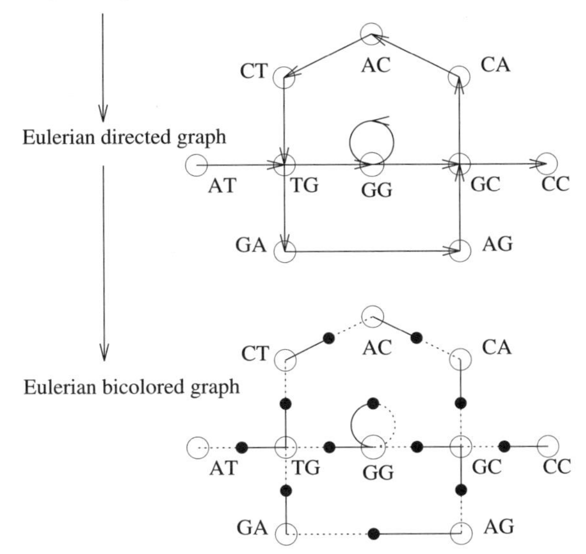

Two alternating Eulerian paths correspond to two sequence reconstructions

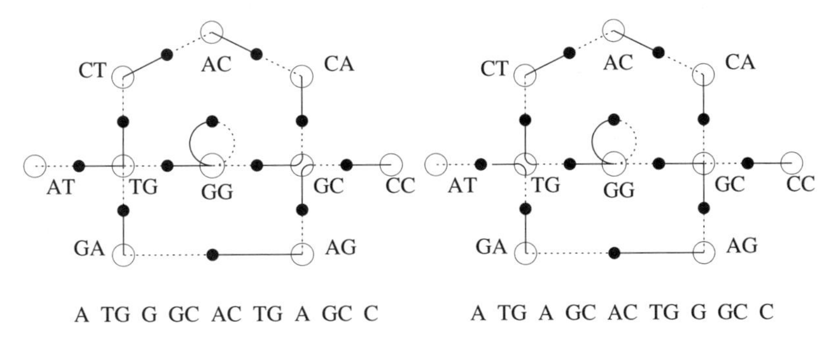

Figure 5.10: Ukkonen's conjecture. The sequence (Eulerian path) on the right is obtained from the sequence on the left by a transposition defined by the interleaving pairs of dinucleotides TG and GC.

We denote an Euler switch of C at x, y as $s = s(x, y)$ and write $C' = s \cdot C$. An Euler switch s of C at x, y transforms four 2-paths in C

$$\overline{v}x\overline{w}, \overline{\overline{v}}y\overline{\overline{w}}, \hat{v}x\hat{w}, \hat{\hat{v}}y\hat{\hat{w}}$$

into four 2-paths in C'

$$\overline{v}x\hat{w}, \hat{\hat{v}}y\overline{\overline{w}}, \hat{v}x\overline{w}, \overline{\overline{v}}y\hat{\hat{w}}.$$

Denote

$$\Delta_x(s) = w(\overline{v}x\hat{w}) + w(\hat{v}x\overline{w}) - w(\overline{v}x\overline{w}) - w(\hat{v}x\hat{w})$$

$$\Delta_y(s) = w(\hat{\hat{v}}y\overline{\overline{w}}) + w(\overline{\overline{v}}y\hat{\hat{w}}) - w(\overline{\overline{v}}y\overline{\overline{w}}) - w(\hat{\hat{v}}y\hat{\hat{w}})$$

and $\Delta(s) = \Delta_x(s) + \Delta_y(s)$. $\Delta(s)$ is the change in the weight of the Eulerian cycle C' as compared to C. An Euler switch s is *increasing* for cycle C if $w(s \cdot C) > w(C)$.

Consider a valid set of 2-paths S containing four 2-paths:

$$\overline{v}x\overline{w}, \overline{\overline{v}}y\overline{\overline{w}}, \hat{v}x\hat{w}, \hat{\hat{v}}y\hat{\hat{w}}.$$

A *switch* of this set at vertices x, y is a new valid set of 2-paths that contains instead of the above four 2-paths the following ones:

$$\overline{v}x\hat{w}, \hat{\hat{v}}y\overline{\overline{w}}, \hat{v}x\overline{w}, \overline{\overline{v}}y\hat{\hat{w}}.$$

If S is an Euler set of 2-paths then a switch of S is called non-Euler if it transforms S into a non-Euler set of 2-paths. For example, a switch at x, y of a set of 2-paths corresponding to the Euler cycle $\ldots\overline{v}x\underline{\overline{w}\ldots\hat{v}}x\hat{w}\ldots\overline{\overline{v}}y\underbrace{\overline{\overline{w}}\ldots\hat{\hat{v}}}y\hat{\hat{w}}\ldots$ is a non-Euler switch.

Gusfield et al., 1998 [147] studied the 2-optimal Eulerian cycle problem in 2-in-2-out directed graphs, i.e., in graphs with maximal degree 2.

Theorem 5.6 *Every two Eulerian cycles in a 2-in-2-out graph can be transformed into each other by means of a sequence of Euler switches $s_1, \ldots s_t$ such that $\Delta(s_1) \geq \Delta(s_2) \geq \ldots \geq \Delta(s_t)$.*

Proof Let $s_1, \ldots, s_t$ be a sequence of Euler switches transforming C into C^* such that the vector $(\Delta(s_1), \Delta(s_2), \ldots, \Delta(s_t))$ is *lexicographically maximal* among all sequences of Euler switches transforming C into C^*. If this vector does not satisfy the condition $\Delta(s_1) \geq \Delta(s_2) \geq \ldots \geq \Delta(s_t)$, then $\Delta(s_i) < \Delta(s_{i+1})$ for $1 \leq i \leq t-1$. Let $C' = s_{i-1} \cdots s_1 C$. If the switch s_{i+1} is an Euler switch in C' (i.e., change of the system of 2-paths in C' imposed by s_{i+1} defines an Eulerian cycle), then $s_1, \ldots, s_{i-1}, s_{i+1}$ is lexicographically larger than $s_1, \ldots, s_{i-1}, s_i$, a contradiction. (Pevzner, 1995 [267] implies that there exists a transformation of C into C^* with the prefix $s_1, \ldots s_{i-1}, s_{i+1}$.) Therefore, the switch s_{i+1} at x, y is not an Euler switch in C'. This implies that the occurrences of x and y in C' are non-interleaving: $C' = \ldots x \ldots x \ldots y \ldots y \ldots$. On the other hand, since the switch s_i at z, u is an Euler switch in C', the occurrences of z and u in C'

are interleaving: $C' = \ldots z \ldots u \ldots z \ldots u \ldots$. We need to consider all kinds of interleaving arrangements of the four vertices x, y, z, and u in C'. The condition that s_{i+1} is an Euler switch in $s_i \cdot C'$ (i.e., s_i moves vertices x and y in such a way that they become interleaving in $s_i \cdot C'$) makes most of these cases invalid (in fact, this is the key idea of the proof). It is easy to see that all valid arrangements of x, y, z, and u in C' are "equivalent" to the arrangement

$$C' = \ldots z \underline{\ldots x \ldots} u \ldots x \ldots y \ldots z \underbrace{\ldots}_{} u \ldots y \ldots$$

In this case s_i "inserts" x between the occurrences of y, thus making x and y interleaving:

$$s_i \cdot C' = \ldots z \underbrace{\ldots}_{} u \ldots x \ldots y \ldots z \underline{\ldots x \ldots} u \ldots y \ldots$$

Note that in this case the switches s' at z, y and s'' at x, u are also Euler switches in C'. Moreover, $\Delta(s_i) + \Delta(s_{i+1}) = \Delta(s') + \Delta(s'')$ Without loss of generality, assume that $\Delta(s') \geq \Delta(s'')$. Then

$$\Delta(s') \geq \frac{\Delta(s') + \Delta(s'')}{2} = \frac{\Delta(s_i) + \Delta(s_{i+1})}{2} > \Delta(s_i)$$

Therefore, the vector $(\Delta(s_1), \ldots, \Delta(s_{i-1}), \Delta(s'))$ is lexicographically larger than the vector $(\Delta(s_1), \ldots, \Delta(s_{i-1}), \Delta(s_i))$, a contradiction. ∎

Theorem 5.6 implies the following:

Theorem 5.7 *(Gusfield et al., 1998 [147]) If C is an Eulerian cycle that is not 2-optimal then there exists an increasing Euler switch of C.*

The proof of theorem 5.6 also implies that in the case when all weights of 2-paths are distinct, a greedy algorithm choosing at every step the switch with maximal weight leads to a 2-optimal Eulerian cycle.

5.8 Positional Sequencing by Hybridization

Although DNA arrays were originally proposed for DNA sequencing, the resolving power of DNA arrays is rather low. With 64-Kb arrays, only DNA fragments as long as 200 bp can be reconstructed in a single SBH experiment. To improve the resolving power of SBH, Broude et al., 1994 [49] suggested *Positional SBH* (PSBH), allowing (with additional experimental work) measurement of approximate positions of every l-tuple in a target DNA fragment. Although this makes the reconstruction less ambiguous, polynomial algorithms for PSBH sequence reconstruction are unknown. PSBH can be reduced to computing Eulerian path with an additional restriction that the position of any edge in the computed Eulerian path should be in the range of positions associated with the edge.

PSBH motivates the *Positional Eulerian Path* Problem. The input to the Positional Eulerian Path Problem is an Eulerian graph $G(V, E)$ in which every edge is associated with a range of integers and the problem is to find an Eulerian path $e_1, \ldots, e_{|E|}$ in G such that the range of e_i contains i:

Positional Eulerian Path Problem Given a directed multigraph $G(V, E)$ and an interval $I_e = \{l_e, h_e\}$, $l_e \leq h_e$ associated with every edge $e \in E$, find an Eulerian path $e_1, \ldots, e_{|E|}$ in G such that $l_{e_i} \leq i \leq h_{e_i}$ for $1 \leq i \leq |E|$.

Hannenhalli et al., 1996 [156] showed that the Positional Eulerian Path Problem is NP-complete. On a positive note, they presented polynomial algorithms to solve a special case of PSBH, where the range of the allowed positions for any edge is bounded by a constant (accurate experimental measurements of positions in PSBH).

Steven Skiena proposed a slightly different formulation of the PSBH problem that models the experimental data more adequately. For this new formulation, the 2-optimal Eulerian path algorithm described in the previous section provides a solution.

Experimental PSBH data provide information about the approximate positions of l-tuples, but usually *do not provide* information about the error range. As a result, instead of an interval $\{l_e, h_e\}$ associated with each edge, we know only the approximate position m_e associated with each edge. In a different and more adequate formulation of the PSBH problem, the goal is to minimize $\sum_{i=0}^{|E|} |(m_{e_{i+1}} - m_{e_i})|$, where $m_{e_0} = 0$ and $m_{e_{|E|+1}} = |E| + 1$. For every pair of consecutive edges e, e' in G, define the weight of the corresponding 2-path as $|m_e - m_{e'}|$. The PSBH problem is to find a 2-optimal Eulerian path of minimal weight.

5.9 Design of DNA Arrays

Since the number of features on a DNA array is fixed, we are interested in the design of a smallest set of probes sufficient to sequence almost all strings of a given length. Suppose that the number of positions m on a DNA array is fixed and the problem is to devise m probes to provide the maximum *resolving power* of a DNA array. It turns out that the *uniform* arrays $C(l)$ containing all l-tuples are rather redundant. Pevzner and Lipshutz, 1994 [271] introduced new arrays with improved resolving power as compared to uniform arrays. These arrays are based on the idea of *pooling* probes into *multiprobes*: synthesizing a set of diverse probes at every address on the array. A multiprobe is a *set* of probes located at a single address of an array. A DNA fragment hybridizes with a multiprobe if it hybridizes with *at least one* probe in the multiprobe. For example, WWS is a multiprobe consisting of eight probes:

$$AAG, AAC, ATG, ATC, TAG, TAC, TTA, TTC$$

(W stands for A or T, while S stands for G or C). RYR is a multiprobe consisting of eight probes:

$$ATA, ATG, ACA, ACG, GTA, GTC, GCA, GCG$$

(R stands for *purines* A or G, while Y stands for *pyrimidines* T or C). TXG is a multiprobe consisting of four probes:

$$TAG, TTG, TGG, TCG$$

(X stands for *any* nucleotide–A, T, G, or C).

An array is now defined as a set of multiprobes C, each multiprobe being a set of probes. The *memory* $|C|$ of the array is the number of multiprobes in C. Each DNA sequence F defines a subset of array C consisting of the multiprobes hybridizing with F (*spectrum* of F in C):

$$F_C = \{p \in C : \text{ multiprobe } p \text{ contains a probe occurring in sequence } \overline{F}\}$$

($\overline{F}$ stands for the sequence complementary to F).

The *binary array* $C_{bin}(l)$ is the array with memory $|C_{bin}(l))| = 2 \cdot 2^l \cdot 4$ composed of all multiprobes of two kinds:

$$\underbrace{\{W,S\},\{W,S\},\ldots,\{W,S\}}_{l},\{N\} \text{ and } \underbrace{\{R,Y\},\{R,Y\},\ldots,\{R,Y\}}_{l},\{N\}$$

where N is a specified nucleotide A,T, G, or C. Each probe is a mixture of 2^l probes of length $l + 1$. For example, the array $C_{bin}(1)$ consists of the 16 multiprobes

$$WA, WC, WG, WT, SA, SC, SG, ST, RA, RC, RG, RT, YA, YC, YG, YT.$$

Each multiprobe is a pool of two dinucleotides (Figure 5.11).

The *gapped array* $C_{gap}(l)$ (Pevzner et al., 1991 [272]) is the array with memory $|C_{gap}(l)| = 2 \cdot 4^l$ composed of all multiprobes of two kinds:

$$N_1N_2\ldots N_l \text{ and } N_1N_2\ldots N_{l-1}\underbrace{XX\ldots X}_{l-1}N_l$$

Binary arrays

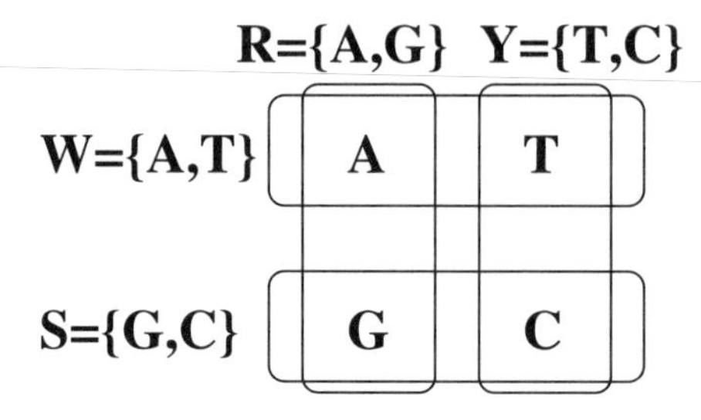

every string of length l in {W,S} or {R,Y} alphabet
is a pool of 2^l strings in {A,T,G,C} alphabet

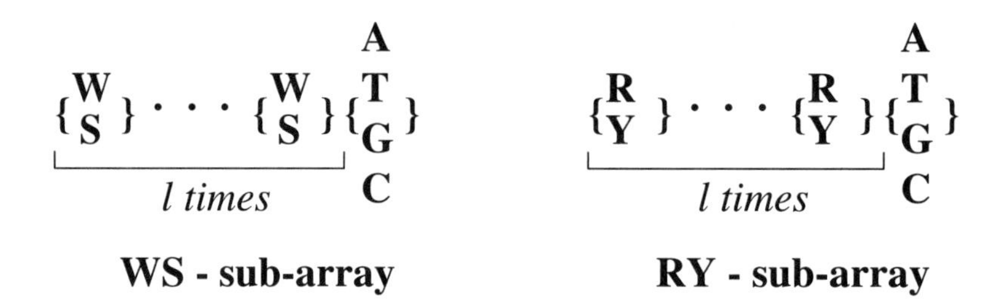

Figure 5.11: Binary arrays.

where N_i is a specified base and X is an unspecified base. Each multiprobe of the first kind consists of the only probe of length l; each multiprobe of the second kind consists of 4^{l-1} probes of length $2l - 1$.

The *alternating array* $C_{alt}(l)$ is the array with memory $|C_{alt}(l)| = 2 \cdot 4^l$ composed of all multiprobes of two kinds:

$$N_1XN_2X \dots N_{l-2}XN_{l-1}XN_l \text{ and } N_1XN_2X \dots N_{l-2}XN_{l-1}N_l.$$

Each multiprobe of the first kind consists of 4^{k-1} probes of length $2k - 1$, while each multiprobe of the second kind consists of 4^{k-2} probes of length $2k - 2$.

5.10 Resolving Power of DNA Arrays

Consider the sequence $F = X_1 \dots X_{m-1}X_mX_{m+1} \dots X_n$ and assume that its prefix $F_m = X_1X_2 \dots X_m$ has already been reconstructed. We will estimate the probability of unambiguously extending the prefix to the right by one nucleotide.

Since F_m is a possible reconstruction of the first m nucleotides of F,

$$(F_m)_C \subset F_C.$$

There are four ways of extending F_m: F_mA, F_mT, F_mG, and F_mC. We define an extension of F_m by a nucleotide N as a *possible extension* if

$$(F_mN)_C \subset F_C. \tag{5.1}$$

We call the sequence F *extendable* after m with respect to array C if the condition (5.1) holds for exactly one of the four nucleotides; otherwise F is called *non-extendable*.

Define $\epsilon(C, F, m)$ as

$$\epsilon(C, F, m) = \begin{cases} 0, & \text{if } F \text{ is extendable after } m \text{ with respect to the array } C \\ 1, & \text{otherwise} \end{cases}$$

The *branching probability* $p(C, n, m)$ is the probability that a random n-sequence is non-extendable after the m-th nucleotide upon reconstruction with array C, i.e.,

$$p(C, n, m) = \frac{1}{4^n} \sum_F \epsilon(C, F, m)$$

where the sum is taken over all 4^n sequences F of length n.

Let us fix m and denote $p(C,n)$=$p(C,n,m)$. Obviously, $p(C,n)$ is an increasing function of n. For a given probability p, the maximum n satisfying the condition $p(C,n) \leq p$ is the maximal sequence length $n_{max}(C,p)$ allowing an *unambiguous extending* with branching probability below p. We demonstrate that for uniform arrays, $n_{max}(C,p) \approx \frac{1}{3} \cdot |C| \cdot p$, while for binary arrays $n_{max}(C,p) \approx \frac{1}{\sqrt{12}} \cdot |C| \cdot \sqrt{p}$. Therefore, the new arrays provide a factor $\sqrt{\frac{3}{4p}}$ improvement in the maximal sequence length as compared to the uniform arrays. For $p = 0.01$ and binary 64-Kb arrays, $n_{max} \approx 1800$, versus $n_{max} \approx 210$ for uniform arrays.

5.11 Multiprobe Arrays versus Uniform Arrays

Consider the sequence $F = X_1 \dots X_{m-1} X_m X_{m+1} \dots X_n$ and assume that its prefix $F_m = X_1 \dots X_m$ has already been reconstructed. Denote the last $(l-1)$-tuple in F_m as $V = X_{m-l+2} \dots X_m$. For the sake of simplicity we suppose $l \leq m$ and $l \ll n \ll 4^l = |C(l)|$.

The sequence F is non-extendable after m using $C(l)$ if the spectrum $F_{C(l)}$ contains a VY_i l-tuple (here, Y_i is an arbitrary nucleotide different from X_{m+1}).

Therefore, $p(C(l), n) = 1 - P\{VY_1, VY_2, VY_3 \notin F_{C(l)}\}$. Assume that the probability of finding each of the 4^l l-tuples at a given position of F is equal to $\frac{1}{4^l}$. The probability that the spectrum of F does not contain VY_1 can be roughly estimated as $(1 - \frac{1}{4^l})^{n-l+1}$ (ignoring potential self-overlaps (Pevzner et al., 1989 [269]) and marginal effects). The probability that the spectrum of F contains neither VY_1, nor VY_2, nor VY_3 can be estimated as $((1 - \frac{1}{4^l})^{(n-l+1)})^3$. Therefore

$$p(C(l), n) = 1 - P\{VY_1, VY_2, VY_3 \notin F_{C(l)}\} \approx$$

$$1 - ((1 - \frac{1}{4^l})^{n-l+1})^3 \approx \frac{3n}{4^l} = \frac{3n}{|C(l)|}. \tag{5.2}$$

Therefore

$$n_{max}(C(l), p) \approx \frac{1}{3} \cdot |C(l)|p.$$

Now we estimate $p(C_{bin}(l), n)$ for $n \ll ||C_{bin}(l)||$. Denote the last l-tuple in the prefix F_m as $V = X_{m-l+1}, \ldots, X_m$, and let V_{WS} and V_{RY} be V written in the $\{W, S\}$ and $\{R, Y\}$ alphabets, respectively. In this case the ambiguity in reconstruction arises when the spectrum $F_{C_{bin}(l)}$ contains *both* a $V_{WS}Y_i$ multiprobe and a $V_{RY}Y_i$ multiprobe for $Y_i \neq X_{m+1}$. Assume that the probability of finding a multiprobe from $C_{bin}(l)$ at a given position of F is $\frac{1}{4 \cdot 2^l}$ and ignore self-overlaps. Then the probability that the spectrum of F does not contain $V_{WS}Y_1$ can be roughly estimated as $(1 - \frac{1}{4 \cdot 2^l})^{n-l}$. Therefore the probability that the spectrum of F contains both $V_{WS}Y_1$ and $V_{RY}Y_1$ is

$$(1 - (1 - \frac{1}{4 \cdot 2^l})^{n-l}) \cdot (1 - (1 - \frac{1}{4 \cdot 2^l})^{n-l}) \approx \frac{n^2}{4 \cdot 2^l \cdot 4 \cdot 2^l}.$$

Similarly to (5.2) we derive the following:

$$p(C_{bin}(l), n) = P\{V_{WS}Y_i \in F_{C_{bin}(l)} \text{ and } V_{RY}Y_i \in F_{C_{bin}(l)}\} \approx$$

$$1 - (1 - \frac{n^2}{4 \cdot 2^l \cdot 4 \cdot 2^l})^3 \approx 3 \cdot \frac{n}{4 \cdot 2^l} \cdot \frac{n}{4 \cdot 2^l} = \frac{12n^2}{|C_{bin}(l)|^2}.$$

Therefore, for $C_{bin}(l)$,

$$n_{max}(C_{bin}(l), p) \approx \frac{1}{\sqrt{12}} \cdot |C_{bin}(l)|\sqrt{p}.$$

Next we estimate the branching probability of gapped arrays $C_{gap}(l)$. Let $m \geq 2l - 1$ and $n \ll ||C_{gap}(l)||$. Denote $U = X_{m-2l+4} \ldots X_{m-l+2}$. In this case the ambiguity arises when the spectrum $F_{C_{gap}(l)}$ contains *both* a VY_i l-multiprobe and a $U\underbrace{XX\ldots X}_{l-1}Y_i$ $(2l-1)$-multiprobe (here, $Y_i \neq X_{m+1}$). Assume that the probability of finding each multiprobe from $C_{gap}(l)$ at a given position of F is $\frac{1}{4^l}$ and ignore self-overlaps. Then the probability that the spectrum of F does not contain VY_1 can be roughly estimated as $(1-\frac{1}{4^l})^{n-l}$. The probability that the spectrum of F does not contain $U\underbrace{XX\ldots X}_{l-1}Y_1$ can be roughly estimated as $(1-\frac{1}{4^l})^{n-(2l-1)+1}$. Therefore, the probability that the spectrum of F contains both VY_1 and $U\underbrace{XX\ldots X}_{l-1}Y_1$ is

$$(1-(1-\frac{1}{4^l})^{n-l})\cdot(1-(1-\frac{1}{4^l})^{n-2l+2}) \approx \frac{n^2}{4^l\cdot 4^l}.$$

Similarly to (5.2), we derive the following:

$$p(C_{gap}(l),n) = P\{VY_i \in F_{C_{gap}(l)} \text{ and } UY_i \in S(C_{gap}(l),F)\} \approx$$

$$1-(1-\frac{n^2}{4^l\cdot 4^l})^3 \approx 3\cdot\frac{n}{4^l}\cdot\frac{n}{4^l} = \frac{12n^2}{|C_{gap}(l)|^2}.$$

Similar arguments demonstrate that

$$n_{max}(C_{alt}(l),p) \approx \frac{1}{\sqrt{12}}\cdot|C_{alt}(l)|\sqrt{p}.$$

5.12 Manufacture of DNA Arrays

DNA arrays can be manufactured with the use of $VLSIPS$, *very large scale immobilized polymer synthesis* (Fodor et al., 1991 [110], Fodor et al., 1993 [109]). In $VLSIPS$, probes are grown one nucleotide at a time through a photolithographic process consisting of a series of chemical steps. Every nucleotide carries a photolabile protection group protecting the probe from further growing. This group can be removed by illuminating the probe with light. In each chemical step, a predefined region of the array is illuminated, thus removing a photolabile protecting group from that region and "activating" it for further nucleotide growth. The entire array is then exposed to a particular nucleotide (which bears its own photolabile protecting group), but reactions only occur in the activated region. Each time the process is repeated, a new region is activated and a single nucleotide is appended to each probe in that region. By appending nucleotides to the proper regions in the appropriate sequence, it is possible to grow a complete set of l-length probes in as few as $4 \cdot l$ steps. The light-directed synthesis allows random access to all positions of the array and can be used to make arrays with any probes at any site.

The proper regions are activated by illuminating the array through a series of masks, like those in Figure 5.12. Black areas of a mask correspond to the region of the array to be illuminated, and white areas correspond to the region to be shadowed. Unfortunately, because of diffraction, internal reflection, and scattering, points that are close to the border between an illuminated region and a shadowed region are often subject to unintended illumination. In such a region, it is uncertain whether a nucleotide will be appended or not. This uncertainty gives rise to probes with unknown sequences and unknown lengths, that may hybridize to a target DNA strand, thus complicating interpretation of the experimental data. Methods are being sought to minimize the lengths of these borders so that the level of uncertainty is reduced. Two-dimensional Gray codes, described below, are optimal $VLSIPS$ masks that minimize the overall border length of all masks.

Figure 5.12 presents two $C(3)$ arrays with different arrangements of 3-tuples and masks for synthesizing the first nucleotide A (only probes with first nucleotide A are shown). The *border length* of the mask at the bottom of Figure 5.12 is significantly smaller than the border length of the mask at the top of Figure 5.12. We are trying to arrange the probes on the array $C(l)$ in such a way that the overall border length of *all* $4 \times l$ masks is minimal. For two l-tuples x and y, let $\delta(x, y)$ be the number of positions in which x and y differ. Clearly, the overall border length of all masks equals $2 \sum \delta(x, y)$, where the sum is taken over all pairs of neighboring probes on the array. This observation establishes the connection between minimization of border length and Gray codes.

An l-bit *Gray* code is defined as a permutation of the binary numbers between 0 and $2^l - 1$ such that neighboring numbers have exactly one differing bit, as do the first and last numbers. For example, the 4-bit *binary reflected Gray code* is shown below:

```
0  1  2  3  4  5  6  7  8  9 10 11 12 13 14 15
-----------------------------------------------
0  0  0  0  0  0  0  0  1  1  1  1  1  1  1  1
0  0  0  0  1  1  1  1  1  1  1  1  0  0  0  0
0  0  1  1  1  1  0  0  0  0  1  1  1  1  0  0
0  1  1  0  0  1  1  0  0  1  1  0  0  1  1  0
```

This Gray code can be generated recursively, starting with the 1-bit Gray code

$$G_1 = \{0, 1\},$$

Masks for VLSIPS

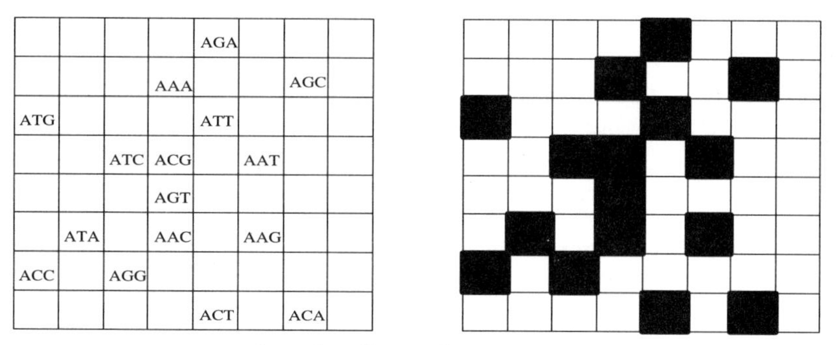

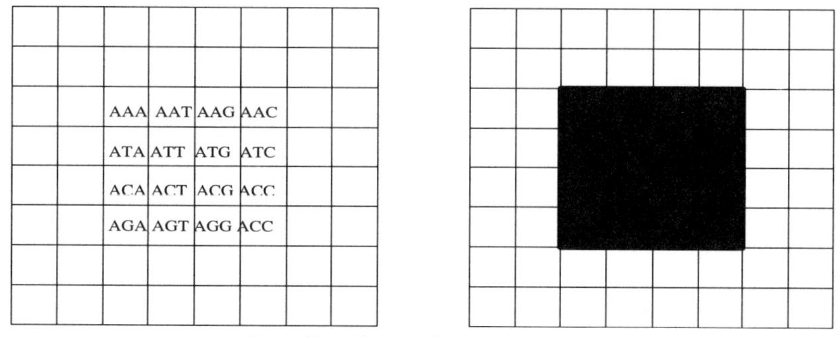

border length=16

Figure 5.12: Two masks with different border lengths.

as follows. For an l-bit Gray code

$$G_l = \{g_1, g_2, ..., g_{2^l-1}, g_{2^l}\},$$

define an $(l+1)$-bit Gray code as follows:

$$G_{l+1} = \{0g_1, 0g_2, ..., g_{2^l-1}, 0g_{2^l}, 1g_{2^l}, 1g_{2^l-1}, ..., 1g_2, 1g_1\}.$$

The elements of G_l are simply copied with 0s added to the front, then reversed with 1s added to the front. Clearly, all elements in G_{l+1} are distinct, and consecutive elements in G_{l+1} differ by exactly one bit.

We are interested in a two-dimensional Gray code composed of strings of length l over a four-letter alphabet. In other words, we would like to generate a 2^l-by-2^l matrix in which each of the 4^l l-tuples is present at a position (i, j), and each pair of adjacent l-tuples (horizontally or vertically) differs in exactly one position. Such a Gray code can be generated from the one-digit two-dimensional Gray code

$$G_1 = \begin{matrix} A & T \\ G & C \end{matrix}$$

as follows. For an l-digit two-dimensional Gray code

$$G_l = \begin{matrix} g_{1,1} & \cdots & g_{1,2^l} \\ \cdots & \cdots & \cdots \\ g_{2^l,1} & \cdots & g_{2^l,2^l} \end{matrix}$$

define the $(l+1)$-digit two-dimensional Gray code as

$$G_{l+1} = \begin{matrix} Ag_{1,1} & \cdots & Ag_{1,2^l} & Tg_{1,2^l} & \cdots & Tg_{1,1} \\ \cdots & \cdots & \cdots & \cdots & \cdots & \cdots \\ Ag_{2^l,1} & \cdots & Ag_{2^l,2^l} & Tg_{2^l,2^l} & \cdots & Tg_{2^l,1} \\ \\ Gg_{2^l,1} & \cdots & Gg_{2^l,2^l} & Cg_{2^l,2^l} & \cdots & Cg_{2^l,1} \\ \cdots & \cdots & \cdots & \cdots & \cdots & \cdots \\ Gg_{1,1} & \cdots & Gg_{1,2^l} & Cg_{1,2^l} & \cdots & Cg_{1,1} \end{matrix}$$

In particular,

$$G_2 = \begin{matrix} AA & AT & TT & TA \\ AG & AC & TC & TG \\ GG & GC & CC & CG \\ GA & GT & CT & CA \end{matrix}$$

The elements of G_l are copied into the upper left quadrant of G_{l+1}, then reflected horizontally and vertically into the three adjacent quadrants. As, Ts, Cs, and Gs are placed in front of the elements in the upper left, upper right, lower right, and lower left quadrant, respectively.

The construction above is one of many possible Gray codes. Two-dimensional Gray codes can be generated from any pair of one-dimensional Gray codes G_1 and G_2 by taking the two-dimensional product $G(i, j) = G_1(i) * G_2(j)$, where $*$ is a *shuffle* operation (an arbitrary fixed shuffling like that of two decks of cards). The simplest shuffling is the concatenation of $G_1(i)$ and $G_2(j)$.

For uniform arrays, Gray-code masks have minimal overall border lengths among all masks, and the ratio of the border length of the Gray-code mask to the border length of the standard mask approaches $\frac{1}{2}$ (Feldman and Pevzner, 1994 [99]).

5.13 Some Other Problems and Approaches

5.13.1 SBH with universal bases

Preparata et al., 1999 [280] pushed the idea of multiprobe arrays further and described the arrays that achieve the information-theoretic lower bound for the number of probes required for unambiguous reconstruction of an arbitrary string of length n. These arrays use universal bases such as inosine that stack correctly without binding and play a role of "don't care" symbols in the probes. The design of these arrays is similar to the design of gapped arrays with more elaborate patterns of gaps. Another approach to pooling in SBH was proposed by Hubbell, 2000 [171].

5.13.2 Adaptive SBH

The idea of adaptive SBH can be explained with the following example. Imagine a super-programmer who implements very complicated software to compute a super-number. After the first run on SuperPC, he learns that a cycle in his program presents a bottleneck, and it will take a hundred years before the super-number is computed. The usual approach to overcome this problem in the software industry is to analyze the time-consuming cycle, localize the bottleneck, and write new faster software (or modify the old). However, if the super-programmer is expensive (he is these days!), this may not be the best approach. A different approach would be to analyze the time-consuming cycle and to build *new hardware* that executes the bottleneck cycle so fast that we could compute the super-number *with the existing software*. Of course, it makes sense only if the cost of building a new SuperPC is lower than the cost of the super-programmer.

For adaptive SBH, a DNA array is the computer and DNA is the program. We are not at liberty to change the program (DNA fragment), but we can build new arrays after we learn about the bottlenecks in sequence reconstruction. Skiena and Sundaram, 1995 [316] and Margaritas and Skiena, 1995 [232] studied error-free adaptive SBH and came up with elegant theoretical bounds for the number of rounds needed for sequence reconstruction. This idea was further developed by Kruglyak, 1998 [210].

5.13.3 SBH-style shotgun sequencing

Idury and Waterman, 1995 [175] suggested using the Eulerian path SBH approach for sequence assembly in traditional DNA sequencing. The idea is simple and elegant: treat every read of length n as $n - l + 1$ l-tuples for sufficiently large l (i.e., $l = 30$). Since most 30-tuples are unique in the human genome, this approach leads to a very efficient sequencing algorithm in the case of error-free data. Idury and Waterman, 1995 [175] also attempted to adapt this algorithm for the case of sequencing errors.

5.13.4 Fidelity probes for DNA arrays

One current approach to quality control in DNA array manufacturing is to synthesize a small set of test probes that detect variation in the manufacturing process. These fidelity probes consist of identical copies of the same probe, but they are deliberately manufactured using different steps of the manufacturing process. A known target is hybridized to these probes, and the hybridization results reflect the quality of the manufacturing process. It is desirable not only to detect variations, but also to analyze the variations that occur, indicating in what step manufacture went wrong. Hubbell and Pevzner, 1999 [172] describe a combinatorial approach that constructs a small set of fidelity probes that not only detect variations, but also point out the erroneous manufacturing steps.

Chapter 6

Sequence Comparison

6.1 Introduction

Mutation in DNA is a natural evolutionary process: DNA replication errors cause substitutions, insertions, and deletions of nucleotides, leading to "editing" of DNA texts. Similarity between DNA sequences can be a clue to common evolutionary origin (as with the similarity between globin genes in humans and chimpanzees) or a clue to common function (as with the similarity between the ν-sys oncogene and the growth-stimulating hormone).

Establishing the link between cancer-causing genes and a gene involved in normal growth and development (Doolittle, 1983 [89], Waterfield, 1983 [353]) was the first success story in sequence comparison. *Oncogenes* are genes in viruses that cause a cancer-like transformation of infected cells. Oncogene ν-sys in the *simian sarcoma virus* causes uncontrolled cell growth and leads to cancer in monkeys. The seemingly unrelated *growth factor* PDGF is a protein that stimulates cell growth. When these genes were compared, significant similarity was found. This discovery confirmed a conjecture that cancer may be caused by a normal growth gene being switched on at the wrong time.

Levenshtein, 1966 [219] introduced the notion of *edit distance* between strings as the minimum number of edit operations needed to transform one string into another, where the edit operations are insertion of a symbol, deletion of a symbol, and substitution of a symbol for another one. Most DNA sequence comparison algorithms still use this or a slightly different set of operations. Levenshtein introduced a definition of edit distance but never described an algorithm for finding the edit distance between two strings. This algorithm has been discovered and re-discovered many times in different applications ranging from speech processing (Vintsyuk, 1968 [347]) to molecular biology (Needleman and Wunsch, 1970 [251]). Although the details of the algorithms are slightly different in different applications, they all are variations of dynamic programming.

Finding differences (edit distance) between sequences is often equivalent to finding similarities between these sequences. For example, if edit operations are limited to insertions and deletions (no substitutions), the edit distance problem is equivalent to the Longest Common Subsequence (LCS) Problem. Mathematicians became interested in the LCS Problem long before the dynamic programming algorithm for sequence comparison was discovered. Studies of the symmetric group revealed surprising connections between representation theory and the problem of finding the LCS between two permutations. The first algorithm for this problem was described by Robinson, 1938 [287]. Robinson's work was forgotten until the 1960s, when Schensted, 1961 [306] and Knuth, 1970 [201] re-discovered the relationships between the LCS and Young tableaux.

Although most algorithmic aspects of sequence comparison are captured by the LCS Problem, biologists prefer using *alignments* for DNA and protein sequence comparison. The alignment of the strings V and W is a two-row matrix such that the first (second) row contains the characters of V (W) in order, interspersed with some spaces. The score of an alignment is defined as the sum of the scores of its columns. The column score is often positive for coinciding letters and negative for distinct letters.

In the early papers on sequence alignment, scientists attempted to find the similarity between *entire* strings V and W, i.e., global alignment. This is meaningful for comparisons between members of the same protein family, such as $globins$, that are very conserved and have almost the same length in organisms ranging from fruit flies to humans. In many biological applications, the score of alignment between substrings of V and W may be larger than the score of alignment between the entire strings. This problem is known as the *local alignment* problem. For example, *homeobox* genes, which regulate embryonic development, are present in a large variety of species. Although homeobox genes are very different in different species, one region of them–called *homeodomain*–is highly conserved. The question arises how to find this conserved area and ignore the areas that show very little similarity. Smith and Waterman, 1981 [320] proposed a clever modification of dynamic programming that solves the local alignment problem.

Back in 1983, it was surprising to find similarities between a cancer-causing gene and a gene involved in normal growth and development. Today, it would be even more surprising *not to find* any similarity between a newly sequenced gene and the huge GenBank database. However, GenBank database search is not as easy now as it was 20 years ago. When we are trying to find the closest match to a gene of length 10^3 in a database of size 10^9 even quadratic dynamic programming algorithms may be too slow. One approach is to use a fast parallel hardware implementation of alignment algorithms; another one is to use fast heuristics that usually work well but are not guaranteed to find the closest match.

Many heuristics for fast database search in molecular biology use the same *filtering* idea. Filtering is based on the observation that a good alignment usually

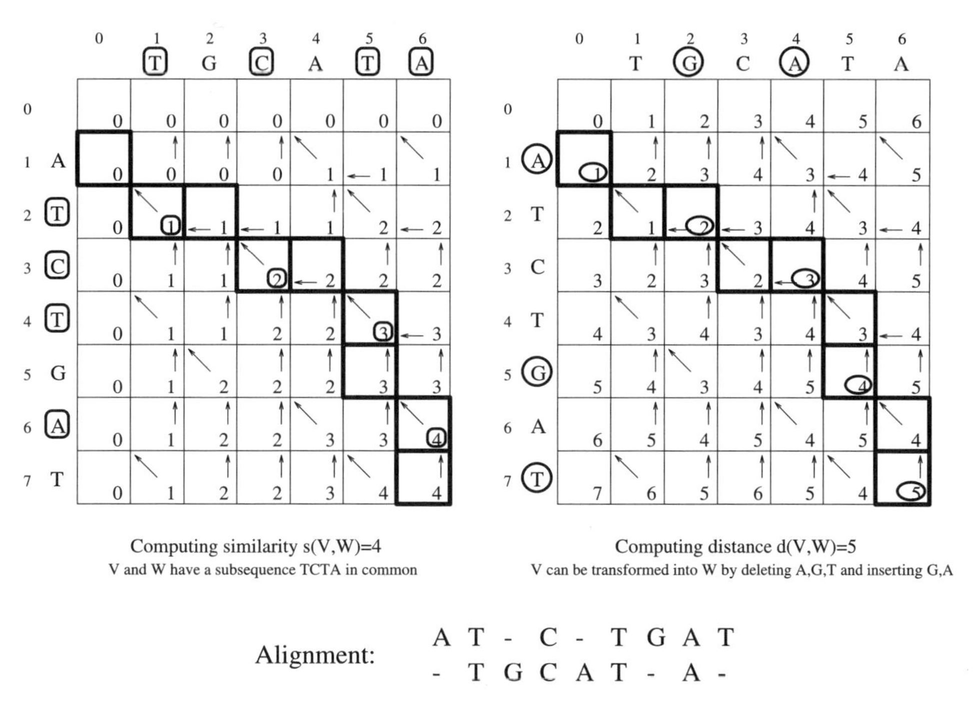

Figure 6.1: Dynamic programming algorithm for computing longest common subsequence.

includes short identical or very similar fragments. Thus one can search for such short substrings and use them as seeds for further analysis. The filtration idea for fast sequence comparison goes back to the early 1970s, well before the popular FASTA and BLAST algorithms were invented. Knuth, 1973 [202] suggested a method for pattern matching with one mismatch based on the observation that strings differing by a single error must match exactly in either the first or the second half. For example, approximate pattern matching of 9-tuples with one error can be reduced to the exact pattern matching of 4-tuples with further extending of the 4-tuple matches into 9-tuple approximate matches. This provides an opportunity for filtering out the positions that do not share common 4-tuples, a large portion of all pairs of positions. The idea of filtration in computational molecular biology was first described by Dumas and Ninio, 1982 [92], and then was taken significantly further by Wilbur and Lipman, 1983 [368] and Lipman and Pearson, 1985 [225] in their FASTA algorithm. It was further developed in BLAST, now a dominant database search tool in molecular biology (Altschul et al., 1990 [5]).

6.2 Longest Common Subsequence Problem

Define a *common subsequence* of strings $V = v_1 \dots v_n$ and $W = w_1 \dots w_m$ as a sequences of indices

$$1 \leq i_1 < \dots < i_k \leq n$$

and a sequences of indices

$$1 \leq j_1 < \dots < j_k \leq m$$

such that

$$v_{i_t} = w_{j_t} \text{ for } 1 \leq t \leq k.$$

Let $s(V, W)$ be the length (k) of a *longest common subsequence* (LCS) of V and W. Clearly, $d(V, W) = n + m - 2s(V, W)$ is the minimum number of insertions and deletions needed to transform V into W. Figure 6.1 presents an LCS of length 4 for the strings $V = ATCTGAT$ and $W = TGCATA$ and a shortest sequence of 2 insertions and 3 deletions transforming V into W.

A simple dynamic programming algorithm to compute $s(V, W)$ has been discovered independently by many authors. Let $s_{i,j}$ be the length of LCS between the i-prefix $V_i = v_1 \dots v_i$ of V and the j-prefix $W_j = w_1 \dots w_j$ of W. Let $s_{i,0} = s_{0,j} = 0$ for all $1 \leq i \leq n$ and $1 \leq j \leq m$. Then, $s_{i,j}$ can be computed by the following recurrency:

$$s_{i,j} = \max \begin{cases} s_{i-1,j} \\ s_{i,j-1} \\ s_{i-1,j-1} + 1, & \text{if } v_i = w_j \end{cases}$$

The first (second) term corresponds to the case when v_i (w_j) is not present in the LCS of V_i and W_j, and the third term corresponds to the case when both v_i and w_j are present in the LCS of V_i and W_j (v_i *matches* w_j). The *dynamic programming* table in Figure 6.1(left) presents the computation of the *similarity* score $s(V, W)$ between V and W, while the table in Figure 6.1(right) presents the computation of *edit distance* between V and W. The edit distance $d(V, W)$ is computed according to the initial conditions $d_{i,0} = i$, $d_{0,j} = j$ for all $1 \leq i \leq n$ and $1 \leq j \leq m$ and the following recurrency:

$$d_{i,j} = \min \begin{cases} d_{i-1,j} + 1 \\ d_{i,j-1} + 1 \\ d_{i-1,j-1} & \text{if } v_i = w_j \end{cases}$$

The length of an LCS between V and W can be read from the element (n, m) of the dynamic programming table. To *construct* an LCS, one has to keep the information on which of the three quantities ($s_{i-1,j}$, $s_{i,j-1}$, or $s_{i-1,j-1} + 1$) corresponds to the maximum in the recurrence for $s_{i,j}$ and *backtrack* through the dynamic programming table. The following algorithm achieves this goal by introducing the pointers $\leftarrow$, $\uparrow$, and $\nwarrow$, corresponding to the above three cases:

LCS (V, W)
for $i \leftarrow 1$ **to** n
  $s_{i,0} \leftarrow 0$
for $i \leftarrow 1$ **to** m
  $s_{0,i} \leftarrow 0$
for $i \leftarrow 1$ **to** n
  for $j \leftarrow 1$ **to** m
    if $v_i = w_j$
      $s_{i,j} \leftarrow s_{i-1,j-1} + 1$
      $b_{i,j} \leftarrow \nwarrow$
    else if $s_{i-1,j} \geq s_{i,j-1}$
      $s_{i,j} \leftarrow s_{i-1,j}$
      $b_{i,j} \leftarrow \uparrow$
    else
      $s_{i,j} \leftarrow s_{i,j-1}$
      $b_{i,j} \leftarrow \leftarrow$
return s and b

The following recursive program prints out the longest common subsequence using backtracking along the path from (n, m) to $(0, 0)$ according to pointers $\leftarrow$, $\uparrow$, and $\nwarrow$. The initial invocation is **PRINT-LCS**(b, V, n, m).

PRINT-LCS (b, V, i, j)
if $i = 0$ or $j = 0$
  return
if $b_{i,j} = \nwarrow$
  PRINT-LCS$(b, V, i-1, j-1)$
  print v_i
else if $b_{i,j} = \uparrow$
  PRINT-LCS$(b, V, i-1, j)$
else
  PRINT-LCS$(b, V, i, j-1)$

6.3 Sequence Alignment

Let $\mathcal{A}$ be a k-letter *alphabet*, and let V and W be two sequences over $\mathcal{A}$. Let $\mathcal{A}' = \mathcal{A} \cup \{-\}$ be an extended alphabet, where $'-'$ denotes *space*. An *alignment* of strings $V = v_1, \ldots v_n$ and $W = w_1, \ldots, w_m$ is a $2 \times l$ matrix A ($l \geq n, m$), such that the first (second) row of A contains the characters of V (W) in order interspersed with $l - n$ ($l - m$) spaces (Figure 6.1(bottom)). We assume that no column of the alignment matrix contains two spaces. The columns of the alignment containing a space are called *indels*, and the columns containing a space in the first (second) row are called *insertions* (*deletions*). The columns containing the same letter in both rows are called *matches*, while the columns containing different letters are called *mismatches*. The score of a column containing symbols x and y from the extended alphabet is defined by a $(k+1) \times (k+1)$ matrix defining the similarity scores $\delta(x, y)$ for every pair of symbols x and y from $\mathcal{A}'$. The score of the alignment is defined as the sum of the scores of its columns. The simplest matrix δ assumes premiums $\delta(x, x) = 1$ for the matches and penalizes every mismatch by $\delta(x, y) = -\mu$ and every insertion or deletion by $\delta(x, -) = \delta(-, x) = -\sigma$. In this case the *score* of the alignment is defined as # matches – μ# mismatches – σ# indels. The Longest Common Subsequence Problem is the alignment problem with the parameters $\mu = \infty$, $\sigma = 0$. The common matrices for protein sequence comparison, *Point Accepted Mutations (PAM)* and *BLOSUM*, reflect the frequency with which amino acid x replaces amino acid y in evolutionary related sequences (Dayhoff et al., 1978 [82], Altschul, 1991 [6], Henikoff and Henikoff, 1992 [158]).

The (global) *sequence alignment* problem is to find the alignment of sequences V and W of maximal score. The corresponding recurrency for the score $s_{i,j}$ of optimal alignment between V_i and W_j is

$$s_{i,j} = \max \begin{cases} s_{i-1,j} + \delta(v_i, -) \\ s_{i,j-1} + \delta(-, w_j) \\ s_{i-1,j-1} + \delta(v_i, w_j) \end{cases}$$

Every alignment of V and W corresponds to a path in the *edit graph* of sequences V and W (Figure 6.2). Therefore, the sequence alignment problem corresponds to the *longest path problem* from the source to the sink in this *directed acyclic graph*.

6.4 Local Sequence Alignment

Frequently, biologically significant similarities are present in certain parts of DNA fragments and are not present in others. In this case biologists attempt to maximize $s(v_i \ldots v_{i'}, w_j \ldots w_{j'})$ where the maximum is taken over all substrings $v_i \ldots v_{i'}$ of V and $w_j \ldots w_{j'}$ of W.

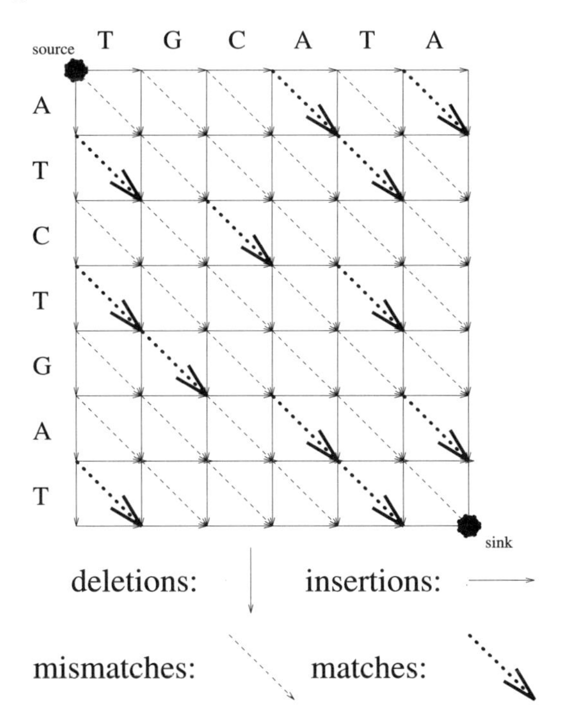

Figure 6.2: Edit graph. The weights of insertion and deletion edges are $-\sigma$, the weights of mismatch edges are $-\mu$, and the weights of match edges are 1.

The global alignment problem corresponds to finding the longest path between vertices $(0, 0)$ and (n, m) in the edit graph, while the local alignment problem corresponds to finding the longest path among paths between arbitrary vertices (i, j) and (i', j') in the edit graph. A straightforward and inefficient approach to this problem is to find the longest path between every pair of vertices (i, j) and (i', j'). Instead of finding the longest path from every vertex (i, j), the local alignment problem can be reduced to finding the longest paths from the source by adding edges of weight 0 from the source to every other vertex. These edges provide a "free" jump from the source to any other vertex (i, j). A small difference in the following recurrency reflects this transformation of the edit graph (Smith and Waterman, 1981 [320]):

$$s_{i,j} = \max \begin{cases} 0 \\ s_{i-1,j} + \delta(v_i, -) \\ s_{i,j-1} + \delta(-, w_j) \\ s_{i-1,j-1} + \delta(v_i, w_j) \end{cases}$$

The largest value of $s_{i,j}$ represents the score of the local alignment of V and W (rather than $s_{n,m}$ for global alignment).

Optimal local alignment reports only the longest path in the edit graph. At the same time several local alignments may have biological significance and the methods are sought to find k best non-overlapping local alignments (Waterman and Eggert, 1987 [359], Huang et al., 1990 [168]). These methods are particularly important for comparison of multi-domain proteins sharing similar blocks that are shuffled in one protein as compared to another. In this case, a single local alignment representing all significant similarities does not exist.

6.5 Alignment with Gap Penalties

Mutations are usually manifestations of errors in DNA replication. Nature frequently deletes or inserts entire substrings as a unit, as opposed to deleting or inserting individual nucleotides. A *gap* in alignment is defined as a continuous sequence of spaces in one of the rows. It is natural to assume that the score of a gap consisting of x spaces is not just the sum of scores of x indels, but rather a more general function. For *affine gap penalties*, the score for a gap of length x is $-(\rho + \sigma x)$, where $\rho > 0$ is the penalty for the introduction of the gap and $\sigma > 0$ is the penalty for each symbol in the gap. Affine gap penalties can be accommodated by introducing long vertical and horizontal edges in the edit graph (e.g., an edge from (i, j) to $(i+x, j)$ of length $-(\rho+\sigma x)$) and further computing the longest path in this graph. Since the number of edges in the edit graph for affine gap penalties increases, at first glance it looks as though the running time for the alignment algorithm also increases from $O(n^2)$ to $O(n^3)$. However, the following three recurrences keep the running time down:

$$\overset{\downarrow}{s}_{i,j} = \max \begin{cases} \overset{\downarrow}{s}_{i-1,j} - \sigma \\ s_{i-1,j} - (\rho + \sigma) \end{cases}$$

$$\vec{s}_{i,j} = \max \begin{cases} \vec{s}_{i,j-1} - \sigma \\ s_{i,j-1} - (\rho + \sigma) \end{cases}$$

$$s_{i,j} = \max \begin{cases} s_{i-1,j-1} + \delta(v_i, w_j) \\ \overset{\downarrow}{s}_{i,j} \\ \vec{s}_{i,j} \end{cases}$$

The variable $\overset{\downarrow}{s}_{i,j}$ computes the score for alignment between V_i and W_j ending with a deletion (i.e., a gap in W), while the variable $\vec{s}_{i,j}$ computes the score for alignment ending with an insertion (i.e., a gap in V). The first term in the recurrences for

$\overset{\downarrow}{s}_{i,j}$ and $\vec{s}_{i,j}$ corresponds to extending the gap, while thc sccond term corresponds to initiating the gap. Although affine gap penalties is the most commonly used model today, some studies indicate that non-linear gap penalties may have some advantages over the affine ones (Waterman, 1984 [356], Gonnet et al., 1992 [133]). Efficient algorithms for alignment with non-linear gap penalties were proposed by Miller and Myers, 1988 [236] and Galil and Giancarlo, 1989 [116].

6.6 Space-Efficient Sequence Alignment

In comparison of long DNA fragments, the limited resource in sequence alignment is not time but space. Hirschberg, 1975 [163] proposed a *divide-and-conquer* approach that performs alignment in linear space for the expense of just doubling the computational time.

The time complexity of the dynamic programming algorithm for sequence alignment is roughly the number of edges in the edit graph, i.e., $O(nm)$. The space complexity is roughly the number of vertices in the edit graph, i.e., $O(nm)$. However, if we only want to compute the score of the alignment (rather than the alignment itself), then the space can be reduced to just twice the number of vertices in a single column of the edit graph, i.e., $O(n)$. This reduction comes from the observation that the only values needed to compute the alignment scores $s_{*,j}$ (column j) are the alignment scores $s_{*,j-1}$ (column $j-1$). Therefore, the alignment scores in the columns before $j-1$ can be discarded while computing alignment scores for columns $j, j+1, \ldots$. In contrast, computing the alignment (i.e., finding the longest path in the edit graph) requires backtracking through the entire matrix $(s_{i,j})$. Therefore, the entire backtracking matrix needs to be stored, thus leading to the $O(nm)$ space requirement.

The longest path in the edit graph connects the *start* vertex $(0,0)$ with the *sink* vertex (n,m) and passes through an (unknown) *middle vertex* $(i, \frac{m}{2})$ (assume for simplicity that m is even). Let's try to find its middle vertex instead of trying to find the entire longest path. This can be done in linear space by computing the scores $s_{*,\frac{m}{2}}$ (lengths of the longest paths from $(0,0)$ to $(i, \frac{m}{2})$ for $0 \leq i \leq n$) and the scores of the paths from $(i, \frac{m}{2})$ to (n,m). The latter scores can be computed as the scores of the paths $s^{reverse}_{*,\frac{m}{2}}$ from (n,m) to $(i, \frac{m}{2})$ in the reverse edit graph (i.e., the graph with the directions of all edges reversed). The value $s_{i,\frac{m}{2}} + s^{reverse}_{i,\frac{m}{2}}$ is the length of the longest path from $(0,0)$ to (n,m) passing through the vertex $(i, \frac{m}{2})$. Therefore, $\max_i(s_{i,\frac{m}{2}} + s^{reverse}_{i,\frac{m}{2}})$ computes the length of the longest path and identifies a middle vertex.

Computing these values requires the time equal to the area of the left rectangle (from column 1 to $\frac{m}{2}$) plus the area of the right rectangle (from column $\frac{m}{2}+1$ to m) and the space $O(n)$ (Figure 6.3). After the middle vertex $(i, \frac{m}{2})$ is found, the problem of finding the longest path from $(0,0)$ to (n,m) can be partitioned into two subproblems: finding the longest path from $(0,0)$ to the middle vertex $(i,\frac{m}{2})$ and finding the longest path from the middle vertex $(i,\frac{m}{2})$ to (n,m). Instead of trying to find these paths, we first try to find the middle vertices in the corresponding rectangles (Figure 6.3). This can be done in the time equal to the area of these rectangles, which is two times smaller than the area of the original rectangle. Computing in this way, we will find the middle vertices of all rectangles in time $area + \frac{area}{2} + \frac{area}{4} + \ldots \leq 2 \times area$ and therefore compute the longest path in time $O(nm)$ and space $O(n)$:

Path $(source, sink)$
if $source$ and $sink$ are in consecutive columns
    output the longest path from the $source$ to the $sink$
else
    $middle \leftarrow$ middle vertex between $source$ and $sink$
    Path $(source, middle)$
    Path $(middle, sink)$

6.7 Young Tableaux

An *increasing subsequence* of a permutation $\pi = x_1x_2\ldots x_n$ is a sequence of indices $1 \leq i_1 < \ldots < i_k \leq n$ such that $x_{i_1} < x_{i_2} \ldots < x_{i_k}$. Decreasing subsequences are defined similarly. Finding the *longest increasing subsequence* (LIS) is equivalent to finding the LCS between π and the identity permutation $12\ldots n$. It is well known that every permutation on n elements has either an increasing or a decreasing subsequence of length at least $\sqrt{n}$. This result is closely related to the non-dynamic programming approach to the LCS that is described below.

A *partition* of integer n is a sequence of positive integers $\lambda_1 \geq \lambda_2 \geq \ldots \geq \lambda_l$ such that $\sum_{i=1}^{i=l} \lambda_i = n$. If $\lambda = (\lambda_1\lambda_2\ldots\lambda_l)$ is a partition of n, then we write $\lambda \vdash n$. Suppose $\lambda = (\lambda_1, \lambda_2, \ldots, \lambda_l) \vdash n$. Then the *Young diagram* or *shape* λ, is an array of n cells into l left-justified rows with row i containing λ_i cells for $1 \leq i \leq l$. The leftmost diagram in Figure 6.4 presents the Young diagram for $(4,2,1,1) \vdash 8$, while the rightmost one presents the Young diagram for $(4,2,2,1) \vdash 9$. A *Young tableau* (or simply *tableau* of shape λ) is an array obtained by replacing the cells of the Young diagram λ with the numbers $1, 2, \ldots, n$ bijectively (rightmost table in Figure 6.4). A tableau λ is *standard* if its rows and columns are increasing sequences. Sagan, 1991 [292] is an excellent introduction to combinatorics of Young tableaux, and our presentation follows this book.

A *bitableau* is a pair of standard Young tableaux with the same Young diagram λ. The Robinson-Schensted-Knuth algorithm (RSK) describes an explicit bijection

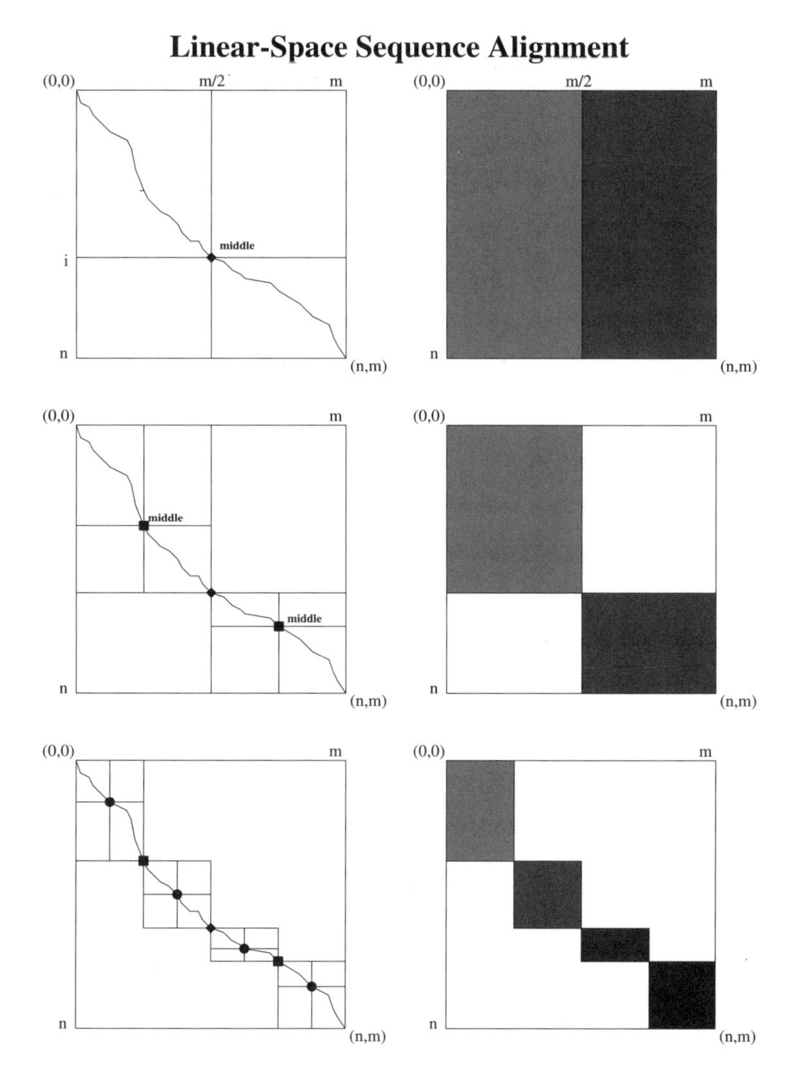

Figure 6.3: Space-efficient sequence alignment. The computational time (area of solid rectangles) decreases by a factor of 2 at every iteration.

between bitableaux with n cells and the symmetric group S_n (all permutations of order n). It was first proposed (in rather fuzzy terms) by Robinson, 1938 [287] in connection with representation theory. Schensted, 1961 [306] re-discovered and clearly described this algorithm on a purely combinatorial basis. Later, Knuth, 1970 [201] generalized the algorithm for the case of the LCS.

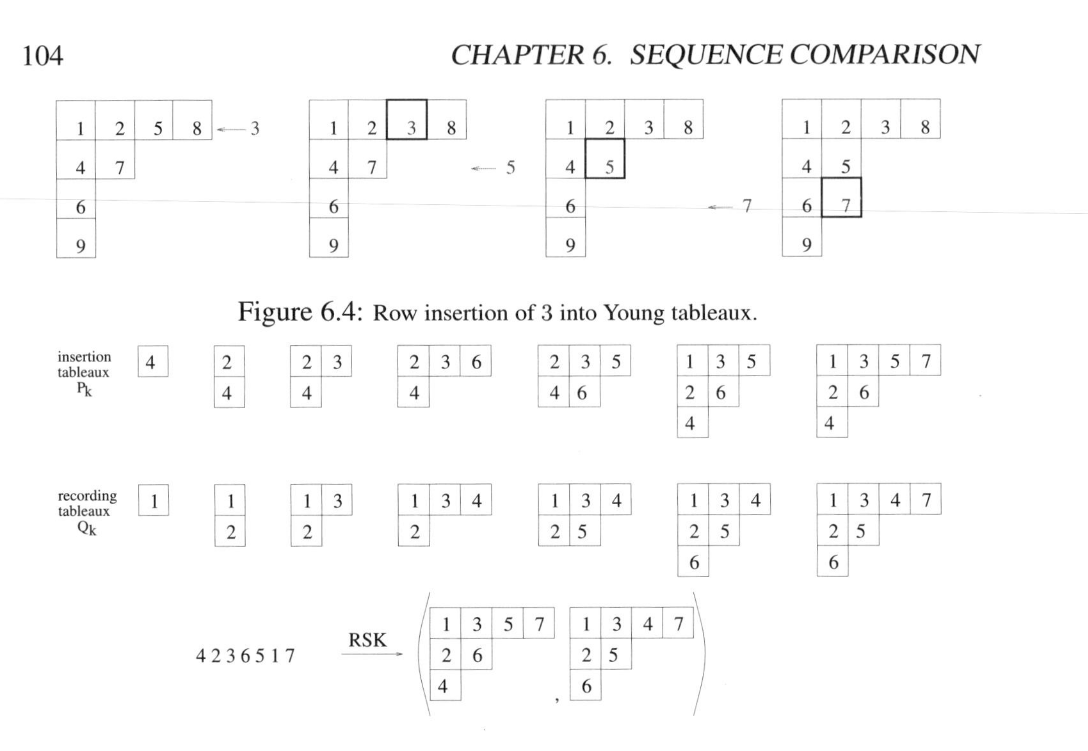

Figure 6.4: Row insertion of 3 into Young tableaux.

Figure 6.5: RSK algorithm for permutation $\pi = 4236517$.

Let P be a *partial tableau*, i.e., a Young diagram with distinct entries whose rows and columns increase. For an array R and an element x, define R_{x+} as the smallest element of R greater than x and R_{x-} as the largest element of R smaller than x. For x not in P, define *row insertion* of x into P by the following algorithm:

$R \leftarrow$ the first row of P
While x is less than some element of row R
  Replace R_{x+} by x in R.
  $x \leftarrow R_{x+}$
  $R \leftarrow$ next row down
Place x at the end of row R.

The result of row inserting x into P is denoted $r_x(P)$. Note that the insertion rules ensure that $r_x(P)$ still has increasing rows and columns (Figure 6.4).

The bijection between permutations and bitaubleaux is denoted as $\pi \overset{RSK}{\rightarrow} (P, Q)$ where $\pi \in S_n$ and P, Q are standard λ-tableaux, $\lambda \vdash n$. For a permutation $\pi = x_1 \dots x_n$ we construct a sequence of tableaux

$$(P_0, Q_0) = (\emptyset, \emptyset), (P_1, Q_1), \dots, (P_n, Q_n) = (P, Q)$$

where $x_1, \dots x_n$ are *inserted* into the $P's$ and $1, \dots, n$ are *placed* in the $Q's$ so that shape of P_i coincides with the shape of Q_i for all i.

Placement of an element in a tableau is even easier than insertion. Suppose that Q is a partial tableau and that (i, j) is an outer corner of Q. If k is greater than every element of Q, then *to place* k in Q at cell (i, j), merely set $Q_{i,j} = k$ (Figure 6.5).

Finally we describe how to build the sequence of bitableaux (P_i, Q_i) from $\pi = x_1 \dots x_n$. Assuming that (P_{k-1}, Q_{k-1}) has already been constructed, define $P_k = r_{x_k}(P_{k-1})$ and Q_k as the result of placement of k into Q_{k-1} at the cell (i, j) where the insertion terminates (Figure 6.5). We call P the *insertion* tableau and Q the *recording* tableau.

Given the rightmost tableau $R_x(P)$ in Figure 6.4 and the position of the last added element (7), can we reconstruct the leftmost tableau P in Figure 6.4? Since element 7 was bumped from the previous row by $R_{7-} = 5$ (see RSK algorithm), we can reconstruct the second tableau in Figure 6.4. Since element 5 was bumped from the first row by the element $R_{5-} = 3$, we can reconstruct the original tableau P. This observation implies the RSK theorem:

Theorem 6.1 *The map* $\pi \overset{RSK}{\rightarrow} (P, Q)$ *is a bijection between elements of* S_n *and pairs of standard tableaux of the same shape* $\lambda \vdash n$.

Proof Construct an inverse bijection $(P, Q) \overset{RSK}{\rightarrow} \pi$ by reversing the RSK algorithm step by step. We begin by defining $(P_n, Q_n) = (P, Q)$. Assuming that (P_k, Q_k) has been constructed, we will find x_k (the k-th element of π) and (P_{k-1}, Q_{k-1}). Find the cell (i, j) containing k in Q_k. Since this is the largest element in Q_k, $P_{i,j}$ must have been the last element to be displaced in the construction of P_k. We can use the following procedure to *delete* $P_{i,j}$ from P. For convenience, we assume the existence of an empty zeroth row above the first row of P_k.

Set $x \leftarrow P_{i,j}$ and erase $P_{i,j}$
$R \leftarrow$ the $(i-1)$-st row of P_k
While R is not the zeroth row of P_k
  Replace R_{x-} by x in R.
  $x \leftarrow R_{x-}$
  $R \leftarrow$ next row up.
$x_k \leftarrow x$.

It is easy to see that P_{k-1} is P_k after the deletion process just described is complete and Q_{k-1} is Q_k with the k removed. Continuing in this way, we eventually recover all the elements of π in reverse order. ■

Lemma 6.1 *If $\pi = x_1 \dots x_n$ and x_k enters P_{k-1} in column j, then the longest increasing subsequence of π ending in x_k has length j.*

Proof We induct on k. The result is trivial for $k = 1$, so suppose it holds for all values up to $k - 1$.

First, we need to show the existence of an increasing subsequence of length j ending in x_k. Let y be the element of P_{k-1} in cell $(1, j-1)$. Then $y < x_k$, since x_k enters in column j. Also, by induction, there is an increasing subsequence of length $j - 1$ ending in y. Combining this subsequence with x_k we get the desired subsequence of length j.

Now we prove that there cannot be a longer increasing subsequence ending in x_k. If such a subsequence exists, let x_i be its element preceding x_k. By induction, when x_i is inserted it enters in some column (weakly) to the right of column j. Thus the element y in cell $(1, j)$ of P_i satisfies $y \leq x_i < x_k$. However, the entries in a given position of a tableau never increase with subsequent insertions (see RSK algorithm). Thus the element in cell $(1, j)$ of P_{k-1} is smaller than x_k, contradicting the fact that x_k displaces it. ■

This lemma implies the following:

Theorem 6.2 *The length of the longest increasing subsequence of permutation π is the length of the first row of $P(\pi)$.*

6.8 Average Length of Longest Common Subsequences

Let $\mathcal{V}$ and $\mathcal{W}$ be two sets of n-letter strings over the same alphabet. Given $\mathcal{V}$, $\mathcal{W}$ and a probability measure p on $\mathcal{V} \times \mathcal{W}$, we are interested in the *average length* of LCS, which is

$$s(n) = \sum_{V \in \mathcal{V}, W \in \mathcal{W}} s(V, W) \cdot p(V, W).$$

where $s(V, W)$ is the length of the LCS between V and W. Two examples of the LCS average length problem are of particular interest.

Longest increasing subsequence in random permutation $s_{per}(n)$. $\mathcal{V}$ contains only the string $(1, \dots, n)$, $\mathcal{W}$ contains all permutations of length n, and $p(V, W) =$

$\frac{1}{n!}$. The problem of finding $s_{per}(n)$ was raised by Ulam, 1961 [341]. Hammersley, 1972 [150] proved that

$$\lim_{n \to \infty} \frac{s_{per}(n)}{\sqrt{n}} = s_{per}$$

where s_{per} is a constant. Even before the convergence had been proven, Baer and Brock, 1968 [15] had conjectured that $s_{per} = 2$ on the basis of extensive computations. Hammersley, 1972 [150] proved that $\frac{\pi}{2} \leq s_{per} \leq e$. Later, Kingman, 1973 [198] improved the bounds to $1.59 < c < 2.49$. Logan and Shepp, 1977 [227] and Vershik and Kerov, 1977 [342] proved that $s_{per} = 2$ by conducting a technically challenging analysis of asymptotics of random Young tableaux and by using theorems 6.1 and 6.2.

Longest common subsequence $s_k(n)$ in a k-letter alphabet. Both $\mathcal{V}$ and $\mathcal{W}$ contain all k^n n-letter words in a k-letter alphabet, and $p(V, W) = \frac{1}{k^n \cdot k^n}$. Chvatal and Sankoff, 1975 [70] noticed that

$$\lim_{n \to \infty} \frac{s_k(n)}{n} = s_k$$

where s_k is a constant. They gave lower and upper bounds for s_k that were later improved by Deken, 1979, 1983 [83, 84] and Chvatal and Sankoff, 1983 [71].

In the 1980s, two conjectures about s_k were stated:

Sankoff-Mainville conjecture [305]: $\lim_{k \to \infty}(s_k \cdot \sqrt{k}) = 2$

Arratia-Steele conjecture [328]: $s_k = \frac{2}{1+\sqrt{k}}$

These conjectures can be formulated as statements about the length of the first row of Young tableaux. Instead of finding the length of the first row, one may try to find a limited shape of Young tableaux yielding simultaneously all characteristics of the Young tableaux, in particular, the length of the first row. At first glance, this problem looks more general and difficult than simply finding the length of the first row. However, the proof $s_{per} = 2$ revealed a paradoxical situation: it may be easier to find a limited shape of the entire Young tableaux than to find the expected length of a particular row using *ad hoc* combinatorial/probability arguments. In particular, an *ad hoc* combinatorial solution of the Ulam problem is still unknown.

Recently, there has been an avalanche of activity in studies of longest increasing subsequences. Aldous and Diaconis, 1995 [2] used interacting particle representation (modeled by the LIS in a sequence of n independent random real numbers uniformly distributed in an interval) to give a different proof that $s_{per} = 2$. Baik et al., 1999 [21] proved that $s_{per}(n) = 2\sqrt{n} - \mu n^{1/6} + o(n^{1/6})$ where $\mu = 1.711...$

Let Λ_n be the set of all shapes with n cells. Denote an insertion tableaux P corresponding to a permutation π as $P(\pi)$ and consider the set of permutations

$\Gamma = \{\pi : P(\pi)$ contains n in the first row$\}$. For a given shape λ with n cells, let $\Gamma_\lambda = \{\pi : P(\pi)$ contains n in the first row and $P(\pi)$ has shape λ $\}$. Given a random permutation $\pi \in S_n$, let p_n be the probability that $P(\pi)$ contains n in the first row.

Lemma 6.2 $p_n \leq \frac{1}{\sqrt{n}}$.

Proof If λ is a shape, we denote as λ_+ the shape derived from λ by adding a new cell to the end of the first row. Observe that the number of standard tableaux of shape $\mu \in \Lambda_n$ with n in the first row equals the number of standard tableaux of shape $\lambda \in \Lambda_{n-1}$ where $\lambda_+ = \mu$. Let f_λ be the number of Young tableaux of shape λ. According to the RSK theorem,

$$|\Gamma| = \sum_{\mu \in \Lambda_n} |\Gamma_\mu| = \sum_{\lambda \in \Lambda_{n-1}} f_\lambda \cdot f_{\lambda_+}$$

where f_λ is the number of Young tableaux of shape λ. It implies the following:

$$p_n = \frac{|\Gamma|}{n!} = \sum_{\lambda \in \Lambda_{n-1}} \frac{f_\lambda \cdot f_{\lambda_+}}{n!} = \sum_{\lambda \in \Lambda_{n-1}} \frac{(n-1)!}{n!} \frac{f_\lambda \cdot f_{\lambda_+}}{f_\lambda \cdot f_\lambda} \frac{f_\lambda \cdot f_\lambda}{(n-1)!} =$$

$$\sum_{\lambda \in \Lambda_{n-1}} \frac{1}{n} \frac{f_{\lambda_+}}{f_\lambda} p(\lambda)$$

According to the RSK theorem, $p(\lambda) = \frac{f_\lambda \cdot f_\lambda}{(n-1)!}$ is the probability that a random permutation $\pi \in S_{n-1}$ corresponds to a shape λ. Denoting $E(X) = p_n$, the mean value of the random variable $X = \frac{1}{n}\frac{f_{\lambda_+}}{f_\lambda}$. Applying the inequality $E(X) \leq \sqrt{E(X^2)}$, we derive

$$p_n^2 \leq \sum_{\lambda \in \Lambda_{n-1}} \frac{1}{n \cdot n} \cdot \frac{f_{\lambda_+} \cdot f_{\lambda_+}}{f_\lambda \cdot f_\lambda} \frac{f_\lambda \cdot f_\lambda}{(n-1)!} = \sum_{\lambda \in \Lambda_{n-1}} \frac{1}{n \cdot n!} f_{\lambda_+} \cdot f_{\lambda_+} =$$

$$\frac{1}{n} \sum_{\lambda \in \Lambda_{n-1}} \frac{f_{\lambda_+} \cdot f_{\lambda_+}}{n!} = \frac{1}{n} \sum_{\lambda \in \Lambda_{n-1}} p(\lambda_+) \leq \frac{1}{n} \sum_{\mu \in \Lambda_n} p(\mu) \leq \frac{1}{n}$$

since λ_+ ranges over all $\mu \in \Lambda_n$ with the length of the first row larger than the length of the second row. ■

The following theorem was proven (and apparently never published) by Vershik and Kerov in late 1970s. The first published proof appeared much later (Pilpel, 1990 [276]).

Theorem 6.3 $s_{per} \leq 2$.

Proof Given $\pi \in S_n$, let $p_k(n)$ be the probability that element k appears in the first row of $P(\pi)$. Notice that $p_k(n) = p_k$ (the elements $1, \ldots, k$ of a random permutation in S_n are evenly distributed over all possible relative orderings). According to lemma 6.2, the expected length of the first row of $P(\pi)$ is

$$r_1 = \sum_{k=1}^{n} p_k(n) = \sum_{k=1}^{n} p_k \leq \sum_{k=1}^{n} \frac{1}{\sqrt{k}}.$$

As $\frac{1}{\sqrt{k}} \leq 2(\sqrt{k} - \sqrt{k-1})$, we derive $r_1 \leq 2\sqrt{n}$. Since the length of the first row of $P(\pi)$ equals the length of the longest increasing subsequence of π, $s_{per} \leq 2$. ■

6.9 Generalized Sequence Alignment and Duality

A *partially ordered set* is a pair $(P, \prec)$ such that P is a set and $\prec$ is a *transitive and irreflexive binary relation* on P, i.e., $p \prec q$ and $q \prec r$ imply $p \prec r$. A *chain* $p_1 \prec p_2 \ldots \prec p_t$ is a subset of P where any two elements are comparable, and an *antichain* is a subset where no two elements are comparable. Partial orders $\prec$ and $\prec^*$ are called *conjugate* if for any two distinct $p_1, p_2 \in P$ the following condition holds:

$$p_1 \text{ and } p_2 \text{ are } \prec\text{-comparable} \iff p_1 \text{ and } p_2 \text{ are } \prec^*\text{-incomparable}$$

We are interested in *longest $\prec$-sequences*, i.e., chains of maximal length in $\prec$ (*generalized sequence alignment*). Let $I = \{1, 2 \ldots, n\}$ and $J = \{1, 2 \ldots, m\}$ and $P \subset I \times J$. Our interest is in the comparison of the two sequences $V = v_1 v_2 \ldots v_n$ and $W = w_1 w_2 \ldots w_m$ with $P = \{(i, j) : v_i = w_j\}$. Let $p_1 = (i_1, j_1)$ and $p_2 = (i_2, j_2)$ be two arbitrary elements in $I \times J$. Denote

$$\mathbf{\Delta}(p_1, p_2) = (\Delta i, \Delta j) = (i_2 - i_1, j_2 - j_1).$$

Consider a few examples of partial orders on $I \times J$ (corresponding chains are shown in Figure 6.6). Partial orders $\prec_1$ and $\prec_4$ as well as partial orders $\prec_2$ and $\prec_3$ are conjugate.

- Common subsequences(CS):

$$p_1 \prec_1 p_2 \Leftrightarrow \Delta i > 0, \Delta j > 0$$

- Common forests(CF):

$$p_1 \prec_2 p_2 \Leftrightarrow \Delta i \geq 0, \Delta j \geq 0$$

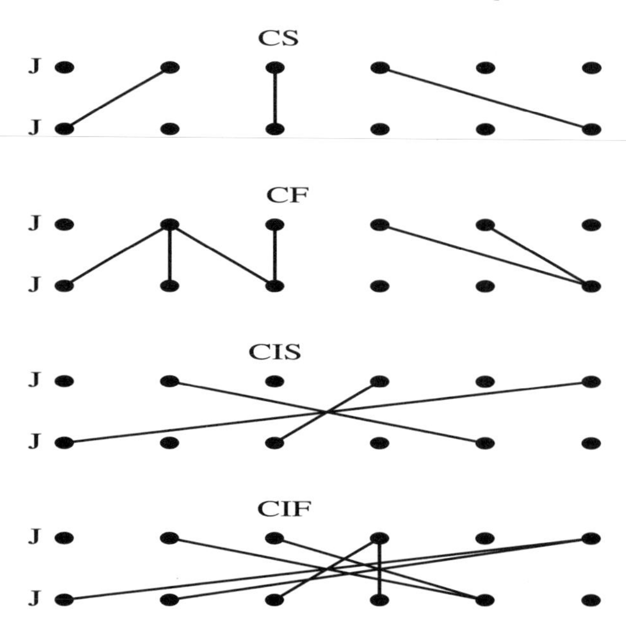

Figure 6.6: Chains in partial orders CS, CF, CIS, and CIF.

- Common inverted subsequences(CIS):

$$p_1 \prec_3 p_2 \Leftrightarrow \Delta i > 0, \Delta j < 0$$

- Common inverted forests(CIF):

$$p_1 \prec_4 p_2 \Leftrightarrow \Delta i \geq 0, \Delta j \leq 0$$

Let $\mathcal{C}$ be a family of subsets of a set P. $\mathcal{C}' \subseteq \mathcal{C}$ is called a *cover* of P if each $p \in P$ is contained in at least one of the subsets $C \in \mathcal{C}'$. The number of elements in $\mathcal{C}'$ is called the *size* of the cover $\mathcal{C}'$, and a cover of minimum size is called a *minimum cover* of P by $\mathcal{C}$. The following theorem was proved by Dilworth, 1950 [86].

Theorem 6.4 *Let P be a partially ordered set. The size of the minimum cover of P by chains is equal to the size of a maximal antichain in P.*

Lemma 6.3 *Let $\prec$ and $\prec^*$ be conjugate partial orders on P. Then the length of a longest $\prec$-sequence in P equals the size of a minimum cover of P by $\prec^*$-sequences.*

Proof According to Dilworth's theorem, the length of a longest antichain in $\prec^*$ equals the size of a minimum cover of P by $\prec^*$-chains. As $\prec$ and $\prec^*$ are conjugate, each antichain in $\prec^*$ is a chain in $\prec$, and each chain in $\prec$ is an antichain in $\prec^*$. Therefore, the length of a longest $\prec$-sequence in P equals the size of a minimum cover of P by $\prec^*$-sequences. ■

Since CS and CIF represent conjugate partial orders, lemma 6.3 implies the following (Pevzner and Waterman, 1993 [273]):

Theorem 6.5 *The length of the longest CS equals the size of a minimum cover by CIF.*

Consider a binary relation on P defined by

$$p_1 \sqsubset p_2 \Longleftrightarrow p_1 \prec p_2 \text{ or } p_1 \prec^* p_2.$$

Lemma 6.4 *$\sqsubset$ is a linear order on P.*

Proof We first prove that $p_1 \sqsubset p_2$ and $p_2 \sqsubset p_3$ implies $p_1 \sqsubset p_3$. If $p_1 \sqsubset p_2$ and $p_2 \sqsubset p_3$, then one of the following conditions holds:

(i) $p_1 \prec p_2$ and $p_2 \prec p_3$,

(ii) $p_1 \prec p_2$ and $p_2 \prec^* p_3$,

(iii) $p_1 \prec^* p_2$ and $p_2 \prec p_3$,

(iv) $p_1 \prec^* p_2$ and $p_2 \prec^* p_3$.

In case (i), $p_1 \prec p_2$ and $p_2 \prec p_3$ imply $p_1 \prec p_3$, and therefore $p_1 \sqsubset p_3$. In case (ii), $p_1 \prec p_2$ and $p_2 \prec^* p_3$ imply neither $p_3 \prec p_1$ nor $p_3 \prec^* p_1$. (In the first case, $p_3 \prec p_1$ and $p_1 \prec p_2$ imply $p_3 \prec p_2$, contradicting $p_2 \prec^* p_3$. In the second case, $p_2 \prec^* p_3$ and $p_3 \prec^* p_1$ imply $p_2 \prec^* p_1$, contradicting $p_1 \prec p_2$). Therefore $p_1 \prec p_3$ or $p_1 \prec^* p_3$, which implies $p_1 \sqsubset p_3$. Notice that cases (iii) and (iv) are symmetric to (ii) and (i) respectively, so we have shown that relation $\sqsubset$ is transitive. The lemma follows from the observation that for each pair p_1, p_2, either $p_1 \sqsubset p_2$ or $p_2 \sqsubset p_1$. ■

6.10 Primal-Dual Approach to Sequence Comparison

Theorem 6.5 reveals a relation between the LCS and the minimum cover problem and leads to the idea of using minimum covers to find LCS. Below we describe a non-dynamic programming algorithm for simultaneous solution of the generalized alignment and minimum cover problems (Pevzner and Waterman, 1993 [273]).

Let $\mathcal{P} = p_1p_2 \ldots p_l$ be an arbitrary ordering of the elements of P, and $\mathcal{P}_i = p_1p_2 \ldots p_i$. Let $\mathcal{C}_i = \{C_1, C_2, \ldots, C_j\}$ be a cover of $\mathcal{P}_i$ by $\prec^*$-sequences, and let $p_1^{max}, p_2^{max}, \ldots, p_j^{max}$ be the $\prec^*$-maximum elements in $C_1, C_2, \ldots, C_j$, correspondingly. Let k be the minimum index ($1 \le k \le j$) fulfilling the following condition:

$$p_k^{max} \prec^* p_{i+1}, \tag{6.1}$$

and if condition (6.1) fails for all k, set $k = j + 1$. For convenience, define C_{j+1} as an empty set. The COVER algorithm constructs a cover $\mathcal{C}_{i+1}$ from $\mathcal{C}_i$ by adding p_{i+1} to C_k. If $k < j + 1$, COVER enlarges C_k:

$$\mathcal{C}_{i+1} = \{C_1, C_2, \ldots, C_{k-1}, C_k \bigcup \{p_{i+1}\}, C_{k+1}, \ldots, C_j\}.$$

If $k = j + 1$, COVER adds $\{p_{i+1}\}$ as a new $\prec^*$-sequence to the cover $\mathcal{C}_{i+1}$:

$$\mathcal{C}_{i+1} = \{C_1, C_2, \ldots, C_j, C_{j+1} = \{p_{i+1}\}\ \}.$$

The algorithm also keeps a *backtrack*:

$$b(p_{i+1}) = \begin{cases} p_{k-1}^{max}, & \text{if } k > 1 \\ \emptyset, & \text{otherwise} \end{cases}$$

Starting with an empty cover $\mathcal{C}_0$, COVER constructs a cover $\mathcal{C}_l$ of P after l iterations, and the size of this cover depends on the ordering of P. The size of the cover $\mathcal{C}_l$ is an upper bound for the length of the longest $\prec$-sequence. The following theorem shows that if $\mathcal{P}$ is the ordering of P in $\sqsubset$, then COVER is a primal-dual algorithm for simultaneous solutions of the longest $\prec$-sequence problem and the minimum cover by $\prec^*$-sequences problem.

Theorem 6.6 *If $\mathcal{P} = p_1p_2 \ldots p_l$ is the ordering of P in $\sqsubset$, then COVER constructs a minimum cover $\mathcal{C}_l = \{C_1, C_2, \ldots, C_t\}$ of P by $\prec^*$-sequences. The backtrack $b(p)$ defines a longest $\prec$-sequence of length t for each $p \in C_t$.*

Proof We show that for each i ($1 \le i \le l$), the cover $\mathcal{C}_i = \{C_1, C_2, \ldots, C_j\}$ satisfies the condition

$$\forall k > 1, \forall p \in C_k : \ \ b(p) \prec p. \tag{6.2}$$

Trivially, this condition holds for $\mathcal{C}_1$. We suppose that it holds for $\mathcal{C}_i$ and prove it for $\mathcal{C}_{i+1}$. Consider two cases:

- *Case 1.* $k < j + 1$ (condition (6.1)). In this case $b(p_{i+1}) = p_{k-1}^{max}$. Since $\mathcal{P}$ is the $\sqsubset$-ordering, $p_{k-1}^{max} \sqsubset p_{i+1}$, and therefore either $p_{k-1}^{max} \prec p_{i+1}$ or $p_{k-1}^{max} \prec^* p_{i+1}$. Since k is the minimum index fulfilling $p_k^{max} \prec^* p_{i+1}$, $p_{k-1}^{max} \prec p_{i+1}$, and therefore condition (6.2) holds for $\mathcal{C}_{i+1}$.

- *Case 2.* $k = j + 1$. In this case $b(p_{i+1}) = p_j^{max}$. Since $\mathcal{P}$ is the $\sqsubset$-ordering, either $p_j^{max} \prec p_{i+1}$ or $p_j^{max} \prec^* p_{i+1}$. Since $k = j + 1$, condition (6.1) fails for each $k \leq j$. Therefore, $p_j^{max} \prec p_{i+1}$, and condition (6.2) holds for $\mathcal{C}_{i+1}$.

Obviously each cover $\mathcal{C}_l = \{C_1, C_2, \ldots, C_t\}$ fulfilling condition (6.2) determines (through the backtrack) a $\prec$-sequence of length t for each $p \in C_t$. According to lemma 6.3, each such sequence is a $\prec$-longest sequence, and $\mathcal{C}_l$ is a minimum cover of P by $\prec^*$-sequences. ■

For the LCS between n-letter sequences, the COVER algorithm can be implemented in $O(nL)$ time, where L is the length of the longest common subsequence, or in $O((l+n)\log n)$ time, where l is the total number of matches between two sequences. Improvements to the classical dynamic programming algorithm for finding LCS have been suggested by Hirschberg, 1977 [164] and Hunt and Szymanski, 1977 [174]. In fact, both the Hirschberg and the Hunt-Szymanski algorithms can be viewed as implementations of the COVER algorithm with various data structures. $\prec^*$-chains in COVER correspond to the *k-candidates* in Hirschberg's algorithm. Maximal elements of $\prec^*$-chains in COVER correspond to the *dominant matches* in the Apostolico, 1986 [9] improvement of Hunt-Szymanski's algorithm.

6.11 Sequence Alignment and Integer Programming

Duality for the LCS is closely related to a new *polyhedral* approach to sequence comparison suggested by Reinert et al., 1997 [283]. They express the alignment problem as an integer linear program and report that this approach to multiple alignment can solve instances that are beyond present dynamic programming approaches. Below we describe the relationship between the LCS problem and integer programming.

Let P be a set with partial order $\prec$, and let $\prec^*$ be a conjugate partial order. Let x_e be the weight associated with $e \in P$. In the case of the LCS of sequences $v_1 \ldots v_n$ and $w_1, \ldots w_m$, P is the set of pairs of positions $e = (i, j)$ such that $v_i = w_j$ and $x_e = 1$ if and only if the positions i and j are matched in the LCS. The LCS problem can be formulated as the following integer programming problem:

$\sum_{x_e \in \alpha} x_e \leq 1$ for every (maximal) antichain α in P
$\sum_{e \in P} x_e \to \max$

Let y_α be a variable associated with a maximal antichain α in P. The *dual* program for the above problem is as follows:

$\sum_{e\in\alpha} y_\alpha \geq 1$ for every $e \in P$
$\sum_\alpha y_\alpha \to \min$

Since antichains in $\prec$ are chains in $\prec^*$, the above program is the *Minimum Path Cover Problem* in a directed acyclic graph representing the conjugate partial order $\prec^*$, which is known to have an integer solution (Cormen et al., 1989 [75]).

6.12 Approximate String Matching

Approximate string matching with k mismatches involves a string $t_1 \ldots t_n$, called the *text*, a shorter string, $q_1 \ldots q_p$, called the *query*, and integers k and m. The *query matching problen* is to find all m-substrings of the query $q_i \ldots q_{i+m-1}$ and the text $t_j \ldots t_{j+m-1}$ that match with at most k mismatches. In the case $p = m$, the *query matching* problem yields the *approximate string matching problem with k-mismatches*.

The approximate string matching problem with k-mismatches has been intensively studied in computer science. For $k = 0$, it reduces to classical string matching, which is solvable in $O(n)$ time (Knuth et al., 1977 [203], Boyer and Moore, 1977 [45]). For $k > 0$, the naive brute-force algorithm for approximate string matching runs in $O(nm)$ time. Linear-time algorithms for approximate string matching were devised by Ivanov, 1984 [177] and Landau and Vishkin, 1985 [213]. For a fixed-size alphabet, the worst-case running time of these algorithms is $O(kn)$.

Although these algorithms yield the best worst-case performance, they are far from being the best in practice (Grossi and Luccio, 1989 [140]). Consequently, several filtration-based approaches have emphasized the expected running time, in contrast to the worst-case running time (Baeza-Yates and Gonnet, 1989 [16], Grossi and Luccio, 1989 [140], Tarhio and Ukkonen, 1990 [334], Baeza-Yates and Perleberg, 1992 [17], Wu and Manber, 1992 [371]).

Using filtration algorithms for approximate string matching involves a two-stage process. The first stage preselects a set of positions in the text that are *potentially* similar to the query. The second stage verifies each potential position, rejecting potential matches with more than k mismatches. Denote as p the number of potential matches found at the first stage of the algorithm. Preselection is usually done in $\alpha n + O(p)$ time, where α is a small constant. If the number of potential matches is small and potential match verification is not too slow, this method yields a significant speed up.

The idea of filtration for the string matching problem first was described by Karp and Rabin, 1987 [189] for the case $k = 0$. For $k > 0$, Owolabi and McGregor, 1988 [258] used the idea of *l-tuple filtration* based on the simple observation that if a query approximately matches a text, then they share at least one l-tuple for sufficiently large l. All l-tuples shared by query and text can be easily found by hashing. If the number of shared l-tuples is relatively small, they can be verified, and all *real* matches with k mismatches can be rapidly located.

l-tuple filtration is based on the following simple observation:

Lemma 6.5 *If the strings* $x_1 \ldots x_m$ *and* $y_1 \ldots y_m$ *match with at most* k *mismatches then they share an* l*-tuple for* $l = \lfloor \frac{m}{k+1} \rfloor$*, i.e.,* $x_i \ldots x_{i+l-1} = y_j \ldots y_{j+l-1}$ *for some* $1 \leq i, j \leq m - l + 1$.

This lemma motivates an l-*tuple filtration* algorithm for query matching with k mismatches:

FILTRATION Algorithm Detection of all m-matches between a query and a text with up to k mismatches.

- *Potential match detection.* Find all matches of l-tuples in both the query and the text for $l = \lfloor \frac{m}{k+1} \rfloor$.
- *Potential match verification.* Verify each potential match by extending it to the left and to the right until either (i) the first $k + 1$ mismatches are found or (ii) the beginning or end of the query or the text is found.

Lemma 6.5 guarantees that FILTRATION finds *all* matches of length m with k or fewer mismatches. Potential match detection in FILTRATION can be implemented by hashing. The running time of FILTRATION is $\alpha n + O(pm)$, where p is the number of potential matches detected at the first stage of the algorithm and α is a small constant. For a Bernoulli text with A equiprobable letters, the expected number of potential matches is roughly $E(p) = \frac{nq}{A^l}$, yielding a fast algorithm for large A and l.

6.13 Comparing a Sequence Against a Database

A *dot-matrix* for sequences V and W is simply a matrix with each entry either 0 or 1, where a 1 at position (i, j) indicates that the l-tuples starting at the i-th position of V and the j-th position of W coincide. A popular protein database search tool, FASTA (Lipman and Pearson, 1985 [225]), uses l-tuple filtration with a usual setup of $l = 2$ (in amino acid alphabet). The positions of l-tuples present in both strings form an (implicit) dot-matrix representation of similarities between the strings. FASTA further assembles ones on the same diagonals of this dot-matrix and attempts to group close diagonals together.

Using shared l-tuples for finding similarities has some disadvantages. BLAST (Altschul et al., 1990 [5]), the dominant database search tool in molecular biology, uses substitution matrices to improve the construction of (implicit) dot-matrices for further analysis of diagonals. Essentially, it attempts to improve the filtration efficiency of FASTA by introducing more stringent rules to locate fewer and better potential matches. Another BLAST feature is the use of Altschul-Dembo-Karlin statistics (Karlin and Altschul, 1990 [186], Dembo and Karlin, 1991 [85]) for estimates of statistical significance. For any two l-tuples $x_1 \ldots x_l$ and $y_1 \ldots y_l$, BLAST defines the *segment score* as $\sum_{i=1}^{l} \delta(x_i, y_i)$, where $\delta(x, y)$ is the similarity score between amino acids x and y. A *maximal segment pair* (MSP) is a pair of l-tuples with the maximum score over all segment pairs in two sequences. A molecular biologist may be interested in all conserved segments, not only in their highest scoring pair. A segment pair is *locally maximal* if its score cannot be improved either by extending or by shortening both segments.

BLAST attempts to find all locally maximal segment pairs in the query sequence and the database with scores above some set threshold. The choice of the threshold is guided by Altschul-Dembo-Karlin statistics, which allows one to identify the lowest value of segment score that is unlikely to happen by chance. BLAST reports sequences that either have a segment score above the threshold or that do not have a segment score above a threshold but do have several segment pairs that in combination are statistically significant.

BLAST abandons the idea of l-tuple filtration and uses a different strategy to find potential matches. It finds all l-tuples that have scores above a threshold with some l-tuple in the query. This can be done either directly—by finding all approximate occurrences of substrings from the query in the database—or in a more involved way. For example, if the threshold is high enough, then the set of such strings is not too large, and the database can be searched for exact occurrences of the strings from this set. This is a well-studied combinatorial pattern matching problem, and the fast Aho and Corasick, 1975 [1] algorithm locates the occurrences of *all* these strings in the database. After the potential matches are located, BLAST attempts to extend them to see whether the resulting score is above the threshold. Altschul et al., 1997 [7] further improved BLAST by allowing insertions and deletions and combining matches on the same and close diagonals.

6.14 Multiple Filtration

For Bernoulli texts with A equiprobable letters, define the *filtration efficiency* of a filtration algorithm as the ratio $\frac{E(r)}{E(p)}$ of the expected number of matches with k mismatches $E(r)$ to the expected number of potential matches $E(p)$. For example, for $k = 1$, the efficiency of the l-tuple filtration, $\approx \frac{A-1}{A^{\lceil \frac{m}{2} \rceil}}$, decreases rapidly as m and A increase. This observation raises the question of devising a filtration method with increased filtration efficiency. The larger the efficiency ratio, the shorter the running time of the verification stage of filtration algorithm.

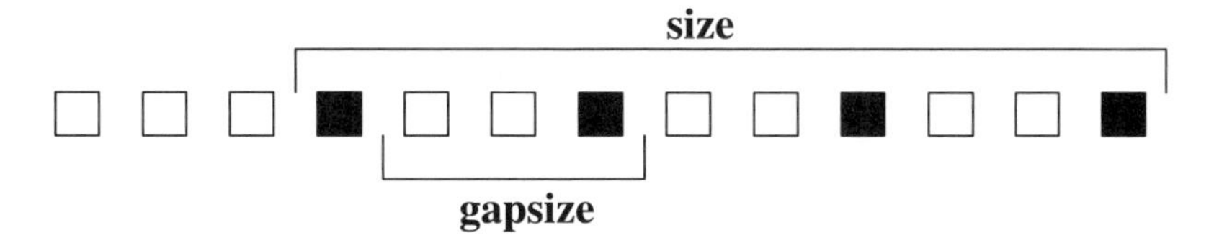

Gapped 4-tuple with gapsize 3 and size 10 starting at position 4

Figure 6.7: A gapped 4-tuple.

Pevzner and Waterman, 1995 [274] described an algorithm that allows the exponential reduction of the number of potential matches at the expense of a linear increase in the filtration time. This significantly reduces the time of the verification stage of the FILTRATION algorithm for the cost of linearly increased time at the detection stage. Taking into account that verification is frequently more time-consuming than detection, the technique provides a trade-off for an optimal choice of filtration parameters.

A set of positions $i, i+t, i+2t, \ldots, i+jt, \ldots, i+(l-1)t$ is called a *gapped l-tuple* with *gapsize* t and *size* $1+t(l-1)$ (Figure 6.7). Continuous l-tuples can be viewed as gapped l-tuples with gapsize 1 and size l. If an l-tuple shared by a pattern and a text starts at position i of the pattern and position j of the query, we call (i,j) the *coordinate* of the l-tuple. Define the *distance* $d(v_1, v_2)$ between l-tuples with coordinates (i_1, j_1) and (i_2, j_2) as

$$d(v_1, v_2) = \begin{cases} i_1 - i_2, & \text{if } i_1 - i_2 = j_1 - j_2 \\ \infty, & \text{otherwise.} \end{cases}$$

Multiple filtration is based on the following observation:

Lemma 6.6 *Let strings $x_1 \ldots x_m$ and $y_1 \ldots y_m$ match with at most k mismatches and $l = \lfloor \frac{m}{k+1} \rfloor$. Then these strings share both a continuous l-tuple and a gapped l-tuple with gapsize $k+1$, with distance d between them satisfying the condition $-k \leq d \leq m-l$.*

This lemma is the basis of a *double-filtration* algorithm for query matching with k mismatches:

DOUBLE-FILTRATION Algorithm Detection of all m-matches between a query and a text with up to k mismatches.

- *Potential match detection.* Find all continuous l-tuple matches between the query and the text that are within the distance $-k \leq d \leq m-l$ from a gapped l-tuple with gapsize $k+1$ match.

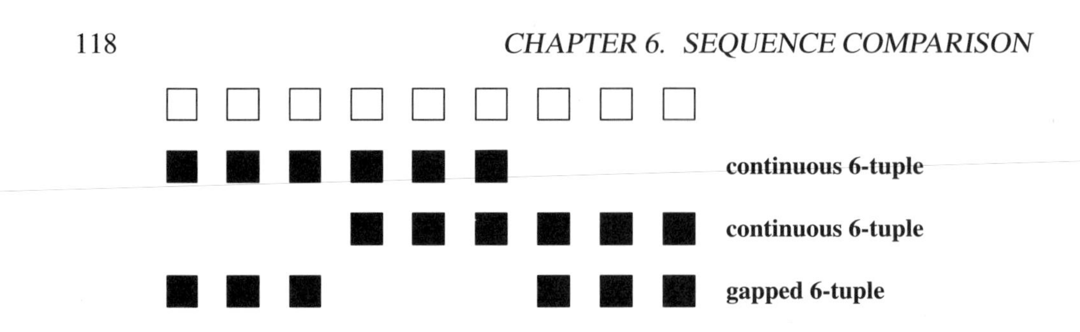

Figure 6.8: Every 9-match with one error contains either a continuous 6-match or a gapped 6-match.

- *Potential match verification.* Verify each potential match by extending it to the left and to the right until either (i) the first $k+1$ mismatches are found or (ii) the beginning or end of the query or the text is found.

Lemma 6.6 guarantees that DOUBLE-FILTRATION finds *all* matches with k mismatches. The efficiency of double filtration is approximately $\frac{A^{l-\delta}}{m(1-\frac{1}{A})}$ times greater than the efficiency of l-tuple filtration for a wide range of parameters (here, $\delta = \lceil \frac{l}{k+1} \rceil$).

Approximate pattern matching of 9-tuples with one error can be reduced to exact pattern matching of 4-tuples. We call it a reduction from a (9,1) pattern matching to a (4,0) pattern matching. Can we reduce a (9,1) pattern matching to a (6,0) pattern matching, thus improving the efficiency of the filtration? The answer is yes if we consider gapped 6-tuples as in Figure 6.8. Another question is whether we can achieve a speedup through reduction from (m, k) pattern matching to (m', k') matching for $0 < k' < k$. Exploring this problem led to the development of sublinear pattern matching algorithms (Chang and Lawler, 1994 [62], Myers, 1994 [245]).

6.15 Some Other Problems and Approaches

6.15.1 Parametric sequence alignment

Sequence alignment is sensitive to the choice of insertion/deletion and substitution penalties, and incorrect choice of these parameters may lead to biologically incorrect alignments. Understanding the influence of the parameters on the resulting alignment and choosing appropriate alignments are very important for biological applications. In the simplest model, when the alignment score is defined as # matches – μ# mismatches – σ# indels, different values of (μ, σ) correspond to different optimal alignments. However, some regions in (μ, σ)-space correspond to the same optimal alignment. In fact, (μ, σ)-space can be decomposed into convex polygons such that any two points in the same polygon correspond to the same optimal alignment (Fitch and Smith, 1983 [107]). Waterman et al., 1992 [360], Gusfield et al., 1994 [146] and Zimmer and Lengauer, 1997 [379] described efficient algorithms for computing a polygonal decomposition of the parameter space.

6.15.2 Alignment statistics and phase transition

Arratia and Waterman, 1989 [13] studied the statistical properties of the local alignment score of two random sequences. It turned out that the statistics of alignment heavily depends on the choice of indel and mismatch penalties. The local alignment score grows first logarithmically and then linearly with the length of the sequences, depending on gap penalties. Those two regions of growth are referred to as the "log region" and as "linear region," and the curve between these regions is called the *phase transition* curve.

Let $v_1 \dots v_n$ and $w_1 \dots w_n$ be two random i.i.d. strings. Let $S_n = S_n(\mu, \delta)$ and $H_n = H_n(\mu, \delta)$ be random variables corresponding to the score (# matches – μ# mismatches – σ# indels) of the global and local alignments of these strings.

Arratia and Waterman, 1994 [14] showed that

$$a = a(\mu, \delta) = \lim_{n \to \infty} \frac{S_n}{n}$$

exists in probability. Moreover, $\{a = 0\} = \{(\mu, \delta) : a(\mu, \delta) = 0\}$ is a continuous phase transition curve separating $[0, \infty]^2$ into two components: $\{a < 0\} = \{(\mu, \delta) : a(\mu, \delta) < 0\}$ and $\{a > 0\} = \{(\mu, \delta) : a(\mu, \delta) > 0\}$.

For the case of $(\mu, \delta) \in \{a > 0\}$, Arratia and Waterman, 1994 [14] showed, that $\lim_{n \to \infty} \frac{H_n(\mu, \delta)}{n} = a(\mu, \delta)$. For the case of $(\mu, \delta) \in \{a < 0\}$, they introduced a constant $b = b(\mu, \delta)$ such that

$$\lim_{n \to \infty} \mathbf{P}\{(1 - \epsilon)b < \frac{H_n(\mu, \delta)}{\log(n)} < (2 + \epsilon)b\} = 1.$$

The problem of computing $a(\mu, \delta)$ for $(\mu, \delta) \in \{a > 0\}$ is difficult. In particular, computing $a(0, 0)$ would solve the Steele-Arratia conjecture. See Vingron and Waterman, 1994 [346] and Bundschuh and Hwa, 1999 [52] for estimates of alignment significance and parameter choice in both the "log region" and the "linear region", and see Waterman and Vingron, 1994 [365] for a fast algorithm to compute the probability that a local alignment score is the result of chance alone.

6.15.3 Suboptimal sequence alignment

The optimal alignment is sometimes not the biologically correct one, and ' methods are sought to generate a set of Δ-suboptimal alignments whose deviation from the optimal is at most Δ (Waterman, 1983 [355]). The problem is equivalent to finding suboptimal paths in the edit graph (Chao, 1994 [63], Naor and Brutlag, 1994 [250]).

6.15.4 Alignment with tandem duplications

Although most alignment algorithms consider only insertions, deletions, and substitutions, other mutational events occur. *Tandem duplication* is a mutational event in which a stretch of DNA is duplicated to produce one or more new copies, each copy following the preceding one in a contiguous fashion. This mutation is a rather common one, making up an estimated 10% of the human genome. Tandem repeats have been implicated in a number of inherited human diseases, including Huntington's disease. Benson, 1997 [30] suggested an efficient algorithm for sequence alignment with tandem duplications. Algorithms for *detecting* tandem repeats were proposed by Landau and Schmidt, 1993 [212], Milosavljevic and Jurka, 1993 [237], and Benson, 1998 [31].

6.15.5 Winnowing database search results

In database searches matches to biologically important regions are frequently obscured by other matches. A large number of matches in one region of the sequence may hide lower-scoring but important matches occurring elsewhere. Since database search programs often report a fixed number of top matches and truncate the output, rules are needed to select a subset of the matches that reveal all important results. The problem is modeled by a list of intervals (alignment regions) with associated alignment scores. If interval I is contained in interval J with a higher score, then I is *dominated* by J. The *winnowing problem* is to identify and discard intervals that are dominated by a fixed number of other intervals. Berman et al., 1999 [34] implemented a version of BLAST that solves the winnowing problem in $O(n \log n)$ time, where n is the number of intervals.

6.15.6 Statistical distance between texts

Let $\mathcal{X}$ be a set of strings—for example, the set of all l-tuples (Blaisdell, 1988 [36]) or gapped l-tuples (Mironov and Alexandrov, 1988 [239]) for a small l. Given a string $x \in \mathcal{X}$ and a text T, define $x(T)$ as the number (or frequency) of occurrences of x in T. Blaisdell, 1988 [36] and Mironov and Alexandrov, 1988 [239] defined the *statistical distance* between texts V and W as

$$d(V,W) = \sqrt{\sum_{x \in \mathcal{X}} (x(V) - x(W))^2}$$

well before Internet search engines started to use related measures to find similar pages on the Web. The efficiency of statistical distance for finding similarities was studied by Pevzner, 1992 [266]. The advantage of the statistical distance method over BLAST is the speed: statistical distance can be computed very fast with database pre-processing. The disadvantage is that statistical distance can miss weak similarities that do not preserve shared l-tuples. As a result, the major application of such algorithms is in "database versus database" comparisons, such as EST clustering. To achieve the very high speed required for large EST databases, the statistical distance approach was recently implemented with suffix arrays (Burkhardt et al., 1999 [55]).

6.15.7 RNA folding

RNAs adopt sophisticated 3-dimensional structures that are important for signal recognition and gene regulation. Pairs of positions in RNA with complementary Watson-Crick bases can form bonds. Bonds (i, j) and (i', j') are interleaving if $i < i' < j < j'$ and non-interleaving otherwise. In a very naive formulation of the RNA folding problem, one tries to find a maximum set of non-interleaving bonds. The problem can be solved by dynamic programming (Nussinov et al., 1978 [253], Waterman, 1978 [354]). In a more adequate model, one attempts to find an RNA fold with minimum energy (Zuker and Sankoff, 1984 [381], Waterman and Smith, 1986 [363], Zuker, 1989 [380]). However, these algorithms are not very reliable. A more promising approach is to derive an RNA fold through multiple alignment of related RNA molecules. Eddy and Durbin, 1994 [95] studied a problem of RNA multiple alignment that takes fold information into account.

Chapter 7

Multiple Alignment

7.1 Introduction

The goal of protein sequence comparison is to discover "biological" (i.e., structural or functional) similarities among proteins. Biologically similar proteins may not exhibit a strong sequence similarity, and one would like to recognize the structural/functional resemblance even when the sequences are very different. If sequence similarity is weak, pairwise alignment can fail to identify biologically related sequences because weak pairwise similarities may fail the statistical test for significance. Simultaneous comparison of many sequences often allows one to find similarities that are invisible in pairwise sequence comparison. To quote Hubbard et al., 1996 [170] "pairwise alignment whispers... multiple alignment shouts out loud."

Straightforward dynamic programming solves the multiple alignment problem for k sequences of length n. Since the running time of this approach is $O((2n)^k)$, a number of different variations and some speedups of the basic algorithm have been devised (Sankoff, 1975 [298], Sankoff, 1985 [299], Waterman et al., 1976 [364]). However, the exact multiple alignment algorithms for large k are not feasible (Wang and Jiang, 1994 [351]), and many heuristics for suboptimal multiple alignment have been proposed.

A natural heuristic is to compute $\binom{k}{2}$ optimal pairwise alignments of the k strings and combine them together in such a way that induced pairwise alignments are close to the optimal ones. Unfortunately, it is not always possible to combine pairwise alignments into multiple alignments, since some pairwise alignments may be incompatible. As a result, many multiple alignment algorithms attempt to combine some compatible subset of optimal pairwise alignments into a multiple alignment. This can be done for some small subsets of all $\binom{k}{2}$ pairwise alignments. The problem is deciding which subset of pairwise alignments to choose for this procedure.

The simplest approach uses pairwise alignment to iteratively add one string to a growing multiple alignment. Feng and Doolittle, 1987 [100] use the pair of strings with greatest similarity and "merge" them together into a new string following the principle "once a gap, always a gap." As a result, the multiple alignment of k sequences is reduced to the multiple alignment of $k-1$ sequences. Many other iterative multiple alignment algorithms use similar strategies (Barton and Sternberg, 1987 [27], Taylor, 1987 [336], Bains, 1986 [22], Higgins et al., 1996 [162]).

Although the Feng and Doolittle, 1987 [100] algorithm works well for close sequences, there is no "performance guarantee" for this method. The first "performance guarantee" approximation algorithm for multiple alignment, with approximation ratio $2-\frac{2}{k}$, was proposed by Gusfield, 1993 [144]. The idea of the algorithm is based on the notion of compatible alignments and uses the principle "once a gap, always a gap."

Feng and Doolittle, 1987 [100] and Gusfield, 1993 [144] use optimal pairwise (2-way) alignments as building blocks for multiple k-way alignments. A natural extension of this approach is to use optimal 3-way (or l-way) alignments as building blocks for k-way alignments. However, this approach faces some combinatorial problems since it is not clear how to define compatible l-way alignments and how to combine them. Bafna et al., 1997 [18] devised an algorithm for this problem with approximation ratio $2 - \frac{l}{k}$.

Biologists frequently depict similarities between two sequences in the form of *dot-matrices*. A dot-matrix is simply a matrix with each entry either 0 or 1, where a 1 at position (i, j) indicates some similarity between the i-th position of the first sequence and the j-th position of the second one. The similarity criteria vary from being purely combinatorial (e.g., a match of length m with at most k mismatches starting at position i of the first sequence and j of the second one) to using correlation coefficients between physical parameters of amino acids. However, no criterion is perfect in its ability to distinguish "real" (biologically relevant) similarities from chance similarities (noise). In biological applications, noise disguises the real similarities, and the problem is determining how to filter noise from dot-matrices.

The availability of several sequences sharing biologically relevant similarities helps in filtering noise from dot-matrices. When k sequences are given, one can calculate $\binom{k}{2}$ pairwise dot-matrices. If all k sequences share a region of similarity, then this region should be visible in all $\binom{k}{2}$ dot-matrices. At the same time, noise is unlikely to occur consistently among all the dot-matrices. The practical problem is to reverse this observation: given the $\binom{k}{2}$ dot-matrices, find similarities shared by all or almost all of the k sequences and filter out the noise. Vingron and Argos, 1991 [344] devised an algorithm for assembling a k-dimensional dot-matrix from 2-dimensional dot-matrices.

7.2 Scoring a Multiple Alignment

Let $\mathcal{A}$ be a finite *alphabet* and let $a_1, \ldots, a_k$ be k sequences (strings) over $\mathcal{A}$. For convenience, we assume that each of these strings contains n characters. Let $\mathcal{A}'$ denote $\mathcal{A} \bigcup \{-\}$, where $'-'$ denotes space. An *alignment* of strings $a_1, ..., a_k$ is specified by a $k \times m$ matrix A, where $m \geq n$. Each element of the matrix is a member of $\mathcal{A}'$, and each row i contains the characters of a_i in order, interspersed with $m - n$ spaces. We also assume that every column of the multiple alignment matrix contains at least one symbol from $\mathcal{A}$. The score of multiple alignment is defined as the sum of scores of the columns and the optimal alignment is defined as the alignment that minimizes the score.

The score of a column can be defined in many different ways. The intuitive way is to assign higher scores to the columns with large variability of letters. For example, in the *multiple shortest common supersequence* problem, the score of a column is defined as the number of different characters from $\mathcal{A}$ in this column. In the *multiple longest common subsequence* problem, the score of a column is defined as -1 if all the characters in the column are the same, and 0 otherwise. In the more biologically adequate *minimum entropy* approach, the score of multiple alignment is defined as the sum of entropies of the columns. The entropy of a column is defined as

$$-\sum_{x \in \mathcal{A}'} p_x \log p_x$$

where p_x is the frequency of letter $x \in \mathcal{A}'$ in a column i. The more variable the column, the higher the entropy. A completely conserved column (as in the multiple LCS problem) would have minimum entropy 0.

The minimal entropy score captures the biological notion of good alignment, but it is hard to efficiently analyze in practice. Below we describe *Distance from Consensus* and *Sum-of-Pairs (SP)* scores, which are easier to analyze.

- *Distance from Consensus.* The consensus of an alignment is a string of the most common characters in each column of the multiple alignment. The *Distance from Consensus* score is defined as the total number of characters in the alignment that differ from the consensus character of their columns.

- *Sum-of-Pairs (SP-score).* For a multiple alignment $A = (a_{ih})$, the induced score of pairwise alignment A_{ij} for sequences a_i and a_j is

  $$s(A_{ij}) = \sum_{h=1}^{m} d(a_{ih}, a_{jh}),$$

  where d specifies the *distance* between elements of $\mathcal{A}'$. The *Sum-of-Pairs score* (SP-score) for alignment A is given by $\sum_{i,j} s(A_{ij})$. In this definition, the score of alignment A is the sum of the scores of *projections* of A onto all pairs of sequences a_i and a_j. We assume the metric properties for distance d, so that $d(x,x) = 0$ and $d(x,z) \leq d(x,y) + d(y,z)$ for all x, y, and z in $\mathcal{A}'$.

7.3 Assembling Pairwise Alignments

Feng and Doolittle, 1987 [100] use the pair of strings with greatest similarity and "merge" them together into a new string following the principle "once a gap, always a gap." As a result, the multiple alignment of k sequences is reduced to the multiple alignment of $k-1$ sequences (one of them corresponds to the merged strings). The motivation for the choice of the closest strings at the early steps of the algorithm is that close strings provide the most reliable information about alignment.

Given an alignment A of sequences $a_1, ..., a_k$ and an alignment A' of some subset of the sequences, we say that A is *compatible* with A' if A aligns the characters of the sequences aligned by A' in the same way that A' aligns them. Feng and Doolittle, 1987 [100] observed that given any tree in which each vertex is labeled with a distinct sequence a_i, and given pairwise alignments specified for each tree edge, there exists a multiple alignment of the k sequences that is compatible with each of the pairwise alignments. In particular, this result holds for a star on k vertices, i.e., a tree with *center* vertex and $k-1$ leaves.

Lemma 7.1 *For any star and any specified pairwise alignments $A_1, \ldots, A_{k-1}$ on its edges, there is an alignment A for the k sequences that is compatible with each of the alignments $A_1, ..., A_{k-1}$.*

Given a star G, define a *star-alignment* A_G as an alignment compatible with optimal pairwise alignments on the edges of this star. The alignment A_G optimizes $k-1$ among $\binom{k}{2}$ pairwise alignments in SP-score. The question of how good the star alignment is remained open until Gusfield, 1993 [144] proved that if the star G is chosen properly, the star alignment approximates the optimal alignment with ratio $2 - \frac{2}{k}$.

Let $G(V,E)$ be an (undirected) graph, and let $\gamma(i,j)$ be a (fixed) shortest path between vertices $i \neq j \in V$ of length $d(i,j)$. For an edge $e \in E$, define the communication cost $c(e)$ as the number of shortest paths $\gamma(i,j)$ in G that use edge e. For example, the communication cost of every edge in a star with k vertices is $k-1$. Define the communication cost of the graph G as $c(G) = \sum_{e \in E} c(e) = \sum_{i \neq j \in V} d(i,j)$. The *complete graph* on k vertices H_k has a minimum communication cost of $c(G) = \frac{k(k-1)}{2}$ among all k-vertex graphs. We call $b(G) = \frac{c(G)}{c(H_k)} = 2\frac{c(G)}{k(k-1)}$ the *normalized communication cost* of G. For a star with

k vertices, $b(G) = 2 - \frac{2}{k}$. Feng and Doolittle, 1987 [100] use a tree to combine pairwise alignments into a multiple alignment, and it turns out that the normalized communication cost of this tree is related to the approximation ratio of the resulting heuristic algorithm.

Define $C(G) = (c_{ij})$ as a $k \times k$ matrix with $c_{ij} = c(e)$ if (i, j) is an edge e in G and $c_{ij} = 0$ otherwise. The *weighted sum-of-pairs score* for alignment A is

$$\sum_{i,j} c_{ij} \cdot s(A_{ij})$$

.

For notational convenience, we use the *matrix dot product* to denote scores of alignments. Thus, letting $S(A) = (s(A_{ij}))$ be the matrix of scores of pairs of sequences, the weighted sum-of-pairs score is $C(G) \odot S(A)$. Letting E be the unit matrix consisting of all 1s except the main diagonal consisting of all 0s, the (unweighted) sum-of-pairs score of alignment A is $E \odot S(A)$.

The pairwise scores of an alignment inherit the triangle inequality property from the distance matrix. That is, for any alignment A, $s(A_{ij}) \leq s(A_{ik}) + s(A_{kj})$, for all i, j, and k. This observation implies the following:

Lemma 7.2 *For any alignment A of k sequences and a star G, $E \odot S(A) \leq C(G) \odot S(A)$.*

7.4 Approximation Algorithm for Multiple Alignments

Let $\mathcal{G}$ be a collection of stars in a k-vertex graph. We say that the collection $\mathcal{G}$ is *balanced* if $\sum_{G \in \mathcal{G}} C(G) = pE$ for some scalar $p > 1$. For example, a collection of k stars with k different center vertices is a balanced collection with $p = 2(k-1)$. Since $C(G)$ is non-zero only at the edges of the star G, and since star alignment A_G induces optimal alignments on edges of G,

$$C(G) \odot S(A_G) \leq C(G) \odot S(A)$$

for any alignment A.

Lemma 7.3 *If $\mathcal{G}$ is a balanced set of stars, then*

$$\min_{G \in \mathcal{G}} C(G) \odot S(A_G) \leq \frac{p}{|\mathcal{G}|} \min_{A} E \odot S(A)$$

Proof We use an averaging argument.

$$\begin{aligned}\min_{G\in\mathcal{G}} C(G) \odot S(A_G) &\leq \tfrac{1}{|\mathcal{G}|} \textstyle\sum_{G\in\mathcal{G}} C(G) \odot S(A_G) \\ &\leq \tfrac{1}{|\mathcal{G}|} \cdot \textstyle\sum_{G\in\mathcal{G}} C(G) \odot S(A) = \tfrac{p}{|\mathcal{G}|} \cdot E \odot S(A)\end{aligned}$$

Here the inequality holds for an arbitrary alignment A, and in particular, it holds for an optimal alignment. ∎

Lemmas 7.2 and 7.3 motivate the **Align** algorithm:

1. Construct a balanced set of stars, $\mathcal{G}$.
2. For each star G in $\mathcal{G}$, assemble a star alignment A_G.
3. Choose a star G such that $C(G) \odot S(A_G)$ is the minimum over all stars in $\mathcal{G}$.
4. Return A_G.

Theorem 7.1 *(Gusfield, 1993 [144]) Given a balanced collection of stars $\mathcal{G}$,* **Align** *returns an alignment with a performance guarantee of $2 - 2/k$ in $O(k \cdot n^2 \cdot |\mathcal{G}|)$ time.*

Proof Note that $\frac{p}{|\mathcal{G}|} = \frac{C(G)\odot E}{E\odot E} = 2-\frac{2}{k}$. **Align** returns the alignment A_G that is optimal for a star $G \in \mathcal{G}$, and for which the smallest score, $\min_{G\in\mathcal{G}} C(G) \odot S(A_G)$, is achieved. Lemmas 7.2 and 7.3 imply that $E \odot S(A_G) \leq C(G) \odot S(A_G) \leq \left(2 - \frac{2}{k}\right) \cdot \min_A E \odot S(A)$. ∎

7.5 Assembling l-way Alignments

An *l-star* $G = (V, E)$ on k vertices is defined by $r = \frac{k-1}{l-1}$ cliques of size l whose vertex sets intersect in only one *center* vertex (Figure 7.1). For a 3-star with $k = 2t + 1$ vertices, $c(G) = (2t - 1)2t + t$ and $b(G) = 2 - \frac{3}{k}$. For an l-star with $k = (l-1)t+1$ vertices, $c(G) = ((l-1)t+1-l+1)(l-1)t+t(\frac{l(l-1)}{2} - l+1)$ and $b(G) = 2 - \frac{l}{k}$. The communication cost of an edge e in an l-star G with center c is

$$c(e) = \begin{cases} k - l + 1, & \text{if } e \text{ is incident to } c \\ 1, & \text{otherwise.} \end{cases}$$

Note that for the communication cost matrix of an l-star G,

$$C(G) \odot E = (k-l+1) \cdot (k-1) + \frac{k-1}{l-1}\binom{l-1}{2} = \binom{k}{2} \cdot \left(2 - \frac{l}{k}\right)$$

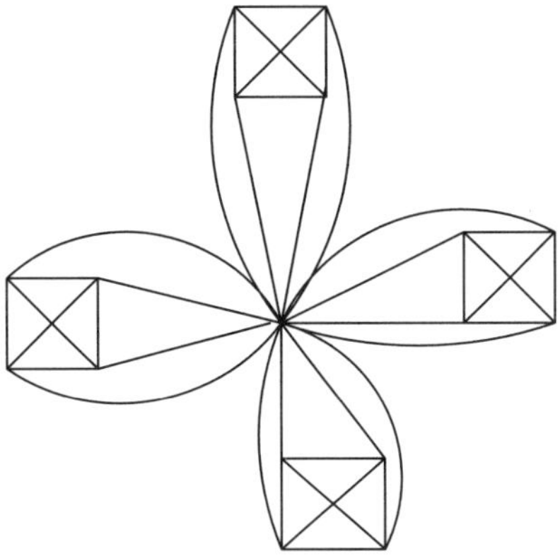

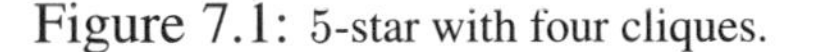

Figure 7.1: 5-star with four cliques.

Let $A_1 \ldots, A_r$ be alignments for the r cliques in the l-star, with each A_i aligning l sequences. A construction similar to Feng and Doolittle, 1987 [100] implies an analog of lemma 7.1:

Lemma 7.4 *For any l-star and any specified alignments $A_1, ..., A_r$ for its cliques, there is an alignment A for the k sequences that is compatible with each of the alignments $A_1, ..., A_r$.*

One can generalize the **Align** algorithm for l-stars and prove analogs of lemmas 7.2 and 7.3. As a result, theorem 7.1 can be generalized for l-stars, thus leading to an algorithm with the performance guarantee equal to the normalized communication cost of l-stars, which is $2 - \frac{l}{k}$. The running time of this algorithm is $O(k(2n)^l|\mathcal{G}|)$. Therefore, the problem is finding a small balanced set of l-stars $\mathcal{G}$.

We have reduced the multiple alignment problem to that of finding an optimal alignment for each clique in each l-star in a balanced set $\mathcal{G}$. How hard is it to find a balanced set $\mathcal{G}$? A trivial candidate is simply the set of all l-stars, which is clearly balanced by symmetry. For $l = 2$, Gusfield, 1993 [144] exploited the fact that there are only k 2-stars to construct a multiple alignment algorithm with an approximation ratio of $2 - \frac{2}{k}$. This is really a special case, as for $l > 2$, the number of l-stars grows exponentially with k, making the algorithm based on generation of all l-stars computationally infeasible. Constructing a *small* balanced set of l-stars is not trivial. Pevzner, 1992 [265] solved the case of $l = 3$ by mapping the problem to maximum matching in graphs. Bafna et al., 1997 [18] further designed a $2 - \frac{l}{k}$ approximation algorithm for arbitrary l.

7.6 Dot-Matrices and Image Reconstruction

Given k sequences, it is easy to generate $\binom{k}{2}$ pairwise dot-matrices. It is much harder to assemble these pairwise dot-matrices into a k-dimensional dot-matrix and to find regions of similarity shared by all or almost all k sequences. To address this problem, Vihinen, 1988 [343] and Roytberg, 1992 [290] proposed "superimposing" pairwise dot-matrices by choosing one reference sequence and relating all others to it. Below we describe the Vingron and Argos, 1991 [344] algorithm for assembling a k-dimensional dot-matrix from 2-dimensional dot-matrices.

We represent the problem of assembling pairwise similarities in a simple geometric framework. Consider M integer points in k-dimensional space,

$$(i_1^1, \ldots, i_k^1), \ldots, (i_1^M, \ldots, i_k^M),$$

for which we do not know the coordinates. Suppose we observe the projections of these points onto each pair of dimensions s and t, $1 \le s < t \le k$:

$$(i_s^1, i_t^1), \ldots, (i_s^M, i_t^M)$$

as well as some other points (*noise*). Suppose also that we cannot distinguish points representing *real* projections from ones representing noise. The *k-dimensional image reconstruction* problem is to reconstruct M k-dimensional points given $\binom{k}{2}$ projections (with noise) onto coordinates s and t for $1 \le s < t \le k$.

In this construction, each similarity (consensus element) shared by k biological sequences corresponds to an integer point $(i_1, \ldots i_k)$ in k-dimensional space, where i_s is the coordinate of the consensus in the s-th sequence, $1 \le s \le k$. In practice, it is hard to find the integer points $(i_1, \ldots, i_k)$ corresponding to consensuses. On the other hand, it is easy to find (though with considerable noise) the projections (i_s, i_t) of all consensuses $(i_1, \ldots i_k)$ onto every pair of coordinates s and t. This observation establishes the link between the multiple alignment problem and the k-dimensional image reconstruction problem.

From the given dots in the side-planes, we propose to keep only those that fulfill the following criterion of consistency: the point (i, j) in projection s, t is called *consistent* if for every other dimension u there exists an integer m such that (i, m) belongs to the projection s, u and (j, m) belongs to the projection t, u (Gotoh, 1990 [135]). Obviously each "real" point, i.e., each one that was generated as a projection of a k-dimensional point, is consistent. In contrast, random points representing noise are expected to be inconsistent. This observation allows one to filter out most (though possibly not all) of the noise and leads to the Vingron and Argos, 1991 [344] algorithm that multiplies and compares dot-matrices.

7.7 Multiple Alignment via Dot-Matrix Multiplication

We model the collection of $\binom{k}{2}$ dot-matrices as a *k-partite graph* $G(V_1 \cup V_2 \cup \ldots \cup V_k, E)$, where V_i is the set of positions in the i-th sequence. We join the vertex $i \in V_s$ with $j \in V_t$ by an (undirected) edge e if there exists a dot at position (i, j) of the dot-matrix comparing sequences s and t. An edge $e \in E$ will be written as $e = (s, i|t, j)$ to indicate that it joins vertices $i \in V_s$ and $j \in V_t$. We denote a *triangle* formed by three edges $(s, i|t, j), (t, j|u, m)$, and $(s, i|u, m)$ as $(s, i|t, j|u, m)$. We now define an edge $(s, i|t, j)$ to be *consistent* if for every $u \neq s, t$, $1 \leq u \leq k$ there exists a triangle $(s, i|t, j|u, m)$ for some m. A subset $E' \subseteq E$ is called *consistent* if for all edges $(s, i|t, j) \in E'$ there exist triangles $(s, i|t, j|u, m)$, $\forall u \neq s, t$, $1 \leq u \leq k$, with all edges of these triangles in E'. The k-partite graph G is defined as *consistent* if its edge-set is consistent. Clearly, if $G'(V, E')$ and $G''(V, E'')$ are consistent graphs, then their *union* $G(V, E' \cup E'')$ is consistent. Therefore, we can associate with any k-partite graph a unique maximal consistent subgraph. Our interest is in the following:

Graph Consistency Problem Find the maximal consistent subgraph of an n-partite graph.

The dot-matrix for sequences s and t is defined as *adjacency* matrix A_{st}:

$$(A_{st})_{ij} = \begin{cases} 1, & \text{if } (s, i|t, j) \in E \\ 0, & \text{otherwise} \end{cases}$$

Each such matrix corresponds to a subset of E, and we will apply the operations $\cup, \cap$, and $\subset$ to the matrices A_{st}. We reformulate the above definition of consistency in terms of boolean multiplication (denoted by "$\circ$") of the adjacency matrices (Vingron and Argos, 1991 [344]). A k-partite graph is consistent if and only if

$$A_{st} \subseteq A_{su} \circ A_{ut} \quad \forall s, t, u : 1 \leq s, t, u \leq k, s \neq t \neq u. \tag{7.1}$$

Characterization (7.1) suggests the following simple procedure to solve the consistency problem: keep only those 1s in the adjacency matrix that are present both in the matrix itself and in all products $A_{su} \circ A_{ut}$. Doing this once for all adjacency matrices will also change the matrices used for the products. This leads to the iterative matrix multiplication algorithm (superscripts distinguish different iterations) that starts with the adjacency matrices $A_{st}^{(0)} := A_{st}$ of the given k-partite graph and defines:

$$A_{st}^{(l+1)} := A_{st}^{(l)} \cap \left(\bigcap_{u \neq s,t} A_{su}^{(l)} \circ A_{ut}^{(l)} \right).$$

Once this is done for all indices s and t, the process is repeated until at some iteration m $A_{st}^{(l+1)} = A_{st}^{(l)}$ for all $1 \leq s, t \leq k$.

The dot-matrix multiplication algorithm (Vingron and Argos, 1991 [344]) converges to the maximal consistent subgraph and requires $O(L^3k^3)$ time per iteration (L is the length of the sequences). Since the number of iterations may be very large, Vingron and Pevzner, 1995 [345] devised an algorithm running in time $O(L^3k^3)$ overall, which is equivalent to the run-time of only one iteration of the matrix-multiplication algorithm. In practical applications the input data for the algorithm are sparse matrices, which makes the algorithm even faster. Expressed in the overall number M of dots in all $\binom{k}{2}$ dot-matrices, the running time is $O(kLM)$.

7.8 Some Other Problems and Approaches

7.8.1 Multiple alignment via evolutionary trees

It often happens that in addition to sequences $a_1, \ldots, a_k$, biologists know (or assume that they know) the evolutionary history (represented by an *evolutionary tree*) of these sequences. In this case, $a_1, \ldots, a_k$ are assigned to the leaves of the tree, and the problem is to reconstruct the ancestral sequences (corresponding to internal vertices of the tree) that minimize the overall number of mutations on the edges of the tree. The score of an edge in the tree is the edit distance between sequences assigned to its endpoints, and the score of the evolutionary tree is the sum of edge scores over all edges of the tree. The optimal multiple alignment for a given evolutionary tree is the assignment of sequences to internal vertices of the tree that produces the minimum score (Sankoff, 1975 [298]). Wang et al., 1996 [352] and Wang and Gusfield, 1996 [350] developed performance guarantee approximation algorithms for evolutionary tree alignment.

7.8.2 Cutting corners in edit graphs

Carrillo and Lipman, 1988 [58] and Lipman et al., 1989 [224] suggested a branch-and-bound technique for multiple alignment. The idea of this approach is based on the observation that if one of the pairwise alignments imposed by a multiple alignment is bad, then the overall multiple alignment won't have a good score. This observation implies that "good" multiple alignment imposes "good" pairwise alignments, thus limiting a search to the vicinity of a main diagonal in a k-dimensional alignment matrix.

Chapter 8

Finding Signals in DNA

8.1 Introduction

Perhaps the first signal in DNA was found in 1970 by Hamilton Smith after the discovery of the Hind II restriction enzyme. The palindromic site of the restriction enzyme is a signal that initiates DNA cutting. Finding the sequence of this site was not a simple problem in 1970; in fact, Hamilton Smith published two consecutive papers on Hind II, one on enzyme purification and the other one on finding the enzyme's recognition signal (Kelly and Smith, 1970 [196]).

Looking back to the early 1970s, we realize that Hamilton Smith was lucky: restriction sites are the simplest signals in DNA. Thirty years later, they remain perhaps the only signals that we can reliably find in DNA. Most other signals (promoters, splicing sites, etc.) are so complicated that we don't yet have good models or reliable algorithms for their recognition.

Understanding gene regulation is a major challenge in computational biology. For example, regulation of gene expression may involve a protein binding to a region of DNA to affect transcription of an adjacent gene. Since protein-DNA binding mechanisms are still insufficiently understood to allow *in silico* prediction of binding sites, the common experimental approach is to locate the approximate position of the binding site. These experiments usually lead to identification of a DNA fragment of length n that contains a binding site (an unknown *magic word*) of length $l \ll n$. Of course, one such experiment is insufficient for finding the binding site, but a sample of experimentally found DNA fragments gives one hope of recovering the magic word.

In its simplest form, the signal finding problem (and the restriction enzyme site problem in particular) can be formulated in the following way. Suppose we are given a sample of K sequences, and suppose there is an (unknown) magic word that appears at different (unknown) positions in those sequences. Can we find the magic word?

A common-sense approach to the magic word problem is to test all words of length l and to find those that appear in all (or almost all) sequences from the sample (Staden, 1989 [326], Wolfertstetter et al., 1996 [370], Tompa, 1999 [338]). If the magic word is the only word that appears that frequently in the sample, then the problem is (probably) solved. Otherwise we should increase l and repeat the procedure.

The described approach usually works fine for short continuous words such as $GAATTC$, the restriction site of EcoRI. However, if the length of sequences in the sample nears 4^6, random words may start competing with the magic word, since some of them may appear in many sequences simply by chance. The situation becomes even more difficult if the nucleotide frequencies in the sample have a skewed distribution.

The problem gets even more complicated when the magic word has gaps, as in $CCAN_9TGG$, the site of the Xcm I restriction enzyme (N stands for any nucleotide, and N_9 indicates a gap of length 9 in the site). Of course, we can try to enumerate all patterns with gaps, but the computational complexity of the problem grows very rapidly, particularly if we allow for patterns with many gaps. Even finding the relatively simple "letter-gap-letter-gap-letter" magic word is not that simple anymore; at least it warranted another pattern-finding paper by Hamilton Smith and colleagues 20 years after the discovery of the first restriction site pattern (Smith et al., 1990 [318]). Another daunting recognition task is to find signals like $Pu^mCN_{40-2000}Pu^mC$, the recognition site of McrBC Endonuclease (Pu stands for A or G).

While the above problems are not simple, the real biological problems in signal finding are much more complicated. The difficulty is that biological signals may be long, gapped, and fuzzy. For example, the magic word for *E.coli* promoters is $TTGACAN_{17}TATAAT$. Enumeration and check of all patterns of this type is hardly possibly due to computational complexity. However, even if we enumerated all patterns of this type, it would be of little help since the pattern above represents an *ideal* promoter but never occurs in known promoter sequences. Rather, it is a consensus of all known promoters: neither consensus bases nor spacing between the two parts of the signal are conserved. In other words, the description of the magic word in this case is something like "12 non-degenerate positions with one gap and a maximum of 4 mismatches." There is no reliable algorithm to find this type of signal yet. The shortcoming of the existing algorithms is that, for subtle signals, they often converge to local minima that represent random patterns rather than a signal.

8.2 Edgar Allan Poe and DNA Linguistics

When William Legrand from Edgar Allan Poe's novel "The Gold-Bug" found a parchment written by the pirate Captain Kidd:

5 3 ‡‡†3 0 5)) 6 * ; 4 8 2 6) 4 ‡.) 4 ‡) : 8 0 6 * ; 4 8 †8 ¶6 0)) 8
5 ; 1 ‡(; : ‡* 8 †8 3 (8 8) 5 * †; 4 6 (8 8 * 9 6 * ? ; 8) * ‡(; 4 8 5
) ; 5 * †2 : * ‡(; 4 9 5 6 * 2 (5 * - - 4) 8 ¶8 * ; 4 0 6 9 2 8 5) ;) 6 †8
) 4 ‡‡; 1 (‡9 ; 4 8 0 8 1 ; 8 : 8 ‡1 ; 4 8 †8 5 ; 4) 4 8 5 †5 2 8 8 0 6
* 8 1 (‡9 ; 4 8 ; (8 8 ; 4 (‡? 3 4 ; 4 8) 4 ‡; 1 6 1 ; : 1 8 8 ; ‡? ;

his friend told him, "Were all the jewels of Golconda awaiting me upon my solution of this enigma, I am quite sure that I should be unable to earn them." Mr. Legrand responded, "It may well be doubted whether human ingenuity can construct an enigma of the kind which human ingenuity may not, by proper application, resolve." He noticed that a combination of three symbols—; 4 8—appeared very frequently in the text. He also knew that Captain Kidd's pirates spoke English and that the most frequent English word is "the." Assuming that ; 4 8 coded for "the," Mr. Legrand deciphered the parchment note and found the pirate treasure (and a few skeletons as well). After this insight, Mr. Legrand had a slightly easier text to decipher:

5 3 ‡‡†3 0 5)) 6 * T H E 2 6) H ‡.) H ‡) : E 0 6 * T H E †E ¶6 0)) E
5 T 1 ‡(T : ‡* E †E 3 (E E) 5 * †T 4 6 (E E * 9 6 * ? T E) * ‡(T H E 5
) T 5 * †2 : * ‡(T H 9 5 6 * 2 (5 * - - H) E ¶E * T H 0 6 9 2 E 5) T) 6 †E
) H ‡‡T 1 (‡9 T H E 0 E 1 T E : E ‡1 T H E †E 5 T H) H E 5 †5 2 E E 0 6
* E 1 (‡9 T H E T (E E T H (‡? 3 H T H E) H ‡T 1 6 1 T : 1 E E T ‡? T

You may try to figure out what codes for "(" and complete the deciphering.

DNA texts are not easy to decipher, and there is little doubt that Nature can construct an enigma of the kind which human ingenuity may not resolve. However, DNA linguistics borrowed Mr. Legrand's scientific method, and a popular approach in DNA linguistics is based on the assumption that frequent or rare words may correspond to signals in DNA. If a word occurs considerably more (or less) frequently than expected, then it becomes a potential "signal," and the question arises as to the "biological" meaning of this word (Brendel et al., 1986 [47], Burge et al., 1992 [53]). For example, Gelfand and Koonin, 1997 [124] showed that the most avoided 6-palindrome in the archaeon *M.jannaschii* is likely to be the recognition site of a restriction-modification system.

DNA linguistics is at the heart of the *pattern-driven* approach to signal finding, which is based on enumerating all possible patterns and choosing the most frequent or the fittest (Brazma et al., 1998 [46]) among them. The fitness measures vary from estimates of the statistical significance of discovered signals to the information content of the fragments that approximately match the signal. The pattern-driven approach includes the following steps:

- Define the fitness measure (e.g., frequency).

- Calculate the fitness of each word with respect to a sample of DNA fragments.

- Report the fittest words as potential signals.

A problem with the pattern-driven approach is efficiency, since the search space for patterns of length l is $|\mathcal{A}|^l$, where $\mathcal{A}$ is the alphabet. To prune the search, one can use the idea behind the Karp-Miller-Rosenberg algorithm (Karp et al., 1972 [188]), which is based on the observation that if a string appears in k sequences, then all of its substrings appear in at least k sequences. Therefore, every frequent string can be assembled from frequent substrings. For example, a simple way to do this is to create a list of all frequent $2l$-tuples from the list of all frequent l-tuples by concatenating every pair of frequent l-tuples and subsequently checking the frequency of these concatenates. Another approach to this problem is to use *suffix trees* (Gusfield, 1997 [145]).

To find frequent and rare words in a text, one has to compute the expected value $E(W)$ and the variance $\sigma^2(W)$ for the number of occurrences (frequency) of every word W. Afterwards, the frequent and rare words are identified as the words with significant deviations from expected frequencies. In many DNA linguistics papers, the variance $\sigma^2(W)$ of the number of occurrences of a word in a text was erroneously assumed to be $E(W)$.

Finding the probability of k occurrences of a word in a text involves an apparatus of generating functions and complex analysis (Guibas and Odlyzko, 1981 [141]). The difficulty is that the probability of a word's occurrence in a text depends not only on the length of the word, but also on the structure of the word overlaps defined by the *autocorrelation polynomial* (Guibas and Odlyzko, 1981 [141]). For example, the distribution of the number of occurrences of AAA (autocorrelation polynomial $1 + x + x^2$) differs significantly from the distribution of the number of occurrences of of ATA (autocorrelation polynomial $1 + x^2$) even in a random Bernoulli text with equiprobable letters (*overlapping words paradox*). Below we discuss the best bet for simpletons explaining the overlapping words paradox.

8.3 The Best Bet for Simpletons

The overlapping words paradox is the basis of *the best bet for simpletons*, studied by John Conway. The best bet for simpletons starts out with two players who select words of length l in a 0-1 alphabet. Player I selects a sequence A of l heads or tails, and Player II, knowing what A is, selects another sequence B of length l. The players then flip a coin until either A or B appears as a block of l consecutive outcomes. The game will terminate with probability 1.

"Funny you don't gamble none, ain't it in your blood?", Shorty said to Smoke (in Jack London's "Smoke Bellew") one night in Elkhorn. Smoke answered, "It is. But the statistics are in my head. I like an even break for my money."

At first glance it looks as though A and B have an even break for their money. Even if somebody realizes that some words are stronger than others in this game, it looks as though A should win after choosing the "strongest" word. The intriguing feature of the game is the fact that if $l \geq 3$, then no matter what A is, Player II can choose a word B that beats A. One more surprise is that the best bet for simpletons is a non-transitive game: A beats B and B beats C does not imply A beats C (remember rock, paper, and scissors?).

Suppose A chooses 00 and B chooses 10. After two tosses either A wins (00), or B wins (10), or the game will continue (01 or 11). However, it makes little sense for A to continue the game since B will win anyway! Therefore, the odds of B over A in this game are 3:1.

The analysis of the best bet for simpletons is based on the notion of a *correlation polynomial* (Guibas and Odlyzko, 1981 [141]). Given two l-letter words A and B, the *correlation* of A and B, to be denoted by $AB = (c_0, \ldots, c_{l-1})$, is an l-letter boolean word (Figure 8.1). The i-th bit of AB is defined to be 1 if the $(n-i)$-prefix (the first $n-i$ letters) of B coincides with the $(n-i)$-suffix (the last $n-i$ letters) of A. Otherwise, the i-th bit of AB is defined to be 0. The *correlation polynomial* of A and B is defined as $K_{AB}(t) = c_0 + c_1 \cdot t^1 + \ldots + c_{l-1} \cdot t^{l-1}$. We also denote $K_{AB} = K_{AB}(\frac{1}{2})$.

John Conway suggested the following elegant formula to compute the odds that B will win over A:

$$\frac{K_{AA} - K_{AB}}{K_{BB} - K_{BA}}$$

Conway's proof of this formula was never published. Martin Gardner, 1974 [118] wrote about this formula:

> I have no idea why it works. It just cranks out the answer as if by magic, like so many of Conway's other algorithms.

The proofs of this formula were given independently by Li, 1980 [222] using martingales and by Guibas and Odlyzko, 1981 [141] using generating functions. In the next section we give a short proof of the Conway equation (Pevzner, 1993 [268]).

8.4 The Conway Equation

Let $AB = (c_0, \ldots, c_{l-1})$ be a correlation of A and B, and let $c_{m_1}, \ldots, c_{m_k}$ be the bits of AB equal to 1. Denote as $\mathcal{H}_{AB}$ the set of k prefixes of $A = a_1 \ldots a_l$ of

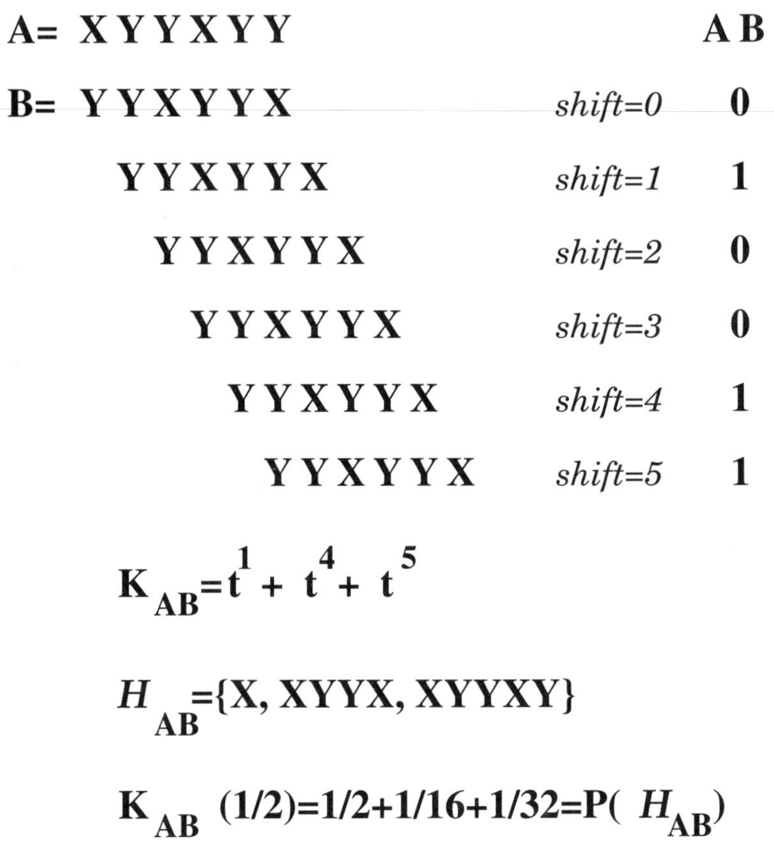

		AB
A= X Y Y X Y Y		
B= Y Y X Y Y X	*shift=0*	**0**
Y Y X Y Y X	*shift=1*	**1**
Y Y X Y Y X	*shift=2*	**0**
Y Y X Y Y X	*shift=3*	**0**
Y Y X Y Y X	*shift=4*	**1**
Y Y X Y Y X	*shift=5*	**1**

$$K_{AB} = t^1 + t^4 + t^5$$

$$H_{AB} = \{X, XYYX, XYYXY\}$$

$$K_{AB}(1/2) = 1/2 + 1/16 + 1/32 = P(H_{AB})$$

Figure 8.1: Correlation polynomial of words A and B ($AB = (010011)$) and the set $\mathcal{H}_{AB}$.

length $m_1, \ldots, m_k$ (Figure 8.1):

$$(a_1 \ldots a_{m_1}),\ (a_1 \ldots a_{m_1} \ldots a_{m_2}),\ \ldots,\ (a_1 \ldots a_{m_1} \ldots a_{m_2} \ldots\ldots a_{m_k}).$$

Given two words X and Y, we denote as $X * Y$ the *concatenation* of X and Y. Given two sets of words $\mathcal{X} = \{X\}$ and $\mathcal{Y} = \{Y\}$, we denote as $\mathcal{X} * \mathcal{Y}$ the set of words $\{X * Y\}$ containing all concatenations of words from $\mathcal{X}$ and words from $\mathcal{Y}$. The set $\mathcal{X} * \mathcal{Y}$ contains $|\mathcal{X}| \cdot |\mathcal{Y}|$ words (perhaps with repeats).

We denote as $P(X) = \frac{1}{2^l}$ the probability of a boolean l-letter word X to represent the result of l head-or-tail trials. For a set of words $\mathcal{X} = \{X\}$, denote

$$P(\mathcal{X}) = \sum_{X \epsilon \mathcal{X}} P(X)$$

We will use the following simple observation:

Lemma 8.1 $K_{AB}(\frac{1}{2}) = P(\mathcal{H}_{AB})$.

A word W is an *A-victory* if it contains A in the end and does not contain B. A word W is an *A-previctory* if $W * A$ is an A-victory. We define $\mathcal{S}_A$ to be the set of all A-previctories. *B-victories, B-previctories*, and the set $\mathcal{S}_B$ of all B-previctories are defined similarly.

The idea of the proof is to consider all *no-victory* words:

$$\mathcal{T} = \{T : T \text{ is neither } A\text{-victory nor } B\text{-victory}\}.$$

Every word $T * A$ for $T \epsilon \mathcal{T}$ corresponds to either an A-victory or a B-victory. If $T * A$ corresponds to an A-victory, then T can be represented in the form A-previctory $* H_{AA}$, where $H_{AA} \epsilon \mathcal{H}_{AA}$ (Figure 8.2a). If $T * A$ corresponds to a B-victory, then T can be represented in the form B-previctory $* H_{BA}$, where $H_{BA} \epsilon \mathcal{H}_{BA}$ (Figure 8.2b). This implies the following representation of no-victories.

Lemma 8.2 $\mathcal{T} = \mathcal{T}_1 = (\mathcal{S}_B * \mathcal{H}_{BA}) \bigcup (\mathcal{S}_A * \mathcal{H}_{AA})$.

Similarly, every word $T * B$ for $T \epsilon \mathcal{T}$ corresponds to either an A-victory or a B-victory. If $T * B$ corresponds to an A-victory, then T can be represented in the form A-previctory $* H_{AB}$, where $H_{AB} \epsilon \mathcal{H}_{AB}$ (Figure 8.2c). If $T * B$ corresponds to a B-victory, then T can be represented in the form B-previctory $* H_{BB}$, where $H_{BB} \epsilon \mathcal{H}_{BB}$ (Figure 8.2d). This implies another representation of no-victories:

Lemma 8.3 $\mathcal{T} = \mathcal{T}_2 = (\mathcal{S}_A * \mathcal{H}_{AB}) \bigcup (\mathcal{S}_B * \mathcal{H}_{BB})$.

Theorem 8.1 *The odds that B wins over A is* $\frac{K_{AA} - K_{AB}}{K_{BB} - K_{BA}}$.

Proof Lemmas 8.1 and 8.3 imply that the overall probability of words in the set $\mathcal{T}_2$ is

$$\begin{aligned} P(\mathcal{T}_2) = P(\mathcal{S}_A * \mathcal{H}_{AB}) + P(\mathcal{S}_B * \mathcal{H}_{BB}) = \\ P(\mathcal{S}_A) \cdot P(\mathcal{H}_{AB}) + P(\mathcal{S}_B) \cdot P(\mathcal{H}_{BB}) = \\ P(\mathcal{S}_A) \cdot K_{AB} + P(\mathcal{S}_B) \cdot K_{BB}. \end{aligned}$$

Similarly, lemmas 8.1 and 8.2 imply

$$P(\mathcal{T}_1) = P(\mathcal{S}_B) \cdot K_{BA} + P(\mathcal{S}_A) \cdot K_{AA}.$$

According to lemmas 8.2 and 8.3, $\mathcal{T}_1$ and $\mathcal{T}_2$ represent the same set $\mathcal{T}$; therefore, $P(\mathcal{T}_1) = P(\mathcal{T}_2)$ and

$$P(\mathcal{S}_A) \cdot K_{AB} + P(\mathcal{S}_B) \cdot K_{BB} = P(\mathcal{S}_B) \cdot K_{BA} + P(\mathcal{S}_A) \cdot K_{AA}.$$

This implies

$$\frac{P(\mathcal{S}_B)}{P(\mathcal{S}_A)} = \frac{K_{AA} - K_{AB}}{K_{BB} - K_{BA}}. \quad \blacksquare$$

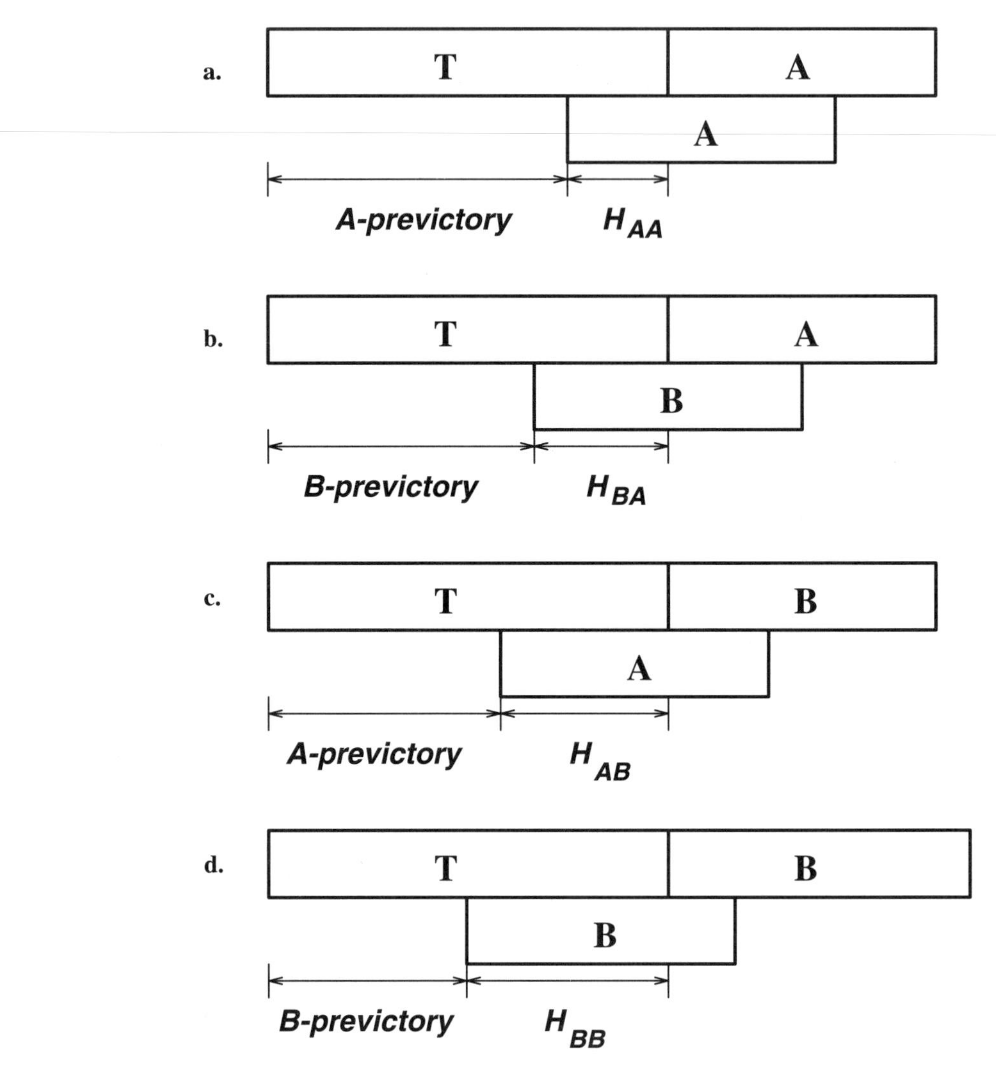

Figure 8.2: Different representations of words from the sets $\mathcal{T} * \{A\}$ and $\mathcal{T} * \{B\}$.

8.5 Frequent Words in DNA

Formulas for the variance $Var(W)$ of the number of occurrences of a word W in a Bernoulli text were given by Gentleman and Mullin, 1989 [126] and Pevzner et al., 1989 [269]. For a Bernoulli text of length n in an l-letter alphabet with equiprobable letters,

$$Var(W) = \frac{n}{l^k} \cdot (2 \cdot K_{WW}(\frac{1}{l}) - 1 - \frac{2k-1}{l^k})$$

where k is the length of word W and $K_{WW}(t)$ is the correlation polynomial of words W and W (*autocorrelation polynomial of W*).

To derive this formula, consider (for the sake of simplicity) a circular text of length n in an l-letter alphabet with the probability of every letter at every position equal to $\frac{1}{l}$. For a fixed k-letter word W, define a random variable x_i as 1 if W starts at the i-th position of the text and as 0 otherwise. Denote the mean of x_i as $p = \frac{1}{l^k}$. The number of occurrences of W in the text is given by the random variable

$$X = \sum_{i=1}^{n} x_i$$

with mean

$$E(X) = \sum_{i=1}^{n} E(x_i) = np$$

and variance

$$Var(X) = E(X^2) - E(X)^2 = \sum_{\{1 \le i,j \le n\}} E(x_i x_j) - E(x_i)E(x_j).$$

Let $d(i,j)$ be the (shortest) distance between positions i and j in a circular text. Then

$$Var(X) = \sum_{\{(i,j):\, d(i,j) \ge k\}} E(x_i x_j) - E(x_i)E(x_j) +$$

$$\sum_{\{(i,j):\, d(i,j)=0\}} E(x_i x_j) - E(x_i)E(x_j) +$$

$$\sum_{\{(i,j):\, 0<d(i,j)<k\}} E(x_i x_j) - E(x_i)E(x_j).$$

Since random variables x_i and x_j are independent for $d(i,j) \ge k$, the first term in the above formula is 0, the second term is simply $n(p - p^2)$, and the last term can be rewritten as follows:

$$\sum_{\{(i,j):\, 0<d(i,j)<k\}} E(x_i x_j) - E(x_i)E(x_j) =$$

$$\sum_{i=1}^{n}\sum_{t=1}^{k-1}\sum_{\{j:\,d(i,j)=t\}} E(x_ix_j)-E(x_i)E(x_j)$$

For a fixed t, $E(x_ix_{i+t})$ equals $p\frac{1}{l^t}$ if the t-th coefficient of the correlation WW equals 1, and 0 otherwise. We can then write $E(x_ix_{i+t}) = c_tp\frac{1}{l^t}$, and since for every i there are exactly two positions j with $d(i,j)=t$,

$$\sum_{t=1}^{k-1}\sum_{\{j:\,d(i,j)=t\}} E(x_ix_j) =$$

$$2p\sum_{t=1}^{k-1} c_t\frac{1}{l^t} = 2p(K_{WW}(\frac{1}{l})-1).$$

Therefore,

$$\sum_{\{(i,j):\,0<d(i,j)<k\}} E(x_ix_j)-E(x_i)E(x_j) = \sum_{i=1}^{n}(2p(K_{WW}(\frac{1}{l})-1)-2(k-1)p^2) =$$

$$np(2K_{WW}(\frac{1}{l})-2-2(k-1)p)$$

and

$$Var(X) = np(2K_{WW}(\frac{1}{l})-1-(2k-1)p).$$

This result demonstrates that the variance of the frequency of occurrences varies significantly between words even for Bernoulli texts. In particular, for a 4-letter alphabet with equal probabilities of letters A, T, G, and C,

$$\frac{Var(AA)}{Var(AT)} = \frac{21}{13} \text{ and } \frac{Var(AAA)}{Var(ATG)} = \frac{99}{59}.$$

Therefore, ignoring the overlapping words paradox leads to about 200% mistakes in estimations of statistical significance while analyzing frequent words. For 2-letter alphabets Pur/Pyr or S/W, ignoring the overlapping words paradox leads to 500%(!) mistakes in estimations of statistical significance ($\frac{Var(SS)}{Var(SW)} = \frac{5}{1}$).

The above formulas allow one to compute the variance for Bernoulli texts. Fousler and Karlin, 1987 [112], Stuckle et al., 1990 [330], and Kleffe and Borodovsky, 1992 [199] presented approximation formulas allowing one to calculate the variance for texts generated by Markov chains. Prum et al., 1995 [281] obtained the normal limiting distribution for the number of word occurrences in the Markovian Model. Finally, Regnier and Szpankowski, 1998 [282] studied approximate word occurrences and derived the exact and asymptotic formulas for the mean, variance, and probability of approximate occurrences.

8.6 Consensus Word Analysis

For a word W and a sample $\mathcal{S}$, denote as $W(\mathcal{S})$ the number of sequences from $\mathcal{S}$ that contain W. If a magic word appears in the sample exactly, then a simple count of $W(\mathcal{S})$ for every l-letter word W would detect the magic word as the most frequent one. The problem gets more complicated when up to k errors (i.e., mismatches) in the magic word are allowed. For this case Waterman et al., 1984 [358] and Galas et al., 1985 [115] suggested *consensus word* analysis, which is essentially an approximate word count. For every word W they defined a neighborhood consisting of all words within distance k from W and counted occurrences of words from the neighborhood in the sample. They also introduced the idea of *weighted occurrences* and assigned higher weight to neighbors with fewer errors. Using consensus word analysis, Galas et al., 1985 [115] were able to detect the $TTGACA$ and $TATAAT$ consensus sequences in the *E. coli* promoter signal.

Let $D_H(s,t)$ be the Hamming distance between two strings s and t of the same length. Mirkin and Roberts, 1993 [238] showed that approximate word count is, in some sense, equivalent to the following:

Consensus String Problem Given a sample $\mathcal{S} = \{s_1, \ldots, s_n\}$ of sequences and an integer l, find a *median* string s of length l and a substring t_i of length l from each s_i, minimizing $\sum_{i=1}^{n} d_H(s, t_i)$.

Li et al., 1999 [221] showed that the Consensus String Problem is NP-hard and gave a polynomial time approximation scheme (PTAS) for this problem. The algorithm is based on the notion of a *majority* string. Given a collection $t_1, \ldots, t_n$ of n strings of length l, the majority string for $t_1, \ldots, t_n$ is the string s whose i-th letter is the most frequent letter among n i-th letters in $t_1, \ldots, t_n$. Li et al., 1999 [221] devised a PTAS for the Consensus String Problem that is based on choosing the majority string for every r length-l substrings $t_{i_1}, \ldots, t_{i_r}$ of $\{s_1, \ldots, s_n\}$.

It is often convenient to concatenate multiple sequences from a sample $\mathcal{S}$ into a single composite sequence, converting the problem of finding the consensus string into the problem of finding the most frequent string in the text. A naive approach to this problem is to find the number of occurrences $W(T)$ of every l-letter string W in text T. Apostolico and Preparata, 1996 [10] devised an efficient algorithm for

String Statistics Problem Given a text T and an integer l, find $W(T)$ for each l-letter string W.

The String Statistics Problem gets difficult if we consider approximate string occurrences. Let $W_k(T)$ be the number of approximate occurrences of W in T with up to k mismatches. We are unaware of an efficient algorithm to solve the following:

Frequent String Problem Given text T and integers l and k, find an l-letter string W maximizing $W_k(T)$ among all l-letter words.

Consensus word analysis is an example of a *sequence-driven* approach to signal finding. A sequence-driven approach does not enumerate all the patterns, but instead considers only the patterns that are present in the sample. Given a collection of frequent words $\mathcal{W}_1$ in a sample $\mathcal{S}_1$ and a collection of frequent words $\mathcal{W}_2$ in a sample $\mathcal{S}_2$, one can intersect $\mathcal{W}_1$ and $\mathcal{W}_2$ to obtain a collection of frequent words in $\mathcal{S}_1 \cup \mathcal{S}_2$. Given a sample of n sequences, one can view it as a set of n samples and start combining sets of frequent words until all the sequences are combined. Particular sequence-driven approaches differ in the way the sets to be combined are chosen and in the way the sets are combined.

8.7 CG-islands and the "Fair Bet Casino"

The most infrequent dinucleotide in many genomes is CG. The reason is that C within CG is typically methylated, and the resulting methyl-C has a tendency to mutate into T. However, the methylation is suppressed around genes in the areas called CG-*islands* where CG appears relatively frequently. The question arises of how to define and find CG-islands in a genome.

Finding CG-islands is not very different from the following "gambling" problem (Durbin et al., 1998 [93]). A dealer in a "Fair Bet Casino" may use either a fair coin or a biased coin that has a probability of $\frac{3}{4}$ of a head. For security reasons, the dealer does not tend to change coins—it happens relatively rarely, with a probability of 0.1. Given a sequence of coin tosses, find out when the dealer used the biased coin and when he used a fair coin.

First, let's solve the problem under the assumption that the dealer never changed the coin. The question is what coin, fair ($p^+(0) = p^+(1) = \frac{1}{2}$) or biased ($p^-(0) = \frac{1}{4}, p^-(1) = \frac{3}{4}$), he used. If the resulting sequence of tosses is $x = x_1 \ldots x_n$, then the probability that it was generated with a fair coin is $P(x|fair\ coin) = \prod_{i=1}^n p^+(x_i) = \frac{1}{2^n}$. The probability that x was generated with a biased coin is $P(x|biased\ coin) = \prod_{i=1}^n p^-(x_i) = \frac{1}{4^{n-k}} \frac{3^k}{4^k} = \frac{3^k}{4^n}$, where k is the number of 1s in x. As a result, when $k < \frac{n}{\log_2 3}$, the dealer most likely used a fair coin, and when $k > \frac{n}{\log_2 3}$, he most likely used a biased coin. We can define the log-odds ratio as follows:

$$\log_2 \frac{P(x|fair\ coin)}{P(x|biased\ coin)} = \sum_{i=1}^{k} \log_2 \frac{p^+(x_i)}{p^-(x_i)} = n - k \log_2 3$$

A naive approach to finding CG-islands is to calculate log-odds ratios for each sliding window of a fixed length. The windows that receive positive scores are potential CG-islands. The disadvantage of such an approach is that we don't know the length of CG-islands in advance. Hidden Markov Models represent a different probabilistic approach to this problem (Churchill, 1989 [69]).

8.8 Hidden Markov Models

A *Hidden Markov Model* (HMM) $\mathcal{M}$ is defined by an alphabet $\sum$, a set of (hidden) states Q, a matrix of state transition probabilities A, and a matrix of emission probabilities E, where

- $\sum$ is an alphabet of symbols.
- Q is a set of states that emit symbols from the alphabet $\sum$.
- $A = (a_{kl})$ is a $|Q| \times |Q|$ matrix of state transition probabilities.
- $E = (e_k(b))$ is a $|Q| \times |\sum|$ matrix of emission probabilities.

Tossing the coin in the "Fair Bet Casino" corresponds to the following HMM:

- $\sum = \{0, 1\}$, corresponding to tail (0) or head (1).
- $Q = \{F, B\}$, corresponding to a fair or biased coin.
- $a_{FF} = a_{BB} = 0.9$, $a_{FB} = a_{BF} = 0.1$
- $e_F(0) = \frac{1}{2}$, $e_F(1) = \frac{1}{2}$, $e_B(0) = \frac{1}{4}$, $e_B(1) = \frac{3}{4}$

A path $\pi = \pi_1 \ldots \pi_n$ in the HMM $\mathcal{M}$ is a sequence of states. For example, if a dealer used the fair coin for the first three and the last three tosses and the biased coin for five tosses in between, the corresponding path is FFFBBBBBFFF. The probability that a sequence x was generated by the path π (given the model $\mathcal{M}$) is

$$P(x|\pi) = \prod_{i=1}^{n} P(x_i|\pi_i) P(\pi_i|\pi_{i+1}) = a_{\pi_0,\pi_1} \cdot \prod_{i=1}^{n} e_{\pi_i}(x_i) \cdot a_{\pi_i,\pi_{i+1}}$$

where for convenience we introduce π_0 and π_{n+1} as the fictitious initial and terminal states *begin* and *end*.

This model defines the probability $P(x|\pi)$ for a given sequence x and a given path π. However, only the dealer knows the real sequence of states π that emitted x. We therefore say that the path of x is hidden and face the following

Decoding Problem Find an optimal path $\pi^* = \arg\max_\pi P(x|\pi)$ for x, such that $P(x|\pi)$ is maximized.

The solution of the decoding problem is provided by the Viterbi, 1967 [348] algorithm, which is a variation of the Bellman, 1957 [29] dynamic programming approach. The idea is that the optimal path for the $(i+1)$-prefix $x_1 \ldots x_{i+1}$ of x uses a path for an i-prefix of x that is optimal among the paths ending in an (unknown) state $\pi_i = k \in Q$.

Define $s_k(i)$ as the probability of the most probable path for the prefix $x_1 \ldots x_i$ that ends with state k ($k \in Q$ and $1 \le i \le n$). Then

$$s_l(i+1) = e_l(x_{i+1}) \cdot \max_{k \in Q} \{s_k(i) \cdot a_{kl}\}.$$

We initialize $s_{begin}(0) = 1$ and $s_k(0) = 0$ for $k \neq begin$. The value of $P(x|\pi^*)$ is

$$P(x|\pi^*) = \max_{k \in Q} s_k(n) a_{k,end}.$$

The Viterbi algorithm runs in $O(n|Q|)$ time. The computations in the Viterbi algorithm are usually done in logarithmic scores $S_l(i) = \log s_l(i)$ to avoid overflow:

$$S_l(i+1) = \log e_l(x_{i+1}) + \max_{k \in Q}\{S_k(i) + \log(a_{kl})\}$$

Given a sequence of tosses x, what is the probability that the dealer had a biased coin at moment i? A simple variation of the Viterbi algorithm allows one to compute the probability $P(\pi_i = k|x)$. Let $f_k(i)$ be the probability of emitting the prefix $x_1 \ldots x_i$ and reaching the state $\pi_i = k$. Then

$$f_k(i) = e_k(x_i) \cdot \sum_{l \in Q} f_l(i-1) \cdot a_{lk}.$$

The only difference between this *forward algorithm* and the Viterbi algorithm is that "max" sign in the Viterbi algorithm changes into the $\sum$ sign in the forward algorithm. Backward probability $b_k(i)$ is defined as the probability of being at state $\pi_i = k$ and emitting the suffix $x_{i+1} \ldots x_n$. The *backward algorithm* uses a similar recurrency:

$$b_k(i) = \sum_{l \in Q} e_l(x_{i+1}) \cdot b_l(i+1) \cdot a_{kl}$$

Finally, the probability that the dealer had a biased coin at moment i is given by

$$P(\pi_i = k|x) = \frac{P(x, \pi_i = k)}{P(x)} = \frac{f_k(i) \cdot b_k(i)}{P(x)}$$

where $P(x, \pi_i = k)$ is the probability of x under the assumption that x_i was produced in state k and $P(x) = \sum_\pi P(x|\pi)$.

8.9 The Elkhorn Casino and HMM Parameter Estimation

The preceding analysis assumed that we knew the state transition and emission probabilities of the HMM. The most difficult problem in applications of HMMs is that these parameters are unknown and need to be estimated. It is easy for an intelligent gambler to figure out that the dealer in the "Fair Bet Casino" is using a biased coin. One way to find this out is to notice that 0 and 1 have different expected frequencies ($\frac{3}{8}$ and $\frac{5}{8}$ correspondingly) and the ratio of 0s to 1s in a day-long sequence of tosses is suspiciously low. However, it is much more difficult to estimate the transition and emission probabilities of the corresponding HMM. Smoke, in Jack London's "Smoke Bellew," made one of the first attempts to figure out the transition probabilities of a roulette wheel in the Elkhorn casino. After long hours and days spent watching the roulette wheel the night came when Smoke proclaimed that he was ready to beat the system. Don't try to do it again in Las Vegas; the gambling technology has changed.

Let Θ be a vector combining the unknown transition and emission probabilities of the HMM $\mathcal{M}$. Given a string x, define $P(x|\Theta)$ as the probability of x given the assignment of parameters Θ. Our goal is to find Θ^* such that

$$\Theta^* = \arg\max_{\Theta} P(x|\Theta).$$

Usually, instead of a single string x, a collection of *training sequences* $x^1, \ldots, x^m$ is given and the goal is to maximize

$$\Theta^* = \arg\max_{\Theta} \prod_{j=1}^{m} P(x^j|\Theta).$$

This is an optimization of a continuous function in multidimensional parameter space Θ. The commonly used algorithms for parameter optimization are heuristics that use a local improvement strategy in the parameter space. If the path $\pi_1 \ldots \pi_n$ corresponding to the observed states $x_1 \ldots x_n$ is known, then we can scan the sequences and compute the empirical estimates for transition and emission probabilities. If A_{kl} is the number of transitions from state k to l and $E_k(b)$ is the number of times b is emitted from state k then, the maximum likelihood estimators are

$$a_{kl} = \frac{A_{kl}}{\sum_{q \in Q} A_{kq}} \qquad e_k(b) = \frac{E_k(b)}{\sum_{\sigma \in \sum} E_k(\sigma)}.$$

Usually, the state sequence $\pi_1 \ldots \pi_n$ is unknown, and in this case, an iterative local improvement strategy called the *Baum-Welch* algorithm is commonly used (Baldi and Brunak, 1997 [24]).

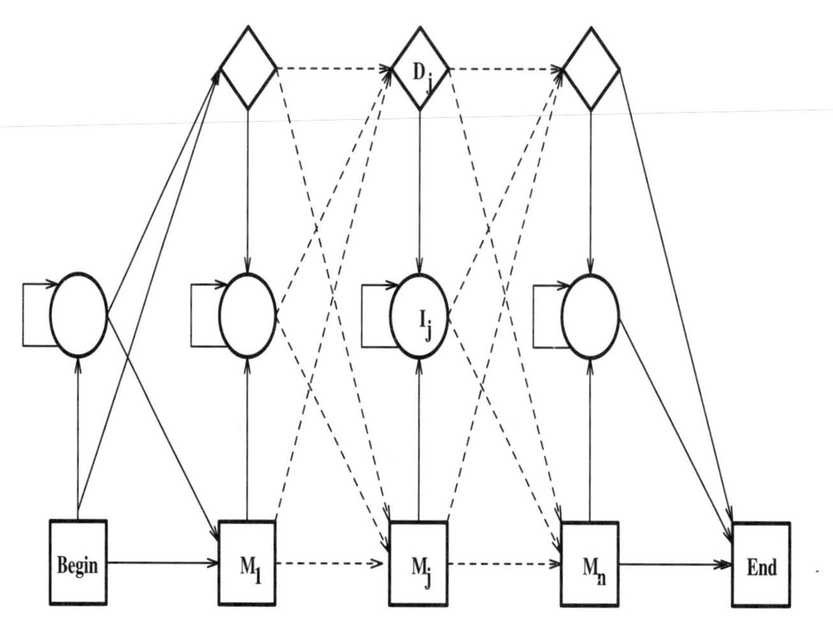

Figure 8.3: Profile HMM.

8.10 Profile HMM Alignment

Given a family of functionally related biological sequences, one can search for new members of the family using pairwise alignments between family members and sequences from a database. However, this approach may fail to find distantly related sequences. An alternative approach is to use the whole set of functionally related sequences for the search.

A *Profile* is the simplest representation of a family of related proteins that is given by multiple alignment. Given an n-column multiple alignment of strings in alphabet $\mathcal{A}$, a profile $\mathcal{P}$ is an $|\mathcal{A}| \times n$ matrix that specifies the frequency $e_i(a)$ of each character a from the alphabet $\mathcal{A}$ in column i (Gribskov et al., 1987 [139]). Profiles can be compared and aligned against each other since the dynamic programming algorithm for aligning two sequences works if both of the input sequences are multiple alignments (Waterman and Perlwitz, 1984 [362]).

HMMs can also be used for sequence comparison (Krogh et al., 1994 [208], Sonnhammer et al., 1997 [324]), in particular for aligning a sequence against a profile. The simplest HMM for a profile $\mathcal{P}$ contains n sequentially linked *match* states $M_1, \ldots, M_n$ with emission probabilities $e_i(a)$ taken from the profile P (Fig 8.3). The probability of a string $x_1 \ldots x_n$ given the profile $\mathcal{P}$ is $\prod_{i=1}^{n} e_i(x_i)$. To model insertions and deletions we add *insertion* states $I_0, \ldots, I_n$ and *deletion* states $D_1, \ldots, D_n$ to the HMM and assume that

$$e_{I_j}(a) = p(a)$$

where $p(a)$ is the frequency of the occurrence of the symbol a in all the sequences. The transition probabilities between matching and insertion states can be defined in the affine gap penalty model by assigning a_{MI}, a_{IM}, and a_{II} in such a way that $\log(a_{MI}) + \log(a_{IM})$ equals the gap creation penalty and $\log(a_{II})$ equals the gap extension penalty. The (silent) deletion states do not emit any symbols.

Define $v_j^M(i)$ as the logarithmic likelihood score of the best path for matching $x_1 \ldots x_i$ to profile HMM $\mathcal{P}$ ending with x_i emitted by the state M_j. Define $v_j^I(i)$ and $v_j^D(i)$ similarly. The resulting dynamic programming recurrency is, of course, very similar to the standard alignment recurrency:

$$v_j^M(i) = \log \frac{e_{M_j}(x_i)}{p(x_i)} + \max \begin{cases} v_{j-1}^M(i-1) + \log(a_{M_{j-1},M_j}) \\ v_{j-1}^I(i-1) + \log(a_{I_{j-1},M_j}) \\ v_{j-1}^D(i-1) + \log(a_{D_{j-1},M_j}) \end{cases}$$

The values $v_j^I(i)$ and $v_j^D(i)$ are defined similarly.

8.11 Gibbs Sampling

Lawrence et al., 1993 [217] suggested using *Gibbs sampling* to find patterns in sequences. Given a set of sequences $x^1, \ldots, x^m$ in an alphabet $\mathcal{A}$ and an integer w, the problem is to find a substring of length w in each x^i in such a way that the similarity between m substrings is maximized.

Let $a^1, \ldots, a^m$ be the starting indices of the chosen substrings in $x^1, \ldots, x^m$, respectively. Denote as q_{ij} the frequency with which the symbol i occurs at the j-th position of the substrings.

Gibbs sampling is an iterative procedure that at each iteration discards one sequence from the alignment and replaces it with a new one. Gibbs sampling starts with randomly choosing substrings of length w in each of m sequences $x^1, \ldots, x^m$ and proceeds as follows.

- At the beginning of every iteration, a substring of length w in each of m sequences $x^1, \ldots, x^m$ is chosen.

- Randomly choose one of the sequences x^r uniformly at random.

- Create a frequency matrix (q_{ij}) from the remaining $m - 1$ substrings.

- For each position i in x^r, calculate the probability $p_i = \prod_{j=0}^{w} q_{x^r_{i+j},j}$ that the substring starting at this position is generated by profile (q_{ij}) (x^r_{i+j} denotes a symbol at position $i + j$ of sequence x^r).

- Choose the starting position i of the new substring in x^r randomly, with probability proportional to p_i.

Although Gibbs sampling is known to work in specific cases, it may, similarly to the Baum-Welch algorithm, converge to a local maximum. Since the described procedure (Lawrence et al., [217]) does not allow for insertions and deletions, Rocke and Tompa, 1998 [288] generalized this method for handling gaps in a pattern.

8.12 Some Other Problems and Approaches

8.12.1 Finding gapped signals

Rigoutsos and Floratos, 1998 [285] addressed the problem of finding gapped signals in a text. A gapped string is defined as a string consisting of an arbitrary combination of symbols from the alphabet and "don't care" symbols. A gapped string P is called an $< l, w >$-string if every substring of P of length l contains at least w symbols from the alphabet. The TERESIAS algorithm (Rigoutsos and Floratos, 1998 [285]) finds all *maximal* $< l, w >$-patterns that appear in at least K sequences in the sample in a two-stage approach. At the first *scanning* stage, it finds all short strings of length l with at least w symbols from the alphabet that appear at least K times in the sample. At the second *convolution* stage, it assembles these short strings into maximal (l, w)-strings.

8.12.2 Finding signals in samples with biased frequencies

The magic word problem becomes difficult if the signal is contained in only a fraction of all sequences and if the background nucleotide distribution in the sample is skewed. In this case, searching for a signal with the maximum number of occurrences may lead to the patterns composed from the most frequent nucleotides. These patterns may not be biologically significant. For example, if A has a of frequency 70% and T, G, and C have frequencies of 10%, then poly(A) may be the most frequent word, thus disguising the real magic word.

To find magic words in biased samples, many algorithms use *relative entropy* to highlight the magic word among the words composed from frequent nucleotides. Given a magic word of length k, the relative entropy is defined as

$$\sum_{j=1}^{k} \sum_{r=A,T,G,C} p_{rj} \log_2 \frac{p_{rj}}{b_r}$$

where p_{rj} is the frequency of nucleotide r in position j among magic word occurrences and b_r is the background frequency of r.

Relative entropy is a good measure for comparing two magic words that have the same number of occurrences in the sample, but not a good measure if the words appear in vastly different numbers of sequences. Hertz and Stormo, 1999 [159] and Tompa, 1999 [338] addressed this problem by designing criteria that account for both number of occurrences and background distribution.

8.12.3 Choice of alphabet in signal finding

Karlin and Ghandour, 1985 [187] observed that there is no way to know beforehand which choice of alphabet is good for revealing signals in DNA or proteins. For example, $WSWS \ldots WSWS$ may be a very strong signal, that is hard to find in the standard A,T, G, C alphabet. This problem was addressed by Sagot et al., 1997 [293].

Chapter 9

Gene Prediction

9.1 Introduction

In the 1960s, Charles Yanofsky, Sydney Brenner, and their collaborators showed that a gene and its protein product are colinear structures with direct correlation between triplets of nucleotides in the gene and amino acids in the protein. However, the concept of the gene as a synthetic string of nucleotides did not live long. Overlapping genes and genes-within-genes were discovered in the late 1960s. These studies demonstrated that the computational problem of gene prediction is far from simple. Finally, the discovery of split human genes in 1977 created a computational gene prediction puzzle.

Eukaryotic genomes are larger and more complex than prokaryotic genomes. This does not come as a surprise since one would expect to find more genes in humans than in bacteria. However, the genome size of many eukaryotes does not appear to be related to genetic complexity; for example, the salamander genome is 10 times larger than the human genome. This paradox was resolved by the discovery that eukaryotes contain not only genes but also large amounts of DNA that do not code for any proteins ("junk" DNA). Moreover, most human genes are interrupted by junk DNA and are broken into pieces called exons. The difference in the sizes of the salamander and human genomes thus reflects larger amounts of junk DNA and repeats in the genome of salamander.

Split genes were first discovered in 1977 independently by the laboratories of Phillip Sharp and Richard Roberts during studies of the adenovirus (Berget et al., 1977 [32], Chow et al., 1977 [67]). The discovery was such a surprise that the paper by Richard Roberts' group had an unusually catchy title for the academic *Cell* magazine: "An amazing sequence arrangement at the 5' end of adenovirus 2 messenger RNA." Berget et al., 1977 [32] focused their experiments on an mRNA that encodes a viral protein known as the hexon. To map the hexon mRNA on viral genome, mRNA was hybridized to adenovirus DNA and the hybrid molecules were analyzed by electron microscopy. Strikingly, the mRNA-DNA hybrids formed in this experiment displayed three loop structures, rather than the continuous duplex segment suggested by the classical "continuous gene" model. Further hybridization experiments revealed that the hexon mRNA is built from four separate fragments of the adenovirus genome. These four exons in the adenovirus genome are separated by three "junk" fragments called *introns*. The discovery of split genes (*splicing*) in the adenovirus was quickly followed by evidence that mammalian genes also have split structures (Tilghman et al., 1978 [337]). These experimental studies raised a computational gene prediction problem that is still unsolved: human genes comprise only 3% of the human genome, and no existing *in silico* gene recognition algorithm provides reliable gene recognition.

After a new DNA fragment is sequenced, biologists try to find genes in this fragment. The traditional statistical way to attack this problem has been to look for features that appear frequently in genes and infrequently elsewhere. Many researchers have used a more biologically oriented approach and attempted to recognize the locations of splicing signals at exon-intron junctions. The goal of such an approach is characterization of sites on RNA where proteins and ribonucleoproteins involved in splicing apparatus bind/interact. For example, the dinucleotides AG and GT on the left and right sides of exons are highly conserved. The simplest way to represent a signal is to give a consensus pattern consisting of the most frequent nucleotide at each position of an alignment of specific signals. Although catalogs of splice sites were compiled in the early 1980s, the consensus patterns are not very reliable for discriminating true sites from pseudosites since they contain no information about nucleotide frequencies at different positions. Ed Trifonov invented an example showing another potential pitfall of consensus:

MELON
MANGO
HONEY
SWEET
COOKY
—
MONEY

The frequency information is captured by *profiles* (or *Position Weight Matrices*) that assign frequency based scores to each possible nucleotide at each position of the signal. Unfortunately, using profiles for splice site prediction has had limited success, probably due to cooperation between multiple binding molecules. Attempts to improve the accuracy of gene prediction led to applications of neural networks and Hidden Markov Models for gene finding.

Large-scale sequencing projects have motivated the need for a new generation of algorithms for gene recognition. The similarity-based approach to gene prediction is based on the observation that a newly sequenced gene has a good chance of having an already known relative in the database (Bork and Gibson, 1996 [41]). The flood of new sequencing data will soon make this chance even greater. As a result, the trend in gene prediction in the late 1990s shifted from statistics-based approaches to similarity-based and EST-based algorithms. In particular, Gelfand et al., 1996 [125] proposed a combinatorial approach to gene prediction, that uses related proteins to derive the exon-intron structure. Instead of employing statistical properties of exons, this method attempts to solve a combinatorial puzzle: to find a set of substrings in a genomic sequence whose concatenation (splicing) fits one of the known proteins.

After predictions are made, biologists attempt to experimentally verify them. This verification usually amounts to full-length mRNA sequencing. Since this process is rather time-consuming, *in silico* predictions find their way into databases and frequently lead to annotation errors. We can only guess the amount of incorrectly annotated sequences in GenBank, but it is clear that the number of genes that have been annotated without full-length mRNA data (and therefore are potentially erroneous) may be large. The problems of developing an "annotation-ready" gene prediction algorithm and correcting these errors remain open.

9.2 Statistical Approach to Gene Prediction

The simplest way to detect potential coding regions is to look at *Open Reading Frames (ORFs)*. An ORF is a sequence of codons in DNA that starts with a Start codon, ends with a Stop codon, and has no other Stop codons inside. One expects to find frequent Stop codons in non-coding DNA simply because 3 of 64 possible codons are translation terminators. The average distance between Stop codons in "random" DNA is $\frac{64}{3} \approx 21$, much smaller than the number of codons in an average protein (roughly 300). Therefore, long ORFs point out potential genes (Fickett, 1996 [105]), although they fail to detect short genes or genes with short exons.

Many gene prediction algorithms rely on recognizing the diffuse regularities in protein coding regions, such as bias in *codon usage*. Codon usage is a 64-mer vector giving the frequencies of each of 64 possible *codons* (triples of nucleotides) in a window. Codon usage vectors differ between coding and non-coding windows, thus enabling one to use this measure for gene prediction (Fickett, 1982 [104], Staden and McLachlan, 1982 [327]). Gribskov et al., 1984 [138] use a likelihood ratio approach to compute the conditional probabilities of the DNA sequence in a window under a coding and under a non-coding random sequence hypothesis. When the window slides along DNA, genes are often revealed as peaks of the likelihood ratio plots. A better coding sensor is the *in-frame hexamer count*, which is similar to three fifth-order Markov models (Borodovsky and McIninch, 1993 [42]). Fickett and Tung, 1992 [106] evaluated many such coding measures and came to the conclusion that they give a rather low-resolution picture of coding-region boundaries, with many false positive and false negative assignments. Moreover, application of these techniques to eukaryotes is complicated by the exon-intron structure. The average length of exons in vertebrates is 130 bp, and thus exons are often too short to produce peaks in the sliding window plot.

Codon usage, amino acid usage, periodicities in coding regions and other statistical parameters (see Gelfand, 1995 [123] for a review) probably have nothing in common with the way the splicing machinery recognizes exons. Many researchers have used a more biologically oriented approach and attempted to recognize the locations of splicing signals at exon-intron junctions (Brunak et al., 1991 [50]). There exists a (weakly) conserved sequence of eight nucleotides at the boundary of an exon and an intron (5' or *donor* splice site) and a sequence of four nucleotides at the boundary of intron and exon (3' or *acceptor* splice site). Unfortunately, profiles for splice site prediction have had limited success, probably due to cooperation between multiple binding molecules. Profiles are equivalent to a simple type of neural network called perceptron. More complicated neural networks (Uberbacher and Mural, 1991 [339]) and Hidden Markov Models (Krogh et al., 1994 [209], Burge and Karlin, 1997 [54]) capture the statistical dependencies between sites and improve the quality of predictions.

Many researchers have attempted to combine coding region and splicing signal predictions into a signal framework. For example, a splice site prediction is more believable if signs of a coding region appear on one side of the site but not the other. Because of the limitations of individual statistics, several groups have developed gene prediction algorithms that combine multiple pieces of evidence into a single framework (Nakata et al., 1985 [249], Gelfand, 1990 [121], Guigo et al., 1992 [142], Snyder and Stormo, 1993 [321]). Practically all of the existing statistics are used in the Hidden Markov Model framework of GENSCAN (Burge and Karlin, 1997 [54]). This algorithm not only merges splicing site, promoter, polyadenylation site, and coding region statistics, but also takes into account their non-homogeneity. This has allowed the authors to exceed the milestone of 90% accuracy for statistical gene predictions. However, the accuracy decreases significantly for genes with many short exons or with unusual codon usage.

9.3 Similarity-Based Approach to Gene Prediction

The idea of a similarity-based approach to gene detection was first stated in Gish and States, 1993 [129]. Although similarity search was in use for gene *detection* (i.e., answering the question of whether a gene is present in a given DNA fragment) for a long time, the potential of similarity search for gene *prediction* (i.e., not only for detection but for detailed prediction of the exon-intron structure as well) remained largely unexplored until the mid-1990s. Snyder and Stormo, 1995 [322] and Searls and Murphy, 1995 [313] made the first attempts to incorporate similarity analysis into gene prediction algorithms. However, the computational complexity of exploring all exon assemblies on top of sequence alignment algorithms is rather high.

Gelfand et al., 1996 [125] proposed a spliced alignment approach to the exon assembly problem, that uses related proteins to derive the exon-intron structure. Figure 9.1a illustrates the spliced alignment problem for the "genomic" sequence

It was brilliant thrilling morning and the slimy hellish lithe doves

gyrated and gambled nimbly in the waves

whose different blocks make up the famous Lewis Carroll line:

't was brillig, and the slithy toves did gyre and gimble in the wabe

The Gelfand et al., 1996 [125] approach is based on the following idea (illustrated by Oksana Khleborodova). Given a genomic sequence (Figure 9.2), they first find a set of *candidate blocks* that contains all *true* exons (Figure 9.3). This can be done by selecting all blocks between potential *acceptor* and *donor* sites (i.e., between AG and GT dinucleotides) with further *filtering* of this set (in a way that does not lose the actual exons). The resulting set of blocks can contain many false exons, of course, and currently it is impossible to distinguish all actual exons from this set by a statistical procedure. Instead of trying to find the actual exons, Gelfand et al., 1996 [125] select a related *target* protein in GenBank (Figure 9.4) and explore all possible block assemblies with the goal of finding an assembly with the highest similarity score to the target protein (Figure 9.5). The number of different block assemblies is huge (Figures 9.6, 9.7, and 9.8), but the *spliced alignment* algorithm, which is the key ingredient of the method, scans all of them in polynomial time (Figure 9.9).

9.4 Spliced Alignment

Let $G = g_1 \dots g_n$ be a string, and let $B = g_i \dots g_j$ and $B' = g_{i'} \dots g_{j'}$ be substrings of G. We write $B \prec B'$ if $j < i'$, i.e., if B ends before B' starts. A sequence $\Gamma = (B_1, \dots, B_p)$ of substrings of G is a *chain* if $B_1 \prec B_2 \prec \dots \prec B_p$. We denote the *concatenation* of strings from the chain Γ as $\Gamma^* = B_1 * B_2 * \dots * B_p$. Given two strings G and T, $s(G, T)$ denotes the score of the *optimal alignment* between G and T.

Let $G = g_1 \dots g_n$ be a string called *genomic sequence*, $T = t_1 \dots t_m$ be a string called *target sequence*, and $\mathcal{B} = \{B_1, \dots B_b\}$ be a set of substrings of G called *blocks*. Given G, T, and $\mathcal{B}$, the *spliced alignment problem* is to find a chain Γ of strings from $\mathcal{B}$ such that the score $s(\Gamma^*, T)$ of the alignment between

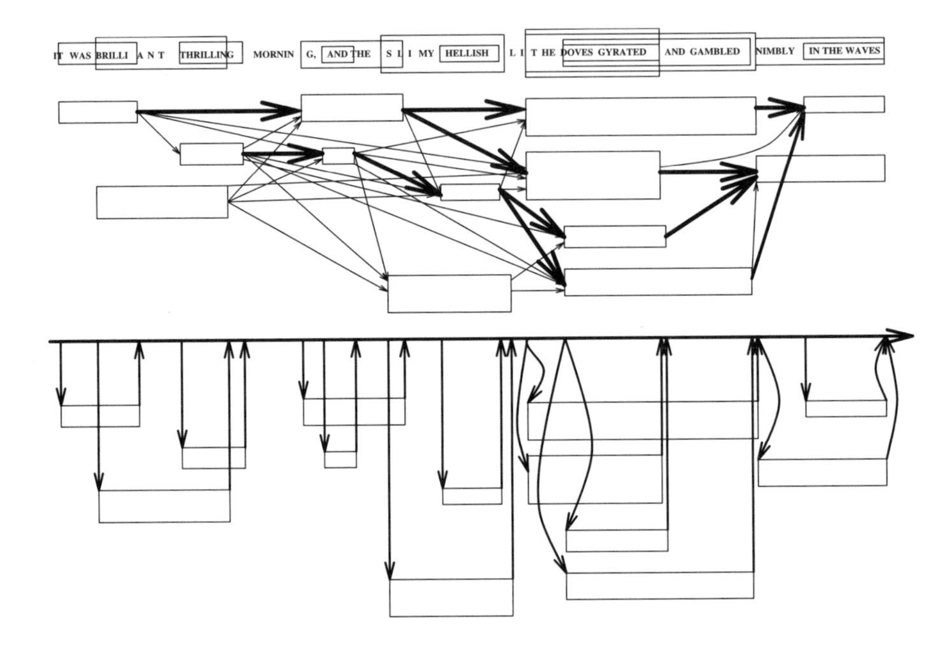

Figure 9.1: Spliced alignment problem: a) block assemblies with the best fit to Lewis Carroll's line, b) corresponding alignment network, and c) equivalent transformation of the alignment network.

the concatenation of these strings and the target sequence is maximum among all chains of blocks from $\mathcal{B}$.

A naive approach to the spliced alignment problem is to detect all relatively high similarities between each block and the target sequence and to assemble these similarities into an optimal subset of compatible similar fragments. The shortcoming of this approach is that the number of blocks is typically very large and the endpoints of the similarity domains are not well defined.

Gelfand et al., 1996 [125] reduced the exon assembly problem to the search of a path in a directed graph (Figure 9.1b). Vertices in this graph correspond to the blocks, edges correspond to potential transitions between blocks, and the path weight is defined as the weight of the optimal alignment between the concatenated

Figure 9.2: Studying genomic sequence.

blocks of this path and the target sequence. Note that the exon assembly problem is different from the standard minimum path problem (the weights of vertices and edges in the graph are not even defined).

Let $B_k = g_m \ldots g_i \ldots g_l$ be a substring of G containing a position i. Define the *i-prefix* of B_k as $B_k(i) = g_m \ldots g_i$. For a block $B_k = g_m \ldots g_l$, let $first(k) = m$, $last(k) = l$, and $size(k) = l - m + 1$. Let $\mathcal{B}(i) = \{k : \ last(k) < i\}$ be the set of blocks ending (strictly) before position i in G. Let $\Gamma = (B_1, \ldots, B_k, \ldots, B_t)$ be a chain such that some block B_k contains position i. Define $\Gamma^*(i)$ as a string

Figure 9.3: Filtering candidate exons.

$\Gamma^*(i) = B_1 * B_2 * \ldots * B_k(i)$. Let

$$S(i, j, k) = \max_{\text{all chains } \Gamma \text{ containing block } B_k} s(\Gamma^*(i), T(j)).$$

The following recurrence computes $S(i, j, k)$ for $1 \leq i \leq n$, $1 \leq j \leq m$, and $1 \leq k \leq b$. For the sake of simplicity we consider sequence alignment with *linear* gap penalties and define $\delta(x, y)$ as a similarity score for every pair of amino acids x and y and δ_{indel} as a penalty for insertion or deletion of amino acids.

Figure 9.4: Finding a target protein.

$$S(i,j,k) = \max \begin{cases} S(i-1,j-1,k) + \delta(g_i,t_j), & \text{if } i \neq first(k) \\ S(i-1,j,k) + \delta_{indel}, & \text{if } i \neq first(k) \\ \max_{l \in \mathcal{B}(first(k))} S(last(l), j-1, l) + \delta(g_i,t_j), & \text{if } i = first(k) \\ \max_{l \in \mathcal{B}(first(k))} S(last(l), j, l) + \delta_{indel}, & \text{if } i = first(k) \\ S(i,j-1,k) + \delta_{indel} & \end{cases} \tag{9.1}$$

After computing the 3-dimensional table $S(i,j,k)$, the score of the optimal spliced alignment is

$$\max_k S(last(k), m, k).$$

Figure 9.5: Using the target protein as a template for exon assembly.

The spliced alignment problem also can be formulated as a *network* alignment problem (Kruskal and Sankoff, 1983 [211]). In this formulation, each block B_k corresponds to a path of length $size(k)$ between vertices $first(k)$ and $last(k)$, and paths corresponding to blocks B_k and B_t are joined by an edge $(last(k), first(t))$ if $B_k \prec B_t$ (Figure 9.1b). The network alignment problem is to find a path in the network with the best alignment to the target sequence.

Gelfand et al., 1996 [125] reduced the number of edges in the spliced alignment graph by making equivalent transformations of the described network, leading to a reduction in time and space. Define

$$P(i, j) = \max_{l \in \mathcal{B}(i)} S(last(l), j, l).$$

Figure 9.6: Assembling.

Then (9.1) can be rewritten as

$$S(i,j,k) = \max \begin{cases} S(i-1,j-1,k) + \delta(g_i,t_j), & \text{if } i \neq first(k) \\ S(i-1,j,k) + \delta_{indel}, & \text{if } i \neq first(k) \\ P(first(k),j-1) + \delta(g_i,t_j), & \text{if } i = first(k) \\ P(first(k),j) + \delta_{indel}, & \text{if } i = first(k) \\ S(i,j-1,k) + \delta_{indel} \end{cases} \qquad (9.2)$$

where

$$P(i,j) = \max \begin{cases} P(i-1,j) \\ \max_{k:\ last(k)=i-1} S(i-1,j,k) \end{cases} \qquad (9.3)$$

Figure 9.7: And assembling...

The network corresponding to (9.2) and (9.3) has a significantly smaller number of edges (Figure 9.1c), thus leading to a practical implementation of the spliced alignment algorithm.

The simplest approach to the construction of blocks $\mathcal{B}$ is to generate all fragments between potential splicing sites represented by AG (acceptor site) and GT (donor site), with the exception of blocks with stop codons in all three frames. However, this approach creates a problem since it generates many short blocks. Experiments with the spliced alignment algorithm have revealed that incorrect predictions for distant targets are frequently associated with the *mosaic effect* caused by very short potential exons. The problem is that these short exons can be easily combined together to fit any target protein. It is easier to "make up" a given sen-

Figure 9.8: And assembling.............

tence from a thousand random short strings than from the same number of longer strings. For example, with high probability, the phrase "filtration of candidate exons" can be made up from a sample of a thousand random two-letter strings ("fi," "lt," "ra," etc. are likely to be present in this sample). The probability that the same phrase can be made up from a sample of the same number of random five-letter strings is close to zero (even finding a string "filtr" in this sample is unlikely). This observation explains the mosaic effect: if the number of short blocks is high, chains of these blocks can replace actual exons in spliced alignments, thus leading to predictions with an unusually large number of short exons. To avoid the mosaic effect, the candidate exons are subjected to some (weak) filtering procedure; for example, only exons with high coding potential may be retained.

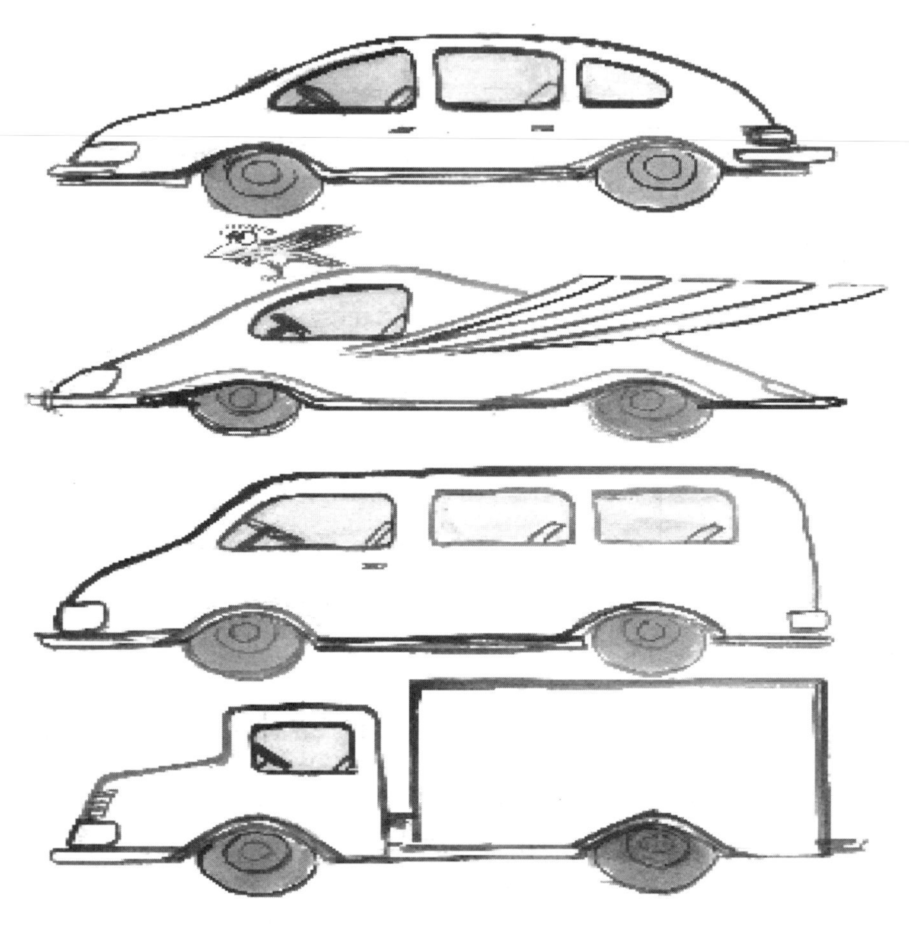

Figure 9.9: Selecting the best exon assembly.

After the optimal block assembly is found, the hope is that it represents the correct exon-intron structure. This is almost guaranteed if a protein sufficiently similar to the one encoded in the analyzed fragment is available: 99% correlation with the actual genes can be obtained from targets with distances of up to 100 PAM (40% similarity). The spliced alignment algorithm provides very accurate predictions if even a distantly related protein is available: predictions at 160 PAM (25% similarity) are still reliable (95% correlation). If a related mammalian protein for an analyzed human gene is known, the accuracy of gene predictions in this fragment is as high as $97\% - 99\%$, and it is 95%, 93%, and 91% for related plant, fungal, and prokaryotic proteins, respectively (Mironov et al., 1998 [242]). Further progress in gene prediction has been achieved by using EST data for similarity-based gene

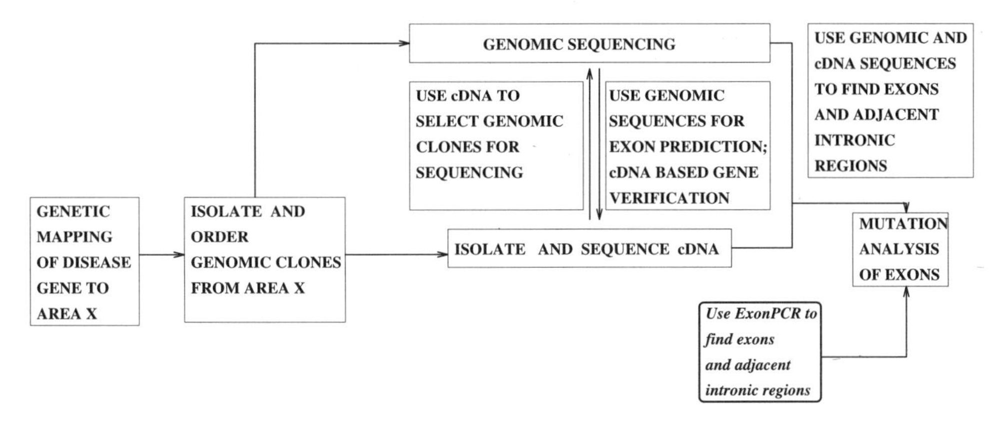

Figure 9.10: Positional cloning and ExonPCR.

prediction. In particular, using EST assemblies, Mironov et al., 1999 [240] found a large number of alternatively spliced genes.

9.5 Reverse Gene Finding and Locating Exons in cDNA

Gene finding often follows the *positional cloning* paradigm:

$$\text{genomic DNA sequencing} \rightarrow \text{exon structure} \rightarrow \text{mRNA} \rightarrow \text{protein}$$

In positional cloning projects, genomic sequences are the primary sources of information for gene prediction, mutation detection, and further search for disease-causing genes. The shift from positional cloning to the *candidate gene library* paradigm is reversing the traditional gene-finding pathway into the following:

$$\text{protein/mRNA} \rightarrow \text{exon structure} \rightarrow \text{limited genomic DNA sequencing}$$

Consequently, modern gene discovery efforts are shifting from single-gene positional cloning to analysis of polygenic diseases with candidate gene libraries of hundred(s) of genes. The genes forming a candidate gene library may come from different sources, e.g., expression analysis, antibody screening, proteomics, etc. Of course, hundred(s) of positional cloning efforts are too costly to be practical.

A positional cloning approach to finding a gene responsible for a disease starts with genetic mapping and proceeds to the detection of disease-related mutations. A multitude of steps are required that include genomic cloning of large DNA fragments, screening cDNA libraries, cDNA isolation, subcloning of the large genomic clones for sequencing, etc. (Figure 9.10). In many gene-hunting efforts, the major motivation for the genomic subcloning and sequencing steps is to determine the gene's exon-intron boundaries. This step is often critical to searches for mutations (or polymorphisms) associated with a disease gene. It requires the design of intronic PCR primers flanking each exon. Traditionally, the exon boundaries are obtained by comparing the cDNA and genomic sequences. The whole process can be time-consuming and may involve multiple subcloning steps and extensive sequencing.

ExonPCR (Xu et al., 1998 [372]) is an alternative experimental protocol that explores the "reverse" gene-finding pathway and provides a fast transition from finding cDNA to mutation detection (Figure 9.10). ExonPCR finds the "hidden" exon boundaries in cDNA (rather than in genomic DNA) and does not require sequencing of genomic clones. In the first step, ExonPCR locates the approximate positions of exon boundaries in cDNA by PCR on genomic DNA using primers designed from the cDNA sequence. The second step is to carry out ligation-mediated PCR to find the flanking intronic regions. As a consequence, the DNA sequencing effort can be vastly reduced.

The computational approaches to finding exon boundaries in cDNA (Gelfand, 1992 [122], Solovyev et al., 1994 [323]) explored *splicing shadows* (i.e., parts of the splicing signals present in exons). However, since the splicing shadow signals are very weak, the corresponding predictions are unreliable. ExonPCR is an experimental approach to finding exon boundaries in cDNA that uses PCR primers in a series of adaptive rounds. Primers are designed from the cDNA sequence and used to amplify genomic DNA. Each pair of primers serves as a query asking the question whether, in the genomic DNA, there exists an intron or introns between the primer sequences. The answer to this query is provided by comparison of the length of PCR products in the cDNA and genomic DNA. If these lengths coincide, the primers belong to the same exon; otherwise, there exists an exon boundary between the corresponding primers. Each pair of primers gives a yes/no answer without revealing the exact positions of exon boundaries. The goal is to find the positions of exon boundaries and to minimize both the number of primers and the number of rounds. Different types of strategies may be used, and the problem is similar to the "Twenty Questions" game with genes. The difference with a parlor game is that genes have a "no answer" option and sometimes may give a false answer and restrict the types of possible queries. This is similar to the "Twenty Questions Game with a Liar" (Lawler and Sarkissian, 1995 [216]) but involves many additional constraints including lower and upper bounds on the length of queries (distance between PCR primers).

ExonPCR attempts to devise a strategy that minimizes the total number of PCR primers (to reduce cost) and at the same time minimizes the number of required rounds of PCR experiments (to reduce time). However, these goals conflict with each other. A minimum number of primer pairs is achieved in a sequential "dichotomy" protocol where only one primer pair is designed in every round based on the results of earlier rounds of experiments. This strategy is unrealistic since it leads to an excessive number of rounds. An alternative, "single-round" protocol designs all possible primer pairs in a single round, thus leading to an excessively large number of primers. Since these criteria are conflicting, ExonPCR searches for a trade-off between the dichotomy strategy and the single-round strategy.

9.6 The Twenty Questions Game with Genes

In its simplest form, the problem can be formulated as follows: given an (unknown) set I of integers in the interval $[1, n]$, reconstruct the set I by asking the minimum number of queries of the form "does a given interval contain an integer from I?" In this formulation, interval $[1, n]$ corresponds to cDNA, I corresponds to exon boundaries in cDNA, and the queries correspond to PCR reactions defined by a pair of primers. A non-adaptive (and trivial) approach to this problem is to generate n single-position queries: does an interval $[i, i]$ contain an integer from I? In an adaptive approach, queries are generated in rounds based on results from all previous queries (only one query is generated in every round).

For the sake of simplicity, consider the case when the number of exon boundaries k is known. For $k = 1$, the optimal algorithm for this problem requires at least $\lg n$ queries and is similar to Binary Search (Cormen et al., 1989 [75]). For $k > 1$, it is easy to derive the lower bound on the number of queries used by any algorithm for this problem, which utilizes the decision tree model. The decision tree model assumes sequential computations using one query at a time. Assume that every vertex in the decision tree is associated with all k-point sets (k-sets) that are consistent with all the queries on the path to this vertex. Since every leaf in the decision tree contains only one k-set, the number of leaves is $\binom{n}{k}$. Since every tree of height h has at most 2^h leaves, the lower bound on the height of the (binary) decision tree is $h \geq \lg \binom{n}{k}$. In the biologically relevant case $k << n$, the minimum number of queries is approximately $k \lg n - k \lg k$. If a biologist tolerates an error Δ in the positions of exon boundaries, the lower bound on the number of queries is approximately $k \lg \frac{n}{\Delta} - k \lg k$. The computational and experimental tests of ExonPCR have demonstrated that it comes close to the theoretical lower bound and that about 30 primers and 3 rounds are required for finding exon boundaries in a typical cDNA sequence.

9.7 Alternative Splicing and Cancer

Recent studies provide evidence that oncogenic potential in human cancer may be modulated by alternative splicing. For example, the progression of prostate cancer from an androgen-sensitive to an androgen-insensitive tumor is accompanied by a change in the alternative splicing of fibroblast growth factor receptor 2 (Carstens et al., 1997 [59]). In another study, Heuze et al., 1999 [160] characterized a prominent alternatively spliced variant for Prostate Specific Antigene, the most important marker available today for diagnosing and monitoring patients with prostate cancer. The questions of what other important alternatively spliced variants of these and other genes are implicated in cancer remains open. Moreover, the known alternative variants of genes implicated in cancer were found by chance in a case-by-case fashion.

Given a gene, how can someone find *all* alternatively spliced variants of this gene? The problem is far from simple since alternative splicing is very frequent in human genes (Mironov et al., 1999 [240]), and computational methods for alternative splicing prediction are not very reliable.

The first systematic attempt to elucidate the splicing variants of genes implicated in (ovarian) cancer was undertaken by Hu et al., 1998 [167]. They proposed long RT-PCR to amplify full-length mRNA and found a new splicing variant for the human multidrug resistance gene MDR1 and the major vault protein (MVP). This method is well suited to detecting a few prominent variants using fixed primers but will have difficulty detecting rare variants (since prominent variants are not suppressed). It also may fail to identify prominent splicing variants that do not amplify with the selected primer pair.

The computational challenges of finding all alternatively spliced variants (*an Alternative Splicing Encyclopedia* or *ASE*) can be explained with the following example. If a gene with three exons has an alternative variant that misses an intermediate exon, then some PCR products in the cDNA library will differ by the length of this intermediate exon. For example, a pair of primers, one from the middle of the first exon and another from the middle of the last exon, will give two PCR products that differ by the length of the intermediate exon. This will lead to detection of both alternatively spliced variants.

Of course, this is a simplified and naive description that is used for illustration purposes only. The complexity of the problem can be understood if one considers a gene with 10 exons with one alternative sliding splicing site per exon. In this case, the number of potential splicing variants is at least 2^{10}, and it is not clear how to find the variants that are present in the cell. The real problem is even more complicated, since some of these splicing variants may be rare and hard to detect by PCR amplification.

Figure 9.11 illustrates the problem of building an ASE for the "genomic" sequence

'twas brilliant thrilling morning and the slimy hellish lithe doves

gyrated and gambled nimbly in the waves

whose alternatively spliced variants "make up" different mRNAs that are similar to the Lewis Carroll's famous "mRNA":

't was brillig, and the slithy toves did gyre and gimble in the wabe

The "exon assembly" graph (Figure 9.11) has an exponential number of paths, each path representing a potential splicing variant. The problem is to figure out which paths correspond to real splicing variants. For example, one can check whether there exists a splicing variant that combines the potential exons X and Y represented by $\boxed{T\ \ WAS\ \ BRILLI}$ and $\boxed{G,\ \ AND\ \ THE\ \ SL}$ with a *spanning primer* XY that spans both X and Y (for example, $BRILLIG,\ \ AND\ T$). In practice, an XY-primer is constructed by concatenation of the last 10 nucleotides of exon X with first 10 nucleotides of exon Y. Pairing XY with another primer (e.g., one taken from the end of exon Y) will confirm or reject the hypothesis about the existence of a splicing variant that combines exons X and Y. Spanning primers allow one to trim the edges in the exon assembly graph that are not supported by experimental evidence. Even after some edges of the graph are trimmed, this approach faces the difficult problem of deciding which triples, quadruples, etc. of exons may appear among alternatively spliced genes. Figure 9.11 presents a relatively simple example of an already trimmed exon assembly graph with just five potential exons and five possible paths: ABCDE, ACDE, ABDE, ABCE, and ACE. The only spanning primers for the variant ACE are AC and CE. However, these spanning primers (in pairs with some other primers) do not allow one to confirm or rule out the existence of the ACE splicing variant. The reason is that the presence of a PCR product amplified by a primer pair involving, let's say, AC, does not guarantee the presence of the ACE variant, since this product may come from the ACBD alternative variant. Similarly, the CE primer may amplify an ABCE splicing variant. If we are lucky, we can observe a relatively short ACE PCR product, but this won't happen if ACE is a relatively rare variant. The solution would be given by forming a pair of spanning primers involving *both* AC and CE. This primer pair amplifies ACE but does not amplify any other splicing variants in Figure 9.11.

The pairs of primers that amplify variant X but do not amplify variant Y are called X+Y- pairs. One can use X+Y- pairs to detect some rare splicing variant X in the background of a prominent splicing variant Y. However, the problem of designing a reliable experimental and computational protocol for finding all alternative variants remains unsolved.

9.8 Some Other Problems and Approaches

9.8.1 Hidden Markov Models for gene prediction

The process of breaking down a DNA sequence into genes can be compared to the process of parsing a sentence into grammatical parts. This naive parsing metaphor was pushed deeper by Searls and Dong, 1993 [312], who advocated a linguistic approach to gene finding. This concept was further developed in the Hidden Markov Models approach for gene prediction (Krogh et al., 1994 [209]) and culminated in

'T WAS BRILLIG, AND THE SLITHY TOVES DID GYRE AND GIMBLE IN THE WABE
T WAS BRILLIG, AND THE SL THE DOVES GYRATED AND GAMBLED IN THE WAVE
T WAS BRILLIG, AND THE SL THE DOVES GYRATED NIMBLY IN THE WAVE
T HRILLING AND HEL LISH DOVES GYRATED AND GAMBLED IN THE WAVE
T HRILLING AND HEL LISH DOVES GYRATED NIMBLY IN THE WAVE

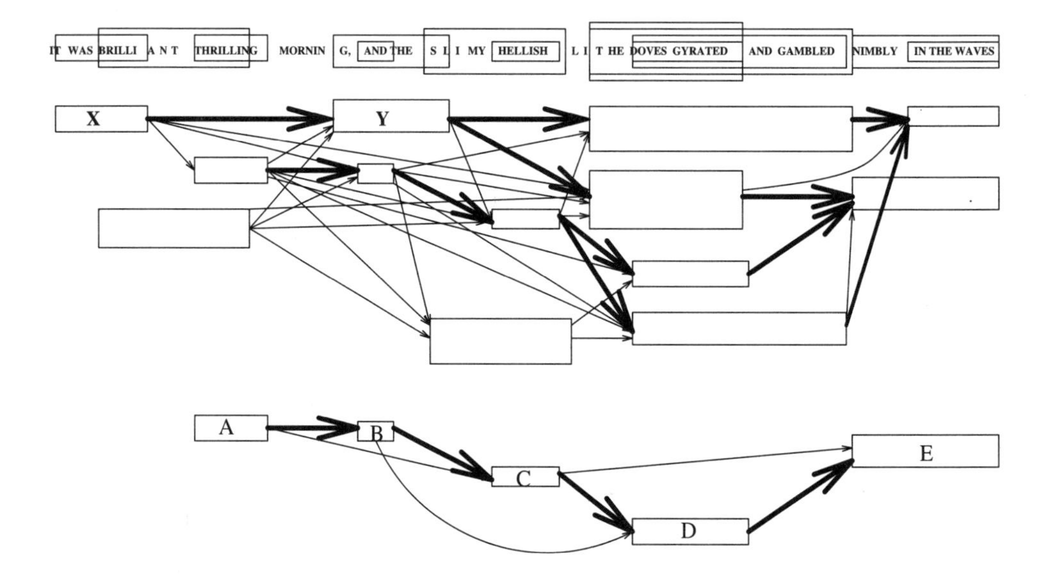

Figure 9.11: Constructing an Alternative Splicing Encyclopedia (ASE) from potential exons. Four different splicing variants (above) correspond to four paths (shown by bold edges) in the exon assembly graph. The overall number of paths in this graph is large, and the problem is how to identify paths that correspond to real splicing variants. The graph at the bottom represents the trimmed exon assembly graph with just five potential splicing variants (paths).

the program GENSCAN (Burge and Karlin, 1997 [54]). HMMs for gene finding consist of many blocks, with each block recognizing a certain statistical feature. For example, profile HMMs can be used to model acceptor and donor sites. Codon statistics can be captured by a different HMM that uses Start codons as *start* state, codons as intermediate states, and Stop codon as *end* state. These HMMs can be combined together as in the Burge and Karlin, 1997 [54] GENSCAN algorithm. In a related approach, Iseli et al., 1999 [176] developed the ESTScan algorithm for gene prediction in ESTs.

9.8.2 Bacterial gene prediction

Borodovsky et al., 1986 [43] were the first to apply Markov chains for bacterial gene prediction. Multiple bacterial sequencing projects created the new computational challenge of *in silico* gene prediction in the absence of any experimental analysis. The problem is that in the absence of experimentally verified genes, there are no positive or negative test samples from which to learn the statistical parameters for coding and non-coding regions. Frishman et al., 1998 [113] proposed the "similarity-first" approach, which first finds fragments in bacterial DNA that are closely related to fragments from a database and uses them as the initial training set for the algorithm. After the statistical parameters for genes that have related sequences are found, they are used for prediction of other genes in an iterative fashion. Currently, GenMark (Hayes and Borodovsky, 1998 [157]), Glimmer (Salzberg et al., 1998 [295]), and Orpheus (Frishman et al., 1998 [113]) combine the similarity-based and statistics-based approaches.

Chapter 10

Genome Rearrangements

10.1 Introduction

Genome Comparison versus Gene Comparison In the late 1980s, Jeffrey Palmer and his colleagues discovered a remarkable and novel pattern of evolutionary change in plant organelles. They compared the mitochondrial genomes of *Brassica oleracea* (cabbage) and *Brassica campestris* (turnip), which are very closely related (many genes are 99% identical). To their surprise, these molecules, which are almost identical in gene *sequences*, differ dramatically in gene *order* (Figure 10.1). This discovery and many other studies in the last decade convincingly proved that genome rearrangements represent a common mode of molecular evolution.

Every study of genome rearrangements involves solving a combinatorial "puzzle" to find a series of *rearrangements* that transform one genome into another. Three such rearrangements "transforming" cabbage into turnip are shown in Figure 10.1. Figure 1.5 presents a more complicated *rearrangement scenario* in which mouse X chromosome is transformed into human X chromosome. Extreme conservation of genes on X chromosomes across mammalian species (Ohno, 1967 [255]) provides an opportunity to study the evolutionary history of X chromosome independently of the rest of the genomes. According to Ohno's law, the gene content of X chromosomes has barely changed throughout mammalian development in the last 125 million years. However, the order of genes on X chromosomes has been disrupted several times.

It is not so easy to verify that the six evolutionary events in Figure 1.5 represent a *shortest* series of *reversals* transforming the mouse gene order into the human gene order on the X chromosome. Finding a shortest series of reversals between the gene order of the mitochondrial DNAs of worm *Ascaris suum* and humans presents an even more difficult computational challenge (Figure 10.2).

In cases of genomes consisting of a small number of "conserved blocks," Palmer and his co-workers were able to find the most parsimonious rearrangement

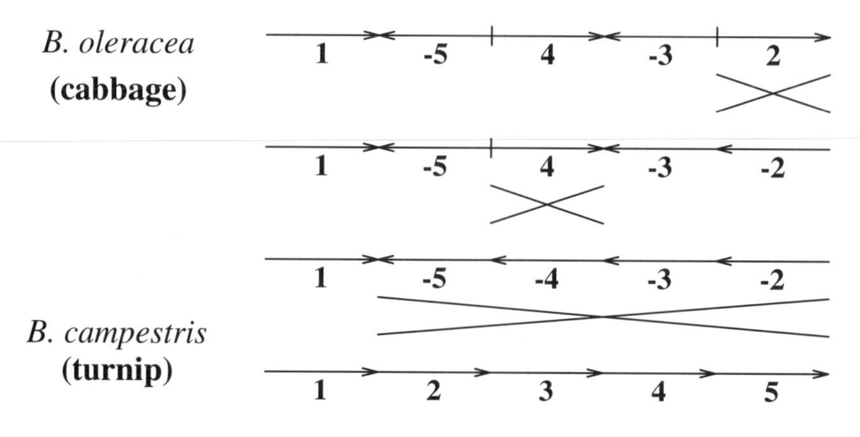

Figure 10.1: "Transformation" of cabbage into turnip.

scenarios. However, for genomes consisting of more than 10 blocks, exhaustive search over all potential solutions is far beyond the capabilities of "pen-and-pencil" methods. As a result, Palmer and Herbon, 1988 [259] and Makaroff and Palmer, 1988 [229] overlooked the most parsimonious rearrangement scenarios in more complicated cases such as turnip versus black mustard or turnip versus radish.

The traditional molecular evolutionary technique is a *gene* comparison, in which phylogenetic trees are being reconstructed based on point mutations of a single gene (or a small number of genes). In the "cabbage and turnip" case, the gene comparison approach is hardly suitable, since the rate of point mutations in cabbage and turnip mitochondrial genes is so low that their genes are almost identical. *Genome comparison* (i.e., comparison of gene orders) is the method of choice in the case of very slowly evolving genomes. Another example of an evolutionary problem for which genome comparison may be more conclusive than gene comparison is the evolution of rapidly evolving viruses.

Studies of the molecular evolution of herpes viruses have raised many more questions than they've answered. Genomes of herpes viruses evolve so rapidly that the extremes of present-day phenotypes may appear quite unrelated; the similarity between many genes in herpes viruses is so low that it is frequently indistinguishable from background noise. Therefore, classical methods of sequence comparison are not very useful for such highly diverged genomes; ventures into the quagmire of the molecular phylogeny of herpes viruses may lead to contradictions, since different genes give rise to different evolutionary trees. Herpes viruses have from 70 to about 200 genes; they all share seven conserved blocks that are rearranged in the genomes of different herpes viruses. Figure 10.3 presents different arrangements of these blocks in Cytomegalovirus (CMV) and Epstein-Barr virus (EBV) and a shortest series of reversals transforming CMV gene order into EBV gene

12 31 34 28 26 17 29 4 9 36 18 35 19 1 16 14 32 33 22 15 11 27 5 20 13 30 23 10 6 3 24 21 8 25 2 7
20 5 27 11 15 22 33 32 14 16 1 19 35 18 36 9 4 29 17 26 28 34 31 12 13 30 23 10 6 3 24 21 8 25 2 7
1 16 14 32 33 22 15 11 27 5 20 19 35 18 36 9 4 29 17 26 28 34 31 12 13 30 23 10 6 3 24 21 8 25 2 7
1 16 15 22 33 32 14 11 27 5 20 19 35 18 36 9 4 29 17 26 28 34 31 12 13 30 23 10 6 3 24 21 8 25 2 7
1 16 15 36 18 35 19 20 5 27 11 14 32 33 22 9 4 29 17 26 28 34 31 12 13 30 23 10 6 3 24 21 8 25 2 7
1 16 15 14 11 27 5 20 19 35 18 36 32 33 22 9 4 29 17 26 28 34 31 12 13 30 23 10 6 3 24 21 8 25 2 7
1 16 15 14 31 34 28 26 17 29 4 9 22 33 32 36 18 35 19 20 5 27 11 12 13 30 23 10 6 3 24 21 8 25 2 7
1 26 28 34 31 14 15 16 17 29 4 9 22 33 32 36 18 35 19 20 5 27 11 12 13 30 23 10 6 3 24 21 8 25 2 7
1 26 28 18 36 32 33 22 9 4 29 17 16 15 14 31 34 35 19 20 5 27 11 12 13 30 23 10 6 3 24 21 8 25 2 7
1 26 28 29 4 9 22 33 32 36 18 17 16 15 14 31 34 35 19 20 5 27 11 12 13 30 23 10 6 3 24 21 8 25 2 7
1 26 28 29 30 13 12 11 27 5 20 19 35 34 31 14 15 16 17 18 36 32 33 22 9 4 23 10 6 3 24 21 8 25 2 7
1 26 11 12 13 30 29 28 27 5 20 19 35 34 31 14 15 16 17 18 36 32 33 22 9 4 23 10 6 3 24 21 8 25 2 7
1 26 27 28 29 30 13 12 11 5 20 19 35 34 31 14 15 16 17 18 36 32 33 22 9 4 23 10 6 3 24 21 8 25 2 7
1 26 27 28 29 30 31 34 35 19 20 5 11 12 13 14 15 16 17 18 36 32 33 22 9 4 23 10 6 3 24 21 8 25 2 7
1 26 27 28 29 30 31 34 35 19 20 9 22 33 32 36 18 17 16 15 14 13 12 11 5 4 23 10 6 3 24 21 8 25 2 7
1 26 27 28 29 30 31 22 9 20 19 35 34 33 32 36 18 17 16 15 14 13 12 11 5 4 23 10 6 3 24 21 8 25 2 7
1 26 27 28 29 30 31 32 33 34 35 19 20 9 22 36 18 17 16 15 14 13 12 11 5 4 23 10 6 3 24 21 8 25 2 7
1 26 27 28 29 30 31 32 33 34 35 36 22 9 20 19 18 17 16 15 14 13 12 11 5 4 23 10 6 3 24 21 8 25 2 7
1 26 27 28 29 30 31 32 33 34 35 36 22 9 24 3 6 10 23 4 5 11 12 13 14 15 16 17 18 19 20 21 8 25 2 7
1 26 27 28 29 30 31 32 33 34 35 36 22 9 8 21 20 19 18 17 16 15 14 13 12 11 5 4 23 10 6 3 24 25 2 7
1 26 27 28 29 30 31 32 33 34 35 36 8 9 22 21 20 19 18 17 16 15 14 13 12 11 5 4 23 10 6 3 24 25 2 7
1 26 27 28 29 30 31 32 33 34 35 36 8 9 22 21 20 19 18 17 16 15 14 13 12 11 5 4 3 6 10 23 24 25 2 7
1 26 27 28 29 30 31 32 33 34 35 36 8 9 22 21 20 19 18 17 16 15 14 13 12 11 5 4 3 2 25 24 23 10 6 7
1 2 3 4 5 11 12 13 14 15 16 17 18 19 20 21 22 9 8 36 35 34 33 32 31 30 29 28 27 26 25 24 23 10 6 7
1 2 3 4 5 11 12 13 14 15 16 17 18 19 20 21 22 9 8 7 6 10 23 24 25 26 27 28 29 30 31 32 33 34 35 36
1 2 3 4 5 6 7 8 9 22 21 20 19 18 17 16 15 14 13 12 11 10 23 24 25 26 27 28 29 30 31 32 33 34 35 36
1 2 3 4 5 6 7 8 9 10 11 12 13 14 15 16 17 18 19 20 21 22 23 24 25 26 27 28 29 30 31 32 33 34 35 36

Figure 10.2: A most parsimonious rearrangement scenario for transformation of worm *Ascaris Suum* mitochondrial DNA into human mitochondrial DNA (26 reversals).

order (Hannenhalli et al., 1995 [152]). The number of such large-scale rearrangements (five reversals) is much smaller than the number of point mutations between CMV and EBV (hundred(s) of thousands). Therefore, the analysis of such rearrangements at the *genome* level may complement the analysis at the *gene* level traditionally used in molecular evolution. Genome comparison has certain merits and demerits as compared to classical gene comparison: genome comparison ignores actual DNA sequences of genes, while gene comparison ignores gene order. The ultimate goal would be to combine the merits of both genome and gene comparison in a single algorithm.

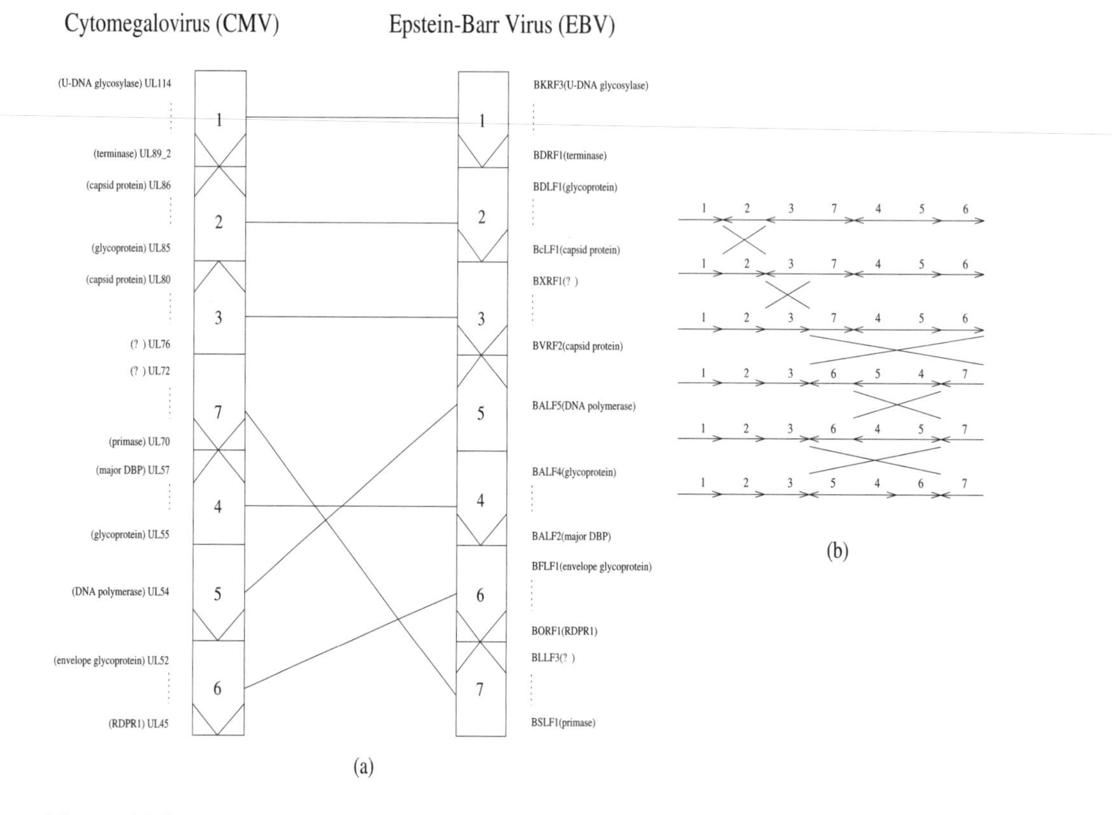

Figure 10.3: Comparative genome organization (a) and the shortest series of rearrangements transforming CMV gene order into EBV gene order (b).

The analysis of genome rearrangements in molecular biology was pioneered in the late 1930s by Dobzhansky and Sturtevant, who published a milestone paper presenting a rearrangement scenario with 17 inversions for the species of *Drosophila* fruit fly (Dobzhansky and Sturtevant, 1938 [87]). With the advent of large-scale mapping and sequencing, the number of *genome comparison* problems is rapidly growing in different areas, including viral, bacterial, yeast, plant, and animal evolution.

Sorting by Reversals A computational approach based on comparison of gene orders was pioneered by David Sankoff (Sankoff et al., 1990, 1992 [302, 304] and Sankoff, 1992 [300]). Genome rearrangements can be modeled by a combinatorial problem of sorting by reversals, as described below. The order of genes in two organisms is represented by permutations $\pi = \pi_1\pi_2\ldots\pi_n$ and $\sigma = \sigma_1\sigma_2\ldots\sigma_n$. A *reversal* $\rho(i,j)$ of an interval $[i,j]$ is the permutation

$$\begin{pmatrix} 1\;2\ldots i-1\;\mathbf{i}\;\mathbf{i+1}\ldots\mathbf{j-1}\;\mathbf{j}\;j+1\ldots n \\ 1\;2\ldots i-1\;\mathbf{j}\;\mathbf{j-1}\ldots\mathbf{i+1}\;\mathbf{i}\;j+1\ldots n \end{pmatrix}$$

Clearly $\rho(i,j)$ has the effect of reversing the order of $\pi_i\pi_{i+1}\ldots\pi_j$ and transforming $\pi_1\ldots\pi_{i-1}\pi_i\ldots\pi_j\pi_{j+1}\ldots\pi_n$ into $\pi\cdot\rho(i,j) = \pi_1\ldots\pi_{i-1}\pi_j\ldots\pi_i\pi_{j+1}\ldots\pi_n$.

Given permutations π and σ, the *reversal distance problem* is to find a series of reversals $\rho_1, \rho_2, \ldots, \rho_t$ such that $\pi\cdot\rho_1\cdot\rho_2\cdots\rho_t = \sigma$ and t is minimal. We call t the *reversal distance* between π and σ. *Sorting* π *by reversals* is the problem of finding the reversal distance $d(\pi)$ between π and the identity permutation $(12\ldots n)$.

Computer scientists have studied a related *sorting by prefix reversals* problem (also known as the *pancake flipping problem*): given an arbitrary permutation π, find $d_{pref}(\pi)$, which is the minimum number of reversals of the form $\rho(1,i)$ sorting π. The pancake flipping problem was inspired by the following "real-life" situation described by Harry Dweigter:

The chef in our place is sloppy, and when he prepares a stack of pancakes they come out all different sizes. Therefore, when I deliver them to a customer, on the way to a table I rearrange them (so that the smallest winds up on top, and so on, down to the largest at the bottom) by grabbing several from the top and flipping them over, repeating this (varying the number I flip) as many times as necessary. If there are n pancakes, what is the maximum number of flips that I will ever have to use to rearrange them?

Bill Gates (an undergraduate student at Harvard in late 1970s, now at Microsoft) and Cristos Papadimitriou made the first attempt to solve this problem (Gates and Papadimitriou, 1979 [120]). They proved that the *prefix reversal diameter* of the symmetric group, $d_{pref}(n) = \max_{\pi\in S_n} d_{pref}(\pi)$, is less than or equal to $\frac{5}{3}n + \frac{5}{3}$, and that for infinitely many n, $d_{pref}(n) \geq \frac{17}{16}n$. The pancake flipping problem still remains unsolved.

The Breakpoint Graph What makes it hard to sort a permutation? In the very first computational studies of genome rearrangements, Watterson et al., 1982 [366] and Nadeau and Taylor, 1984 [248] introduced the notion of a *breakpoint* and noticed some correlations between the reversal distance and the number of breakpoints. (In fact, Sturtevant and Dobzhansky, 1936 [331] implicitly discussed these correlations 60 years ago!) Below we define the notion of a breakpoint.

Let $i \sim j$ if $|i-j| = 1$. Extend a permutation $\pi = \pi_1\pi_2\ldots\pi_n$ by adding $\pi_0 = 0$ and $\pi_{n+1} = n+1$. We call a pair of elements (π_i, π_{i+1}), $0 \leq i \leq n$, of π an *adjacency* if $\pi_i \sim \pi_{i+1}$, and a *breakpoint* if $\pi_i \not\sim \pi_{i+1}$ (Figure 10.4). As the identity permutation has no breakpoints, sorting by reversals corresponds to eliminating breakpoints. An observation that every reversal can eliminate *at most* 2 breakpoints immediately implies that $d(\pi) \geq \frac{b(\pi)}{2}$, where $b(\pi)$ is the number of breakpoints in π. Based on the notion of a breakpoint, Kececioglu and Sankoff, 1995 [194] found an approximation algorithm for sorting by reversals with performance guarantee 2. They also devised efficient bounds, solving the reversal distance problem almost optimally for n ranging from 30 to 50. This range covers the biologically important case of animal mitochondrial genomes.

However, the estimate of reversal distance in terms of breakpoints is very inaccurate. Bafna and Pevzner, 1996 [19] showed that another parameter (size of a maximum cycle decomposition of the breakpoint graph) estimates reversal distance with much greater accuracy.

The *breakpoint graph* of a permutation π is an edge-colored graph $G(\pi)$ with $n+2$ vertices $\{\pi_0, \pi_1, \ldots, \pi_n, \pi_{n+1}\} \equiv \{0, 1, \ldots, n, n+1\}$. We join vertices π_i and π_{i+1} by a *black* edge for $0 \leq i \leq n$. We join vertices π_i and π_j by a *gray* edge if $\pi_i \sim \pi_j$. Figure 10.4 demonstrates that a breakpoint graph is obtained by a superposition of a black path traversing the vertices $0, 1, \ldots, n, n+1$ in the order given by permutation π and a gray path traversing the vertices in the order given by the identity permutation.

A *cycle* in an edge-colored graph G is called *alternating* if the colors of every two consecutive edges of this cycle are distinct. In the following, by cycles we mean alternating cycles. A vertex v in a graph G is called *balanced* if the number of black edges incident to v equals the number of gray edges incident to v. A *balanced graph* is a graph in which every vertex is balanced. Clearly $G(\pi)$ is a balanced graph: therefore, it contains an alternating Eulerian cycle. Therefore, there exists a *cycle decomposition* of $G(\pi)$ into edge-disjoint alternating cycles (every edge in the graph belongs to exactly one cycle in the decomposition). Cycles in an edge decomposition may be self-intersecting. The breakpoint graph in Figure 10.4 can be decomposed into four cycles, one of which is self-intersecting. We are interested in the decomposition of the breakpoint graph into a *maximum* number $c(\pi)$ of edge-disjoint alternating cycles. For the permutation in Figure 10.4, $c(\pi) = 4$.

Cycle decompositions play an important role in estimating reversal distance. When we apply a reversal to a permutation, the number of cycles in a maximum decomposition can change by at most one (while the number of breakpoints can change by two). Bafna and Pevzner, 1996 [19] proved the bound $d(\pi) > n + 1 - c(\pi)$, which is much tighter than the bound in terms of breakpoints $d(\pi) \geq b(\pi)/2$. For most biological examples, $d(\pi) = n + 1 - c(\pi)$, thus reducing the reversal distance problem to the maximal cycle decomposition problem.

Duality Theorem for Signed Permutations Finding a maximal cycle decomposition is a difficult problem. Fortunately, in the more biologically relevant case of *signed permutations,* this problem is trivial. Genes are *directed* fragments of DNA,

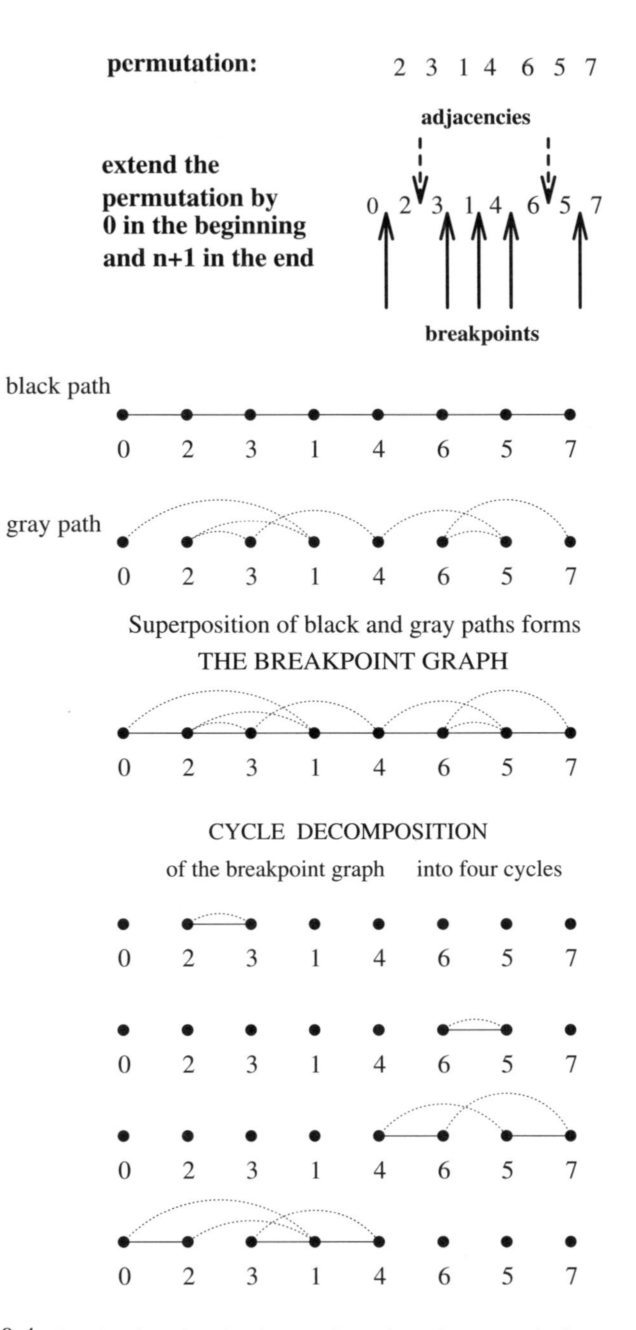

Figure 10.4: Breakpoints, breakpoint graph, and maximum cycle decomposition.

and a sequence of n genes in a genome is represented by a *signed* permutation on $\{1, \ldots n\}$ with a $+$ or $-$ sign associated with every element of π. For example, the gene order for *B. oleracea* presented in Figure 10.1 is modeled by the signed permutation $(+1 - 5 + 4 - 3 + 2)$. In the signed case, every reversal of fragment $[i, j]$ changes both the order and the signs of the elements within that fragment (Figure 10.1). We are interested in the minimum number of reversals $d(\pi)$ required to transform a signed permutation π into the identity signed permutation $(+1 + 2 \ldots + n)$.

Bafna and Pevzner, 1996 [19] noted that the concept of a breakpoint graph extends naturally to signed permutations by mimicking every directed element i by two undirected elements i_a and i_b, which substitute for the tail and the head of the directed element i (Figure 10.5).

For signed permutations, the bound $d(\pi) \geq n + 1 - c(\pi)$ approximates the reversal distance extremely well for both simulated and biological data. This intriguing performance raises the question of whether the bound $d(\pi) \geq n+1-c(\pi)$ overlooks another parameter (in addition to the size of a maximum cycle decomposition) that would allow closing the gap between $d(\pi)$ and $n+1-c(\pi)$. Hannenhalli and Pevzner, 1995 [154] revealed another "hidden" parameter (number of *hurdles* in π) making it harder to sort a signed permutation and showed that

$$n + 1 - c(\pi) + h(\pi) \leq d(\pi) \leq n + 2 - c(\pi) + h(\pi) \tag{10.1}$$

where $h(\pi)$ is the number of hurdles in π. They also proved the duality theorem for signed permutations and developed a polynomial algorithm for computing $d(\pi)$.

Unsigned Permutations and Comparative Physical Mapping Since sorting (unsigned) permutations by reversals is NP-hard (Caprara, 1997 [57]), many researchers have tried to devise a practical approximation algorithm for sorting (unsigned permutations) by reversals.

A *block* of π is an interval $\pi_i \ldots \pi_j$ containing no breakpoints, i.e., (π_k, π_{k+1}) is an adjacency for $0 \leq i \leq k < j \leq n+1$. Define a *strip* of π as a maximal block, i.e., a block $\pi_i \ldots \pi_j$ such that (π_{i-1}, π_i) and (π_j, π_{j+1}) are breakpoints. A strip of one element is called a *singleton*, a strip of two elements is called a *2-strip*, and a strip with more than two elements is called a *long strip*. It turns out that singletons cause a major challenge in sorting unsigned permutations by reversals.

A reversal $\rho(i, j)$ *cuts* a strip $\pi_k \ldots \pi_l$ if either $k < i \leq l$ or $k \leq j < l$. A reversal cutting a strip separates elements that are consecutive in the identity permutation. Therefore, it is natural to expect that for every permutation π there exists an (optimal) sorting of π by reversals that does not cut strips. This, however, is false. Permutation 3412 requires three reversals if we do not cut strips, and yet it can be sorted with two: $3412 \rightarrow 1432 \rightarrow 1234$. Kececioglu and Sankoff, 1993 [192] conjectured that every permutation has an optimal sorting by reversals that does not cut long strips and does not increase the number of breakpoints.

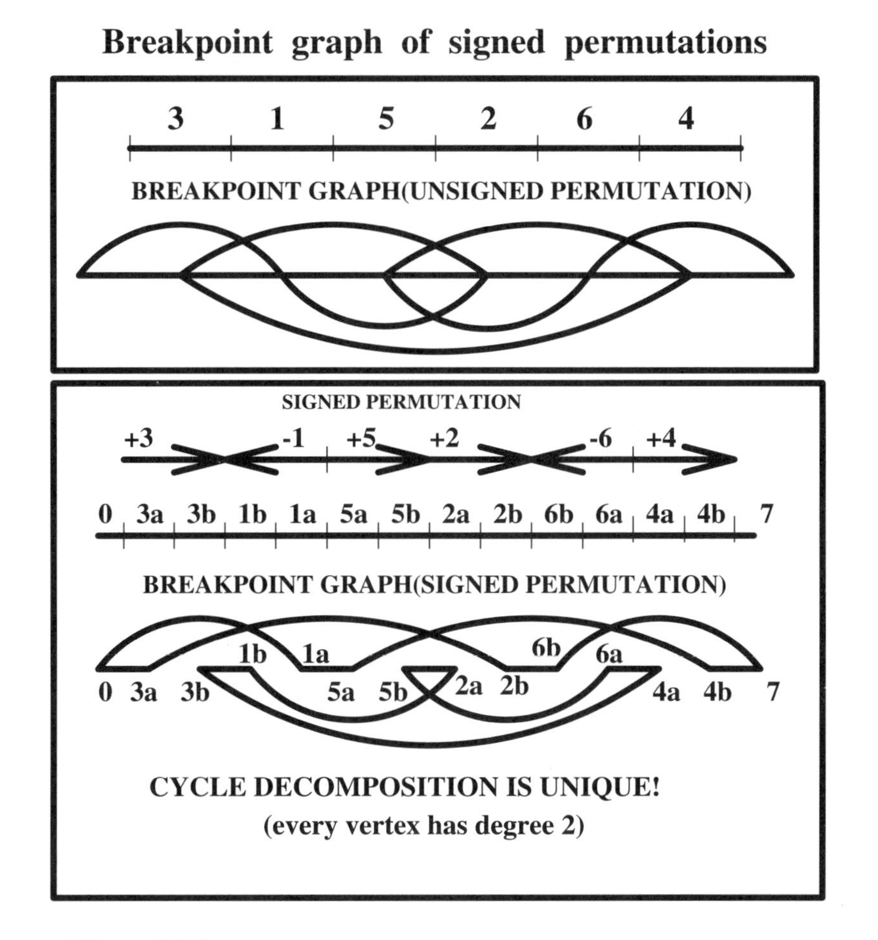

Figure 10.5: Modeling a signed permutation by an unsigned permutation.

Since the identity permutation has no breakpoints, sorting by reversals corresponds to eliminating breakpoints. From this perspective, it is natural to expect that for every permutation there exists an optimal sorting by reversals that never increases the number of breakpoints. Hannenhalli and Pevzner, 1996 [155] proved both the "reversals do not cut long strips" and the "reversals do not increase the number of breakpoints" conjectures by using the duality theorem for signed permutations.

Biologists derive gene orders either by sequencing entire genomes or by using comparative physical mapping. Sequencing provides information about the directions of genes and allows one to represent a genome by a signed permutation. However, sequencing of entire genomes is still expensive, and most currently avail-

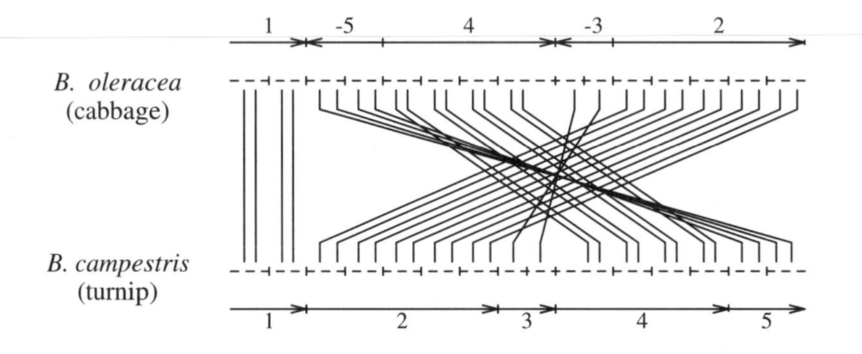

Figure 10.6: Comparative physical map of cabbage and turnip (unsigned permutation) and corresponding signed permutation.

able experimental data on gene orders are based on comparative physical maps. Physical maps usually do not provide information about the directions of genes, and therefore lead to representation of a genome as an *unsigned* permutation π. Biologists try to derive a signed permutation from this representation by assigning a positive (negative) sign to increasing (decreasing) strips of π (Figure 10.6). The "reversals do not cut long strips" property provides a theoretical substantiation for such a procedure in the case of long strips. At the same time, for 2-strips this procedure might fail to find an optimal rearrangement scenario. Hannenhalli and Pevzner, 1996 [155] pointed to a biological example for which this procedure fails and described an algorithm fixing this problem.

Permutations without singletons are called *singleton-free* permutations. The difficulty in analyzing such permutations is posed by an alternative, *"to cut or not to cut"* 2-strips. A characterization of a set of 2-strips *"to cut"* (Hannenhalli and Pevzner, 1996 [155]) leads to a polynomial algorithm for sorting singleton-free permutations and to a polynomial algorithm for sorting permutations with a small number of singletons. The algorithm can be applied to analyze rearrangement scenarios derived from comparative physical maps.

Low-resolution physical maps usually contain many singletons and, as a result, rearrangement scenarios for such maps are hard to analyze. The Hannenhalli and Pevzner, 1996 [155] algorithm runs in polynomial time if the number of singletons is $O(\log n)$. This suggests that $O(\log n)$ singletons is the desired trade-off of resolution for comparative physical mapping in molecular evolution studies. If the number of singletons is large, a biologist might choose additional experiments (i.e., sequencing of some areas) to resolve the ambiguities in gene directions.

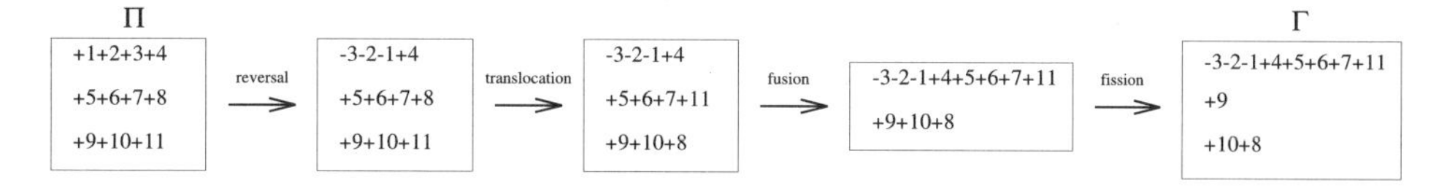

Figure 10.7: Evolution of genome Π into genome Γ.

Rearrangements of Multichromosomal Genomes When the Brothers Grimm described a transformation of a man into a mouse in the fairy tale "Puss in Boots," they could hardly have anticipated that two centuries later humans and mice would be the most genetically studied mammals. Man-mouse comparative physical mapping started 20 years ago, and currently a few thousand pairs of homologous genes are mapped in these species. As a result, biologists have found that the related genes in man and mouse are not chaotically distributed over the genomes, but form "conserved blocks" instead. Current comparative mapping data indicate that both human and mouse genomes are comprised of approximately 150 blocks which are "shuffled" in humans as compared to mice (Copeland et al., 1993 [74]). For example, the chromosome 7 in the human can be viewed as a mosaic of different genes from chromosomes 2, 5, 6, 11, 12, and 13 in the mouse (Fig 1.4). Shuffling of blocks happens quite rarely (roughly once in a million years), thus giving biologists hope of reconstructing a rearrangement scenario of human-mouse evolution. In their pioneering paper, Nadeau and Taylor, 1984 [248] estimated that surprisingly few genomic rearrangements (178 ± 39) have happened since the divergence of human and mouse 80 million years ago.

In the model we consider, every gene is represented by an integer whose *sign* ("+" or "–") reflects the *direction* of the gene. A *chromosome* is defined as a *sequence* of genes, while a *genome* is defined as a *set* of chromosomes. Given two genomes Π and Γ, we are interested in a most parsimonious scenario of *evolution* of Π into Γ, i.e., the shortest sequence of rearrangement events (defined below) required to transform Π into Γ. In the following we assume that Π and Γ contain the same set of genes. Figure 10.7 illustrates four rearrangement events transforming one genome into another.

Let $\Pi = \{\pi(1), \ldots, \pi(N)\}$ be a genome consisting of N chromosomes and let $\pi(i) = (\pi(i)_1 \ldots \pi(i)_{n_i})$, n_i being the number of genes in the i-th chromosome. Every chromosome π can be viewed either from "left to right" (i.e., as $\pi = (\pi_1 \ldots \pi_n)$) or from "right to left" (i.e., as $-\pi = (-\pi_n \ldots - \pi_1)$), leading to two equivalent representations of the same chromosome (i.e., the *directions* of chromosomes are irrelevant). The four most common elementary rearrangement events in multichromosomal genomes are *reversals*, *translocations*, *fusions*, and *fissions*, defined below.

Let $\pi = \pi_1 \dots \pi_n$ be a chromosome and $1 \le i \le j \le n$. A *reversal* $\rho(\pi, i, j)$ on a chromosome π rearranges the genes *inside* $\pi = \pi_1 \dots \pi_{i-1}\pi_i \dots \pi_j\pi_{j+1} \dots \pi_n$ and transforms π into $\pi_1 \dots \pi_{i-1} - \pi_j \dots - \pi_i\pi_{j+1} \dots \pi_n$. Let $\pi = \pi_1 \dots \pi_n$ and $\sigma = \sigma_1 \dots \sigma_m$ be two chromosomes and $1 \le i \le n+1$, $1 \le j \le m+1$. A *translocation* $\rho(\pi, \sigma, i, j)$ exchanges genes *between* chromosomes π and σ and transforms them into chromosomes $\pi_1 \dots \pi_{i-1}\sigma_j \dots \sigma_m$ and $\sigma_1 \dots \sigma_{j-1}\pi_i \dots \pi_n$ with $(i-1)+(m-j+1)$ and $(j-1)+(n-i+1)$ genes respectively. We denote as $\Pi \cdot \rho$ the genome obtained from Π as a result of a rearrangement (reversal or translocation) ρ. Given genomes Π and Γ, the *genomic sorting problem* is to find a series of reversals and translocations $\rho_1, \dots, \rho_t$ such that $\Pi \cdot \rho_1 \cdots \rho_t = \Gamma$ and t is minimal. We call t the *genomic distance* between Π and Γ. The *Genomic distance problem* is the problem of finding the genomic distance $d(\Pi, \Gamma)$ between Π and Γ.

A translocation $\rho(\pi, \sigma, n+1, 1)$ concatenates the chromosomes π and σ, resulting in a chromosome $\pi_1 \dots \pi_n\sigma_1 \dots \sigma_m$ and an *empty* chromosome $\emptyset$. This special translocation, leading to a reduction in the number of (non-empty) chromosomes, is known in molecular biology as a *fusion*. The translocation $\rho(\pi, \emptyset, i, 1)$ for $1 < i < n$ "breaks" a chromosome π into two chromosomes $(\pi_1 \dots \pi_{i-1})$ and $(\pi_i \dots \pi_n)$. This translocation, leading to an increase in the number of (non-empty) chromosomes, is known as a *fission*. Fusions and fissions are rather common in mammalian evolution; for example, the major difference in the overall genome organization of humans and chimpanzees is the fusion of two chimpanzee chromosomes into one human chromosome.

Kececioglu and Ravi, 1995 [191] made the first attempt to analyze rearrangements of multichromosomal genomes. Their approximation algorithm addresses the case in which both genomes contain the same number of chromosomes. This is a serious limitation, since different organisms (in particular humans and mice) have different numbers of chromosomes. From this perspective, every realistic model of genome rearrangements should include fusions and fissions. It turns out that fusions and fissions present a major difficulty in analyzing genome rearrangements. Hannenhalli and Pevzner, 1995 [153] proved the duality theorem for multichromosomal genomes, which computes genomic distance in terms of seven parameters reflecting different combinatorial properties of sets of strings. Based on this result they found a polynomial algorithm for this problem.

The idea of the analysis is to concatenate N chromosomes of Π and Γ into permutations π and γ, respectively, and to mimic genomic sorting of Π into Γ by sorting π into γ by reversals. The difficulty with this approach is that there exist $N!2^N$ different concatenates for Π and Γ, and only some of them, called *optimal concatenates*, mimic an *optimal* sorting of Π into Γ. Hannenhalli and Pevzner, 1995 [153] introduced techniques called *flipping* and *capping* of chromosomes that allow one to find an optimal concatenate.

Of course, gene orders for just two genomes are hardly sufficient to delineate a correct rearrangement scenario. Comparative gene mapping has made possible the generation of comparative maps for many mammalian species (O'Brien and Graves, 1991 [254]). However, the resolution of these maps is significantly lower than thc rcsolution of the human-mouse map. Since comparative physical mapping is rather laborious, one can hardly expect that the tremendous effort involved in obtaining the human-mouse map will be repeated for other mammalian genomes. However, an experimental technique called *chromosome painting* allows one to derive gene order without actually building an accurate "gene-based" map. In the past, the applications of chromosome painting were limited to primates (Jauch et al., 1992 [178]); attempts to extend this approach to other mammals were not successful because of the DNA sequence diversity between distantly related species. Later, Scherthan et al., 1994 [307] developed an improved version of chromosome painting, called *ZOO-FISH*, that is capable of detecting homologous chromosome fragments in distant mammalian species. Using ZOO-FISH, Rettenberger et al., 1995 [284] quickly completed the human-pig chromosome painting project and identified 47 conserved blocks common to human and pig. The success of the human-pig chromosome painting project indicates that gene orders of many mammalian species can be generated with ZOO-FISH inexpensively, thus providing an invaluable new source of data to attack the 100-year-old problem of mammalian evolution.

10.2 The Breakpoint Graph

Cycle decomposition is a rather exotic notion that at first glance has little in common with genome rearrangements. However, the observation that a reversal changes the number of cycles in a maximum decomposition by at most one allows us to bound the reversal distance in terms of maximum cycle decomposition.

Theorem 10.1 *For every permutation π and reversal ρ, $c(\pi\rho) - c(\pi) \leq 1$.*

Proof An arbitrary reversal $\rho(i, j)$ involves four vertices of the graph $G(\pi)$ and leads to replacing two black edges $DEL = \{(\pi_{i-1}, \pi_i), (\pi_j, \pi_{j+1})\}$ by the black edges $ADD = \{(\pi_{i-1}, \pi_j), (\pi_i, \pi_{j+1})\}$.

If these two black edges in ADD belong to the same cycle in a maximum cycle decomposition of $G(\pi\rho)$, then a deletion of that cycle yields a cycle decomposition of $G(\pi)$ with at least $c(\pi\rho) - 1$ cycles. Therefore, $c(\pi) \geq c(\pi\rho) - 1$.

On the other hand, if the black edges in ADD belong to different cycles C_1 and C_2 in a maximum cycle decomposition of $G(\pi\rho)$, then deleting $C_1 \cup C_2$ gives a set of edge-disjoint cycles of size $c(\pi\rho) - 2$ in the graph $G(\pi\rho) \setminus (C_1 \cup C_2)$. Clearly, the set of edges $(C_1 \cup C_2 \cup DEL) \setminus ADD$ forms a balanced graph and must contain at least one cycle. Combining this cycle with the previously obtained $c(\pi\rho) - 2$ cycles, we obtain a cycle decomposition of $G(\pi) = (G(\pi\rho) \setminus (C_1 \cup$

$C_2)) \cup (C_1 \cup C_2 \cup DEL \setminus ADD)$ into at least $c(\pi\rho) - 1$ cycles. ■

Theorem 10.1, together with the observation that $c(\iota) = n + 1$ for the identity permutation ι, immediately implies $d(\pi) \geq c(\iota) - c(\pi) \equiv n + 1 - c(\pi)$:

Theorem 10.2 *For every permutation π, $d(\pi) \geq n + 1 - c(\pi)$.*

10.3 "Hard-to-Sort" Permutations

Define $d(n) = \max_{\pi \in S_n} d(\pi)$ to be the *reversal diameter* of the symmetric group of order n. Gollan conjectured that $d(n) = n - 1$ and that only one permutation γ_n and its inverse permutation γ_n^{-1} require $n - 1$ reversals to be sorted. The *Gollan* permutation, in one-line notation, is defined as follows:

$$\gamma_n = \begin{cases} (3,1,5,2,7,4,\ldots,n-3,n-5,n-1,n-4,n,n-2), & n \text{ even} \\ (3,1,5,2,7,4,\ldots,n-6,n-2,n-5,n,n-3,n-1), & n \text{ odd} \end{cases}$$

Bafna and Pevzner, 1996 [19] proved Gollan's conjecture by showing that $c(\gamma_n) = 2$ and applying theorem 10.2. Further, they demonstrated that the reversal distance between two random permutations is very close to the reversal diameter of the symmetric group, thereby indicating that reversal distance provides a good separation between related and non-related sequences in molecular evolution studies.

We show that the breakpoint graph $G(\gamma_n)$ has at most two disjoint alternating cycles. The subgraph of $G(\gamma_n)$ formed by vertices $\{4, 5, \ldots, n-5, n-4\}$ has a regular structure (Figure 10.8). Direct the black edges of an arbitrary cycle in this subgraph from the bottom to the top and all gray edges from the top to the bottom. Note that in this orientation all edges are directed either $\nwarrow$ or $\swarrow$ or $\downarrow$, and therefore walking along the edges of this cycle slowly but surely leads to the left. How would we return to the initial vertex? we can do so only after reaching one of the "irregular" vertices (1 and 3), which serve as "turning points." The following lemma justifies this heuristic argument.

Lemma 10.1 *Every alternating cycle in $G(\gamma_n)$ contains the vertex 1 or 3.*

Proof Let i be the minimal odd vertex of an alternating cycle X in $G(\gamma_n)$. Consider the sequence i, j, k of consecutive vertices in X, where (i, j) is black and (j, k) is gray. If $i > 5$, then $j = i - 3$ or $j = i - 5$ and $k = j + 1$ or $k = j - 1$ (Figure 10.8), implying that k is odd and $k < i$, a contradiction. If $i = 5$, then $j = 2$ and k is either 1 or 3, a contradiction. Therefore, i is either 1 or 3. ■

Theorem 10.3 *(Gollan conjecture) For every n, $d(\gamma_n) = d(\gamma_n^{-1}) = n - 1$.*

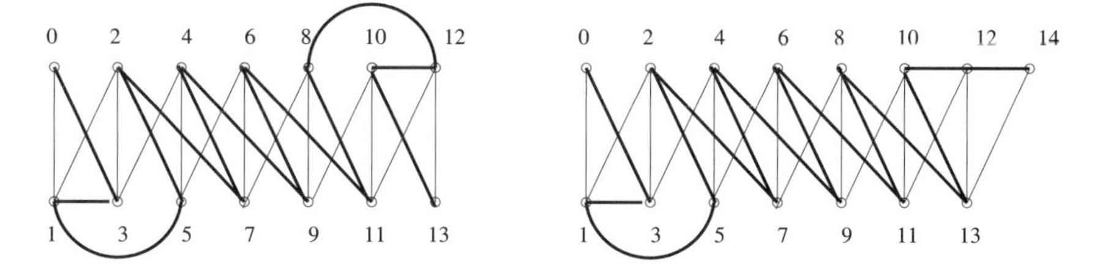

Figure 10.8: $G(\gamma_{12})$ and $G(\gamma_{13})$.

Proof For $n \leq 2$, the claim is trivial. For $n > 2$, partition the vertex set of $G(\gamma_n)$ into $V_l = \{0, 1, 3\}$ and V_r. From lemma 10.1 and the fact that there is no cycle contained in V_l, we see that every alternating cycle must contain at least two edges from the cut (V_l, V_r). As the cut (V_l, V_r) consists of four edges $((1, 2), (1, 5), (3, 2)$, and $(3, 4))$, the maximum number of edge-disjoint alternating cycles in a cycle decomposition of $G(\gamma_n)$ is at most $\frac{4}{2} = 2$.

From theorem 10.2, $d(\gamma_n) \geq n + 1 - c(\gamma_n) \geq n - 1$. On the other hand, $d(\gamma_n) \leq n - 1$, since there exists a simple algorithm sorting every n-element permutation in $n - 1$ steps. Finally note that $d(\gamma_n^{-1}) = d(\gamma_n)$. ■

Bafna and Pevzner, 1996 [19] also proved that γ_n and γ_n^{-1} are the only permutations in S_n with a reversal distance of $n - 1$.

Theorem 10.4 *(strong Gollan conjecture) For every n, γ_n and γ_n^{-1} are the only permutations that require $n - 1$ reversals to be sorted.*

10.4 Expected Reversal Distance

For any permutation $\pi \in S_n$, consider a set of cycles that form a maximum decomposition and partition them by size. Let $c_i(\pi)$ denote the number of alternating cycles of length i in a maximum decomposition, that do not include either vertex 0 or $n + 1$. Let $\delta \leq 2$ be the number of alternating cycles in a maximum decomposition that include either vertex 0 or $n + 1$. Then,

$$c(\pi) = \sum_{i=2}^{2(n+1)} c_i(\pi) + \delta. \tag{10.2}$$

For $k \leq 2(n+1)$, let us consider cycles in the decomposition whose size is at least k. The number of such cycles is $c(\pi) - \sum_{i=2}^{k-1} c_i(\pi) - \delta$. Now, the breakpoint graph of π has exactly $2(n + 1)$ edges. From this and the fact that the cycles are edge disjoint, we have

$$\forall k \leq 2(n+1), \quad c(\pi) - \sum_{i=2}^{k-1} c_i(\pi) - \delta \leq \frac{1}{k}\left(2(n+1) - \sum_{i=2}^{k-1} i c_i(\pi)\right) \quad (10.3)$$

and

$$\forall k \leq 2(n+1), \quad c(\pi) \leq \frac{1}{k}\left(2(n+1) + \sum_{i=2}^{k-1}(k-i)c_i(\pi)\right) + \delta. \quad (10.4)$$

Theorem 10.2, inequality (10.4), and $\delta \leq 2$ imply that for all $k \leq 2(n+1)$, we can bound $d(\pi)$ as

$$d(\pi) \geq \left(1 - \frac{2}{k}\right)(n+1) - \frac{1}{k}\left(\sum_{i=2}^{k-1}(k-i)c_i(\pi)\right) - 2 \quad (10.5)$$

$$\geq \left(1 - \frac{2}{k}\right)(n+1) - \left(\sum_{i=2}^{k-1} c_i(\pi)\right) - 2. \quad (10.6)$$

Consider a permutation π chosen uniformly at random. Denote the expected number of cycles of length i in a maximum cycle decomposition of $G(\pi)$ by $E(c_i(\pi)) = \frac{1}{n!}\sum_{\pi \in S_n} c_i(\pi)$. If we can bound $E(c_i(\pi))$, we can use (10.6) above to get a lower bound on the expected reversal distance. Lemma 10.2 provides such a bound, which is, somewhat surprisingly, independent of n. Note that there is a slight ambiguity in the definition of $E(c_i(\pi))$, which depends on the choice of a maximum cycle decomposition for each $\pi \in S_n$. This does not affect lemma 10.2, however, which holds for an arbitrary cycle decomposition.

Lemma 10.2 $E(c_i(\pi)) \leq \frac{2^i}{i}$.

Proof A cycle of length $i = 2t$ contains t black edges (unordered pairs of vertices) of the form

$$\{(x'_t, x_1), (x'_1, x_2), (x'_2, x_3), \ldots, (x'_{t-1}, x_t)\}, \text{ with } x_j \sim x'_j.$$

Consider the set $x_1, x_2, \ldots, x_t$. First, we claim that in every maximum cycle decomposition, $x_1, x_2, \ldots, x_t$ are all distinct. To see this, consider the case $x_k = x_l$, for some $1 \leq k < l \leq t$. Then, $(x'_k, x_{k+1}), (x'_{k+1}, x_{k+2}), \ldots, (x'_{l-1}, x_l = x_k)$ form an alternating cycle, which can be detached to give a larger decomposition.

We have $\frac{n!}{(n-t)!}$ ways of selecting the ordered set, $x_1, x_2, \ldots, x_t$. Once this is fixed we have a choice of at most two elements for each of the x'_j, giving a bound of $2^t \frac{n!}{(n-t)!}$ on the number of cycles of length $2t$. Note however that we count each $(2t)$-cycle $2t$ times, therefore a tighter bound for the number of cycles of length $2t$ is $\frac{2^t}{2t} \frac{n!}{(n-t)!}$.

Choose an arbitrary $(2t)$-cycle. The number of permutations in which this cycle can occur is no more than the number of ways of permuting the remaining $n - 2t$ elements plus the t pairs that form the cycle. Additionally, each pair can be flipped to give a different order, which gives at most $2^t(n-t)!$ permutations. Let p be the probability that an arbitrary $(2t)$-cycle is present in a random permutation. Then $p \leq \frac{2^t(n-t)!}{n!}$ and

$$E(c_i(\pi)) = E(c_{2t}(\pi)) \leq \sum_{\{\text{all } (2t)-\text{cycles}\}} p \leq \frac{2^{2t}}{2t} = \frac{2^i}{i}.$$

■

Cycles of length 2 correspond to adjacencies in permutation π. There are a total of $2n$ ordered adjacencies. Any such pair occurs in exactly $(n-1)!$ permutations, so the probability that it occurs in a random permutation is $\frac{1}{n}$. Thus, the expected number of adjacencies is $\frac{2n}{n}$ and $E(c_2) = 2$. Note that the expected number of breakpoints in a random permutation is $n - 1$.

We use lemma 10.2 and $E(c_2) = 2$ to get a bound on the expected reversal diameter:

Theorem 10.5 *(Bafna and Pevzner, 1996 [19])* $E(d(\pi)) \geq \left(1 - \frac{4.5}{\log n}\right) n.$

Proof From inequality (10.6), for all $k \leq 2(n+1)$,

$$E(d(\pi)) \geq \left(1 - \frac{2}{k}\right)(n+1) - \sum_{i=2}^{k-1} E(c_i) - 2 \geq \left(1 - \frac{2}{k}\right)(n+1) - \sum_{i=4}^{k-1} 2^i/i - 4 \geq$$

$$\frac{\left(1 - \frac{2}{k}\right)(n+1) - \sum_{i=4}^{k-1} 2^i + 2^4 - 2^4}{4 - 4 \geq n - \frac{2n}{k} - \left(1 - \frac{2}{k}\right) - 2^k + 1 + 8 \geq}$$

$$n - \frac{2n}{k} - 2^k.$$

Choose $k = \log \frac{n}{\log n}$. Then $2^k \leq \frac{n}{k}$, and

$$E(d(\pi)) \geq \left(1 - \frac{3}{\log \frac{n}{\log n}}\right) n \geq \left(1 - \frac{4.5}{\log n}\right) n \quad \text{for } n \geq 2^{16}.$$

Lemma 10.2 and inequality (10.5) for $k = 10$ imply that $E(d(\pi)) \geq \left(1 - \frac{4.5}{\log n}\right) n$ for $19 < n < 2^{16}$. For $1 \leq n \leq 19$, $\left(1 - \frac{4.5}{\log n}\right) n < 1$. ■

10.5 Signed Permutations

Let $\vec{\pi}$ be a *signed* permutation of $\{1, \ldots, n\}$, i.e., a permutation with a $+$ or $-$ sign associated with each element. Define a transformation from a signed permutation $\vec{\pi}$ of order n to an (unsigned) permutation π of $\{1, \ldots, 2n\}$ as follows. To model the signs of elements in $\vec{\pi}$, replace the positive elements $+x$ by $2x - 1, 2x$ and the negative elements $-x$ by $2x, 2x - 1$ (Figure 10.9c). We call the unsigned permutation π the *image* of the signed permutation $\vec{\pi}$. In the breakpoint graph $G(\pi)$, elements $2x - 1$ and $2x$ are joined by both black and gray edges for $1 \leq x \leq n$. Each such pair of a black and a gray edge defines a cycle of length 2 in the breakpoint graph. Clearly, there exists a maximal cycle decomposition of $G(\pi)$ containing all these n cycles of length 2. Define the breakpoint graph $G(\vec{\pi})$ of a signed permutation $\vec{\pi}$ as the breakpoint graph $G(\pi)$ with these $2n$ edges excluded. Observe that in $G(\vec{\pi})$ every vertex has degree 2 (Figure 10.9c), and therefore the breakpoint graph of a signed permutation is a collection of disjoint cycles. Denote the number of such cycles as $c(\vec{\pi})$. We observe that the identity signed permutation of order n maps to the identity (unsigned) permutation of order $2n$, and the effect of a reversal on $\vec{\pi}$ can be mimicked by a reversal on π, thus implying $d(\vec{\pi}) \geq d(\pi)$.

In the following, by a sorting of the image $\pi = \pi_1 \pi_2 \ldots \pi_{2n}$ of a signed permutation $\vec{\pi}$, we mean a sorting of π by reversals $\rho(2i + 1, 2j)$ that "cut" only after even positions of π (between π_{2k-1} and π_{2k} for $1 \leq k \leq n$). The effect of a reversal $\rho(2i + 1, 2j)$ on π can be mimicked by a reversal $\rho(i + 1, j)$ on $\vec{\pi}$, thus implying that $d(\vec{\pi}) = d(\pi)$ if cuts between π_{2i-1} and π_{2i} are forbidden. In the following, all unsigned permutations we consider are images of some signed permutations. For convenience we extend the term "signed permutation" to unsigned permutations $\pi = (\pi_1 \pi_2 \ldots \pi_{2n})$ such that π_{2i-1} and π_{2i} are consecutive numbers for $1 \leq i \leq n$. A reversal $\rho(i, j)$ on π is *legal* if i is odd and j is even. Notice that any reversal on a signed permutation corresponds to a legal reversal on its image, and vice versa. In the following, by reversals we mean legal reversals.

Given an arbitrary reversal ρ, denote $\Delta c = \Delta c(\pi, \rho) = c(\pi\rho) - c(\pi)$ (increase in the size of the cycle decomposition). Theorem 10.1 implies that for every permutation π and reversal ρ, $\Delta c \equiv \Delta c(\pi, \rho) \leq 1$. We call a reversal *proper* if $\Delta c = 1$.

If we were able to find a proper reversal for every permutation, then we would optimally sort a permutation π in $n + 1 - c(\pi)$ steps. However, for a permutation $\pi = +3+2+1$ there is no proper reversal, and therefore, it cannot be sorted in $n + 1 - c(\pi) = 2$ steps (optimal sorting of this permutation is shown in Figure 10.10).

This indicates that besides the number of cycles there exists another "obstacle" to sorting by reversals. The permutation $\pi = +3 + 2 + 1$ contains a *hurdle* that presents such a "hidden" obstacle to sorting by reversal. The notion of a hurdle will be defined in the next section.

We say that a reversal $\rho(i,j)$ *acts* on black edges (π_{i-1}, π_i) and (π_j, π_{j+1}) in $G(\pi)$. $\rho(i,j)$ is a *reversal (acting) on a cycle* C of $G(\pi)$ if the black edges (π_{i-1}, π_i) and (π_j, π_{j+1}) belong to C. A gray edge g is *oriented* if a reversal acting on two black edges incident to g is proper and *unoriented* otherwise. For example, gray edges $(8,9)$ and $(22,23)$ in Fig 10.9c are oriented, while gray edges $(4,5)$ and $(18,19)$ are unoriented.

Lemma 10.3 *Let (π_i, π_j) be a gray edge incident to black edges (π_k, π_i) and (π_j, π_l). Then (π_i, π_j) is oriented if and only if $i - k = j - l$.*

A cycle in $G(\pi)$ is *oriented* if it has an oriented gray edge and unoriented otherwise. Cycles C and F in Figure 10.9c are oriented, while cycles A, B, D, and E are unoriented. Clearly, there is no proper reversal acting on an unoriented cycle.

10.6 Interleaving Graphs and Hurdles

Gray edges (π_i, π_j) and (π_k, π_t) in $G(\pi)$ are *interleaving,* if the intervals $[i,j]$ and $[k,t]$ overlap but neither of them contains the other. For example, edges $(4,5)$ and $(18,19)$ in Figure 10.9c are interleaving, while edges $(4,5)$ and $(22,23)$, as well as $(4,5)$ and $(16,17)$, are non-interleaving. Cycles C_1 and C_2 are *interleaving* if there exist interleaving gray edges $g_1 \in C_1$ and $g_2 \in C_2$.

Let $\mathcal{C}_\pi$ be the set of cycles in the breakpoint graph of a permutation π. Define an *interleaving* graph $H_\pi(\mathcal{C}_\pi, \mathcal{I}_\pi)$ of π with the edge set

$$\mathcal{I}_\pi = \{(C_1, C_2) : \; C_1 \text{ and } C_2 \text{ are interleaving cycles in } G(\pi)\}.$$

Figure 10.9d shows an interleaving graph H_π consisting of three connected components. The vertex set of H_π is partitioned into *oriented* and *unoriented* vertices (cycles in $\mathcal{C}_\pi$). A connected component of H_π is *oriented* if it has at least one oriented vertex and *unoriented* otherwise. For a connected component U, define leftmost and rightmost positions of U as

$$U_{min} = \min_{\pi_i \in C \in U} i \text{ and } U_{max} = \max_{\pi_i \in C \in U}.$$

For example, the component U containing cycles B, C, and D in Figure 10.9c has leftmost vertex $\pi_2 = 6$ and rightmost vertex $\pi_{13} = 17$; therefore, $[U_{min}, U_{max}] = [2, 13]$.

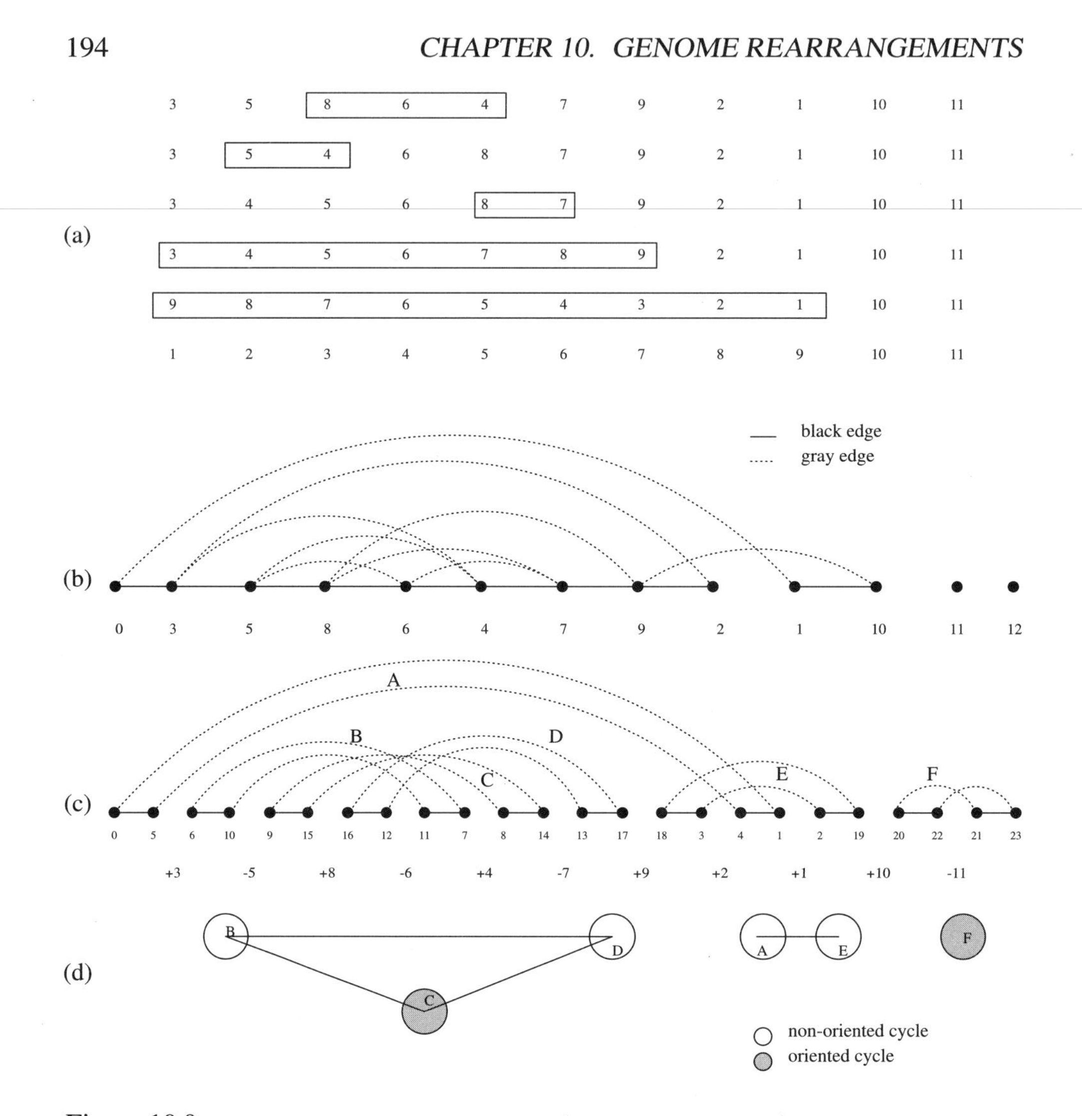

Figure 10.9: (a) Optimal sorting of a permutation (3 5 8 6 4 7 9 2 1 10 11) by five reversals and (b) breakpoint graph of this permutation; (c) Transformation of a signed permutation into an unsigned permutation π and the breakpoint graph $G(\pi)$; (d) Interleaving graph H_π with two oriented and one unoriented component.

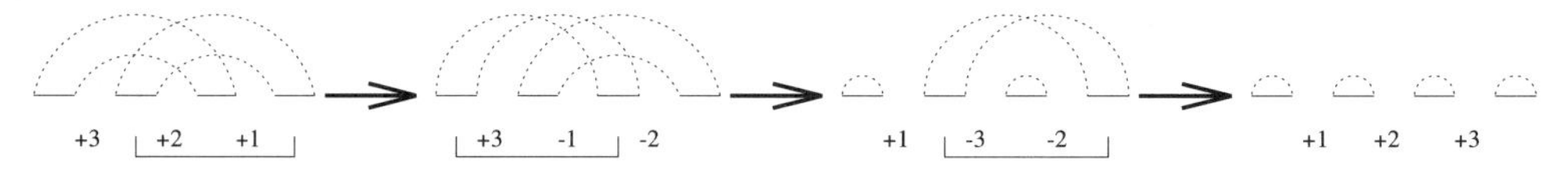

Figure 10.10: Optimal sorting of permutation $\pi = +3 + 2 + 1$ involves a non-proper reversal.

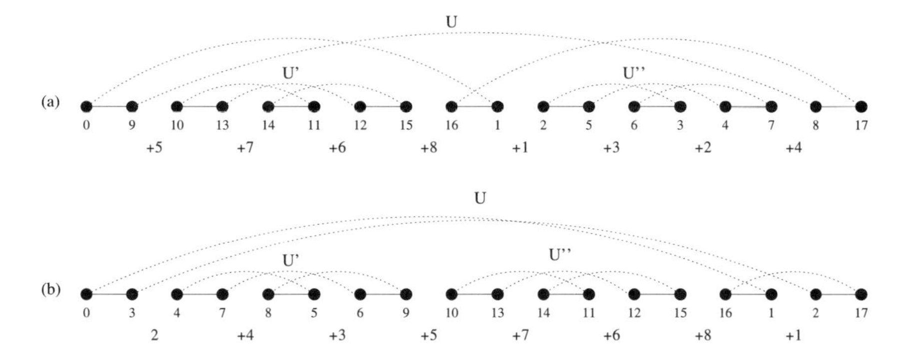

Figure 10.11: (a) Unoriented component U separates U' and U'' by virtue of the edge $(0, 1)$; (b) Hurdle U does not separate U' and U''.

We say that a component U *separates* components U', U'' in π if there exists a gray edge (π_i, π_j) in U such that $[U'_{min}, U'_{max}] \subset [i, j]$, but $[U''_{min}, U''_{max}] \not\subset [i, j]$. For example, the component U in Figure 10.11a separates the components U' and U''.

Let $\prec$ be a partial order on a set P. An element $x \in P$ is called a *minimal* element in $\prec$ if there is no element $y \in P$ with $y \prec x$. An element $x \in P$ is the *greatest* in $\prec$ if $y \prec x$ for every $y \in P$.

Consider the set of unoriented components $\mathcal{U}_\pi$ in H_π, and define the *containment* partial order on this set, i.e., $U \prec W$ if $[U_{min}, U_{max}] \subset [W_{min}, W_{max}]$ for $U, W \in \mathcal{U}_\pi$. A *hurdle* is defined as an unoriented component that is either a minimal hurdle or the greatest hurdle, where a *minimal hurdle* $U \in \mathcal{U}_\pi$ is a minimal element in $\prec$ and the *greatest hurdle* satisfies the following two conditions: (i) U is the greatest element in $\prec$ *and* (ii) U does not separate any two hurdles. Let $h(\pi)$ be the *overall* number of hurdles in π. Permutation π in Figure 10.9c has one unoriented component and $h(\pi) = 1$. Permutation π in Figure 10.11b has two minimal and one greatest hurdle ($h(\pi) = 3$). Permutation π in Figure 10.11a has two minimal and no greatest hurdle ($h(\pi) = 2$), since the greatest unoriented component U in Figure 10.11a separates U' and U''.

The following theorem further improves the bound for sorting signed permutations by reversals:

Theorem 10.6 *For arbitrary (signed) permutation π, $d(\pi) \geq n+1-c(\pi)+h(\pi)$.*

Proof Given an arbitrary reversal ρ, denote $\Delta h \equiv \Delta h(\pi, \rho) = h(\pi\rho) - h(\pi)$. Clearly, every reversal ρ acts on black edges of at most two hurdles, and therefore ρ may "destroy" at most two minimal hurdles. Note that if ρ destroys two minimal hurdles in $\mathcal{U}_\pi$, then ρ cannot destroy the greatest hurdle in $\mathcal{U}_\pi$ (see condition (ii) in the definition of the greatest hurdle). Therefore $\Delta h \geq -2$ for every reversal ρ. ■

Theorem 10.1 implies that $\Delta c \in \{-1, 0, 1\}$. If $\Delta c = 1$, then ρ acts on an oriented cycle and hence does not affect any hurdles in π. Therefore $\Delta h = 0$ and $\Delta(c - h) \equiv \Delta c - \Delta h = 1$. If $\Delta c = 0$, then ρ acts on a cycle and therefore affects at most one hurdle (see condition (ii) in the definition of the greatest hurdle). This implies $\Delta h \geq -1$ and $\Delta(c-h) \leq 1$. If $\Delta c = -1$, then $\Delta(c-h) \leq 1$, since $\Delta h \geq -2$ for every reversal ρ.

Therefore, for an arbitrary reversal ρ, $\Delta(c-h) \leq 1$. This, together with the observation that $c(\iota) = n + 1$ and $h(\iota) = 0$ for the identity permutation ι, implies $d(\pi) \geq (c(\iota) - h(\iota)) - (c(\pi) - h(\pi)) = n + 1 - c(\pi) + h(\pi)$. ■

Hannenhalli and Pevzner, 1995 [154] proved that the lower bound $d(\pi) \geq n + 1 - c(\pi) + h(\pi)$ is very tight. As a first step toward the upper bound $d(\pi) \leq n+1-c(\pi)+h(\pi)+1$, we developed a technique called *equivalent transformations* of permutations.

10.7 Equivalent Transformations of Permutations

The complicated interleaving structure of long cycles in breakpoint graphs poses serious difficulties for analyzing sorting by reversals. To get around this problem we introduce equivalent transformations of permutations, based on the following idea. If a permutation $\pi \equiv \pi(0)$ has a long cycle, transform it into a new permutation $\pi(1)$ by "breaking" this long cycle into two smaller cycles. Continue with $\pi(1)$ in the same manner and form a sequence of permutations $\pi \equiv \pi(0), \pi(1), \ldots, \pi(k) \equiv \sigma$, ending with a *simple* (i.e., having no long cycles) permutation. This section demonstrates that these transformations can be arranged in such a way that every sorting of σ mimics a sorting of π with the same number of reversals. The following sections show how to optimally sort simple permutations. Optimal sorting of the *simple* permutation σ mimics optimal sorting of the *arbitrary* permutation π, leading to a polynomial algorithm for sorting by reversals.

Let $b = (v_b, w_b)$ be a black edge and $g = (w_g, v_g)$ be a gray edge belonging to a cycle $C = \ldots, v_b, w_b, \ldots, w_g, v_g, \ldots$ in the breakpoint graph $G(\pi)$ of a permutation π. A (g, b)-*split* of $G(\pi)$ is a new graph $\hat{G}(\pi)$ obtained from $G(\pi)$ by

- removing edges g and b,
- adding two new vertices v and w,
- adding two new black edges (v_b, v) and (w, w_b),
- adding two new gray edges (w_g, w) and (v, v_g).

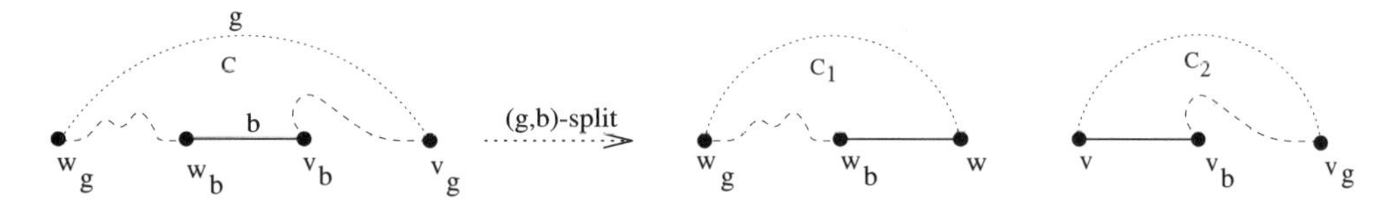

Figure 10.12: Example of a (g, b)-split.

Figure 10.12 shows a (g, b)-split transforming a cycle C in $G(\pi)$ into cycles C_1 and C_2 in $\hat{G}(\pi)$. If $G(\pi)$ is a breakpoint graph of a signed permutation π, then every (g, b)-split of $G(\pi)$ corresponds to the breakpoint graph of a signed *generalized* permutation $\hat{\pi}$ such that $\hat{G}(\pi) = G(\hat{\pi})$. Below we define generalized permutations and describe the *padding* procedure to find a generalized permutation $\hat{\pi}$ corresponding to a (g, b)-split of G.

A generalized permutation $\pi = \pi_1\pi_2 \ldots \pi_n$ is a permutation of arbitrary distinct *reals* (instead of permutations of *integers* $\{1, 2, \ldots, n\}$). In this section, by permutations we mean generalized permutations, and by *identity generalized permutation* we mean a generalized permutation $\pi = \pi_1\pi_2 \ldots \pi_n$ with $\pi_i < \pi_{i+1}$ for $1 \leq i \leq n - 1$. Extend a permutation $\pi = \pi_1\pi_2 \ldots \pi_n$ by adding $\pi_0 = \min_{1 \leq i \leq n} \pi_i - 1$ and $\pi_{n+1} = \max_{1 \leq i \leq n} \pi_i + 1$. Elements π_j and π_k of π are *consecutive* if there is no element π_l such that $\pi_j < \pi_l < \pi_k$ for $1 \leq l \leq n$. Elements π_i and π_{i+1} of π are *adjacent* for $0 \leq i \leq n$. The *breakpoint graph* of a (generalized) permutation $\pi = \pi_1\pi_2 \ldots \pi_n$ is defined as the graph on vertices $\{\pi_0, \pi_1, \ldots, \pi_n, \pi_{n+1}\}$ with black edges between adjacent elements that are not consecutive and gray edges between consecutive elements that are not adjacent. Obviously the definition of the breakpoint graph for generalized permutations is consistent with the notion of the breakpoint graph described earlier.

Let $b = (\pi_{i+1}, \pi_i)$ be a black edge and $g = (\pi_j, \pi_k)$ be a gray edge belonging to a cycle $C = \ldots, \pi_{i+1}, \pi_i, \ldots, \pi_j, \pi_k, \ldots$ in the breakpoint graph $G(\pi)$. Define $\Delta = \pi_k - \pi_j$ and let $v = \pi_j + \frac{\Delta}{3}$ and $w = \pi_k - \frac{\Delta}{3}$. A (g, b)-*padding* of $\pi = (\pi_1\pi_2 \ldots \pi_n)$ is a permutation on $n+2$ elements obtained from π by inserting v and w after the i-th element of π $(0 \leq i \leq n)$:

$$\hat{\pi} = \pi_1\pi_2 \ldots \pi_i v w \pi_{i+1} \ldots \pi_n$$

Note that v and w are both consecutive and adjacent in $\hat{\pi}$, thus implying that if π is (the image of) a signed permutation , then $\hat{\pi}$ is also (the image of) a signed permutation. The following lemma establishes the correspondence between (g, b)-paddings and (g, b)-splits. [†]

Lemma 10.4 $\hat{G}(\pi) = G(\hat{\pi})$.

[†]Of course, a (g, b)-padding of a permutation $\pi = (\pi_1\pi_2 \ldots \pi_n)$ on $\{1, 2, \ldots, n\}$ can be mod-

If g and b are non-incident edges of a *long* cycle C in $G(\pi)$, then the (g, b)-padding breaks C into two *smaller* cycles in $G(\hat{\pi})$. Therefore paddings may be used to transform an arbitrary permutation π into a simple permutation. Note that the number of elements in $\hat{\pi}$ is $\hat{n} = n + 1$ and $c(\hat{\pi}) = c(\pi) + 1$. Below we prove that for every permutation with a long cycle, there exists a padding on non-incident edges of this cycle such that $h(\hat{\pi}) = h(\pi)$, thus indicating that padding provides a way to eliminate long cycles in a permutation without changing the parameter $n + 1 - c(\pi) + h(\pi)$. First we need a series of technical lemmas.

Lemma 10.5 *Let a (g, b)-padding on a cycle C in $G(\pi)$ delete the gray edge g and add two new gray edges g_1 and g_2. If g is oriented, then either g_1 or g_2 is oriented in $G(\hat{\pi})$. If C is unoriented, then both g_1 and g_2 are unoriented in $G(\hat{\pi})$.*

Lemma 10.6 *Let a (g, b)-padding break a cycle C in $G(\pi)$ into cycles C_1 and C_2 in $G(\hat{\pi})$. Then C is oriented if and only if either C_1 or C_2 is oriented.*

Proof Note that a (g, b)-padding preserves the orientation of gray edges in $G(\hat{\pi})$ that are "inherited" from $G(\pi)$ (lemma 10.3). If C is oriented, then it has an oriented gray edge. If this edge is different from g, then it remains oriented in a (g, b)-padding of π, and therefore a cycle (C_1 or C_2) containing this edge is oriented. If $g = (w_g, v_g)$ is the only oriented gray edge in C, then (g, b)-padding adds two new gray edges ((w_g, w) and (v, v_g)) to $G(\hat{\pi})$, one of which is oriented (lemma 10.5). Therefore a cycle (C_1 or C_2) containing this edge is oriented.

If C is an unoriented cycle, then all edges of C_1 and C_2 "inherited" from C remain unoriented. Lemma 10.5 implies that new edges ((w_g, w) and (v, v_g)) in C_1 and C_2 are also unoriented. ■

The following lemma shows that paddings preserve the interleaving of gray edges.

Lemma 10.7 *Let g' and g'' be two gray edges of $G(\pi)$ different from g. Then g' and g'' are interleaving in π if and only if g' and g'' are interleaving in a (g, b)-padding of π.*

This lemma immediately implies

Lemma 10.8 *Let a (g, b)-padding break a cycle C in $G(\pi)$ into cycles C_1 and C_2 in $G(\hat{\pi})$. Then every cycle D interleaving with C in $G(\pi)$ interleaves with either C_1 or C_2 in $G(\hat{\pi})$.*

---

eled as a permutation $\hat{\pi} = (\hat{\pi}_1 \hat{\pi}_2 \dots \hat{\pi}_i v w \hat{\pi}_{i+1} \dots \hat{\pi}_n)$ on $\{1, 2, \dots, n+2\}$ where $v = \pi_j + 1, w = \pi_k + 1$, and $\hat{\pi}_i = \pi_i + 2$ if $\pi_i > \min\{\pi_j, \pi_k\}$ and $\hat{\pi}_i = \pi_i$ otherwise. Generalized permutations were introduced to make the following "mimicking" procedure more intuitive.

Proof Let $d \in D$ and $c \in C$ be interleaving gray edges in $G(\pi)$. If c is different from g, then lemma 10.7 implies that d and c are interleaving in $G(\hat{\pi})$, and therefore D interleaves with either C_1 or C_2. If $c = g$, then it is easy to see that one of the new gray edges in $G(\hat{\pi})$ interleaves with d, and therefore D interleaves with either C_1 or C_2 in $G(\hat{\pi})$. ■

Lemma 10.9 *For every gray edge g there exists a gray edge f interleaving with g in $G(\pi)$.*

Lemma 10.10 *Let C be a cycle in $G(\pi)$ and $g \notin C$ be a gray edge in $G(\pi)$. Then g interleaves with an even number of gray edges in C.*

A (g, b)-padding ϕ transforming π into $\hat{\pi}$ (i.e., $\hat{\pi} = \pi \cdot \phi$) is *safe* if it acts on non-incident edges of a long cycle and $h(\pi) = h(\hat{\pi})$. Clearly, every safe padding breaks a long cycle into two smaller cycles.

Theorem 10.7 *If C is a long cycle in $G(\pi)$, then there exists a safe (g, b)-padding acting on C.*

Proof If C has a pair of interleaving gray edges $g_1, g_2 \in C$, then removing these edges transforms C into two paths. Since C is a long cycle, at least one of these paths contains a gray edge g. Pick a black edge b from another path and consider the (g, b)-padding transforming π into $\hat{\pi}$ (clearly g and b are non-incident edges). This (g, b)-padding breaks C into cycles C_1 and C_2 in $G(\hat{\pi})$, with g_1 and g_2 belonging to different cycles C_1 and C_2. By lemma 10.7, g_1 and g_2 are interleaving, thus implying that C_1 and C_2 are interleaving. Also, this (g, b)-padding does not "break" the component K in H_π containing the cycle C since by lemma 10.8, all cycles from K belong to the component of $H_{\hat{\pi}}$ containing C_1 and C_2. Moreover, according to lemma 10.6, the orientation of this component in H_π and $H_{\hat{\pi}}$ is the same. Therefore the chosen (g, b)-padding preserves the set of hurdles, and $h(\pi) = h(\hat{\pi})$.

If all gray edges of C are mutually non-interleaving, then C is an unoriented cycle. Lemmas 10.9 and 10.10 imply that there exists a gray edge $e \in C'$ interleaving with at least two gray edges $g_1, g_2 \in C$. Removing g_1 and g_2 transforms C into two paths, and since C is a long cycle, at least one of these paths contains a gray edge g. Pick a black edge b from another path and consider the (g, b)-padding of π. This padding breaks C into cycles C_1 and C_2 in $G(\hat{\pi})$, with g_1 and g_2 belonging to different cycles C_1 and C_2. By lemma 10.7, both C_1 and C_2 interleave with C' in $\hat{\pi}$. Therefore, this (g, b)-padding does not break the component K in H_π containing C and C'. Moreover, according to lemma 10.6, both C_1 and C_2 are unoriented, thus implying that the orientation of this component in H_π and $H_{\hat{\pi}}$ is the same. Therefore, the chosen (g, b)-padding preserves the set of hurdles, and hence, $h(\pi) = h(\hat{\pi})$. ■

A permutation π is *equivalent* to a permutation σ ($\pi \rightsquigarrow \sigma$) if there exists a series of permutations $\pi \equiv \pi(0), \pi(1), \ldots, \pi(k) \equiv \sigma$ such that $\pi(i+1) = \pi(i) \cdot \phi(i)$ for a safe (g, b)-padding $\phi(i)$ acting on π_i $(0 \leq i \leq k-1)$.

Theorem 10.8 *For every permutation there exists an equivalent simple permutation.*

Proof Define the *complexity* of a permutation π as $\sum_{C \in \mathcal{C}_\pi} (l(C) - 2)$, where $\mathcal{C}_\pi$ is the set of cycles in $G(\pi)$ and $l(C)$ is the length of a cycle C. The complexity of a simple permutation is 0. Note that every padding on non-incident edges of a long cycle C breaks C into cycles C_1 and C_2 with $l(C) = l(C_1) + l(C_2) - 1$. Therefore

$$(l(C) - 2) = (l(C_1) - 2) + (l(C_2) - 2) + 1,$$

implying that a padding on non-incident edges of a cycle reduces the complexity of permutations. This observation and theorem 10.7 imply that every permutation with long cycles can be transformed into a permutation without long cycles by a series of paddings preserving $b(\pi) - c(\pi) + h(\pi)$. ■

Let $\hat{\pi}$ be a (g, b)-padding of π, and let ρ be a reversal acting on two black edges of $\hat{\pi}$. Then ρ can be mimicked on π by ignoring the padded elements. We need a generalization of this observation. A sequence of generalized permutations $\pi \equiv \pi(0), \pi(1), \ldots, \pi(k) \equiv \sigma$ is called a *generalized sorting* of π if σ is the identity (generalized) permutation and $\pi(i+1)$ is obtained from $\pi(i)$ either by a reversal or by a padding. Note that reversals and paddings in a generalized sorting of π may interleave.

Lemma 10.11 *Every generalized sorting of π mimics a (genuine) sorting of π with the same number of reversals.*

Proof Ignore padded elements. ■

In the following, we show how to find a generalized sorting of a permutation π by a series of paddings and reversals containing $d(\pi)$ reversals. Lemma 10.11 implies that this generalized sorting of π mimics an optimal (genuine) sorting of π.

10.8 Searching for Safe Reversals

Recall that for an arbitrary reversal, $\Delta(c - h) \leq 1$ (see the proof of theorem 10.6). A reversal ρ is *safe* if $\Delta(c - h) = 1$. The first reversal in Figure 10.10 is not proper but it is safe (since $\Delta c = 0$ and $\Delta h = -1$). Figure 10.14 presents examples of safe ($\Delta c = 1, \Delta h = 0$) and unsafe ($\Delta c = 1, \Delta h = 1$) reversals. In the following, we prove the existence of a safe reversal acting on a cycle in an oriented component by analyzing actions of reversals on simple permutations. In this section, by cycles we mean *short* cycles and by permutations we mean simple permutations.

Denote the set of all cycles interleaving with a cycle C in $G(\pi)$ as $V(C)$ (i.e., $V(C)$ is the set of vertices adjacent to C in H_π). Define the sets of edges in the subgraph of H_π induced by $V(C)$

$$E(C) = \{(C_1, C_2) : C_1, C_2 \in V(C) \text{ and } C_1 \text{ interleaves with } C_2 \text{ in } \pi\}$$

and its complement

$$\overline{E}(C) = \{(C_1, C_2) : C_1, C_2 \in V(C) \text{ and } C_1 \text{ does not interleave with } C_2 \text{ in } \pi\}.$$

A reversal ρ acting on an oriented (short) cycle C "destroys" C (i.e., removes the edges of C from $G(\pi)$) and transforms every other cycle in $G(\pi)$ into a corresponding cycle on the same vertices in $G(\pi\rho)$. As a result, ρ transforms the interleaving graph $H_\pi(\mathcal{C}_\pi, \mathcal{I}_\pi)$ of π into the interleaving graph $H_{\pi\rho}(\mathcal{C}_\pi \setminus C, \mathcal{I}_{\pi\rho})$ of $\pi\rho$. This transformation results in complementing the subgraph induced by $V(C)$, as described by the following lemma (Figure 10.13). We denote $\overline{\mathcal{I}}_\pi = \mathcal{I}_\pi \setminus \{(C, D) : D \in V(C)\}$.

Lemma 10.12 *Let ρ be a reversal acting on an oriented (short) cycle C. Then*

- $\mathcal{I}_{\pi\rho} = (\overline{\mathcal{I}}_\pi \setminus E(C)) \cup \overline{E}(C)$, *i.e., ρ removes edges $E(C)$ and adds edges $\overline{E}(C)$ to transform H_π into $H_{\pi\rho}$, and*
- *ρ changes the orientation of a cycle $D \in \mathcal{C}_\pi$ if and only if $D \in V(C)$.*

Lemma 10.12 immediately implies the following:

Lemma 10.13 *Let ρ be a reversal acting on a cycle C, and let A, and B be non-adjacent vertices in $H_{\pi\rho}$. Then (A, B) is an edge in H_π if and only if $A, B \in V(C)$.*

Let K be an oriented component of H_π, and let $\mathcal{R}(K)$ be a set of reversals acting on oriented cycles from K. Assume that a reversal $\rho \in \mathcal{R}(K)$ "breaks" K into a number of connected components $K_1(\rho), K_2(\rho), \ldots$ in $H_{\pi\rho}$ and that the first m of these components are unoriented. If $m > 0$, then ρ may be unsafe, since some of the components $K_1(\rho), \ldots, K_m(\rho)$ may form new hurdles in $\pi\rho$, thus increasing $h(\pi\rho)$ as compared to $h(\pi)$ (Figure 10.14). In the following, we show that there is flexibility in choosing a reversal from the set $\mathcal{R}(K)$, allowing one to substitute a safe reversal σ for an unsafe reversal ρ.

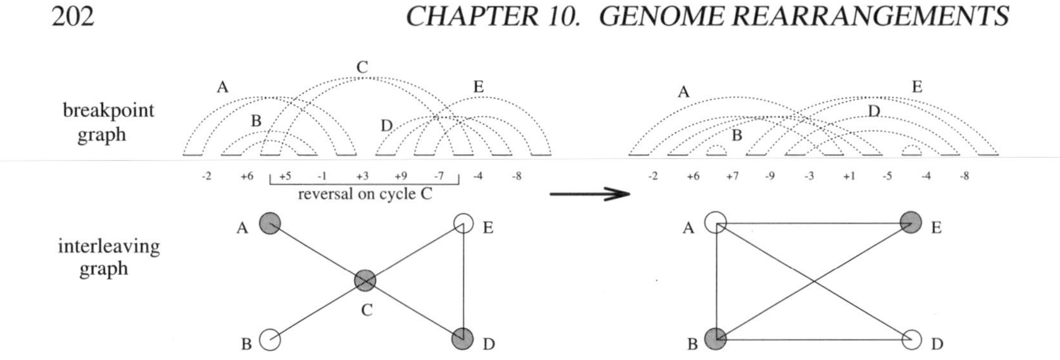

Figure 10.13: Reversal on a cycle C complements the edges between the neighbors of C and changes the orientation of each cycle neighboring C in the interleaving graph.

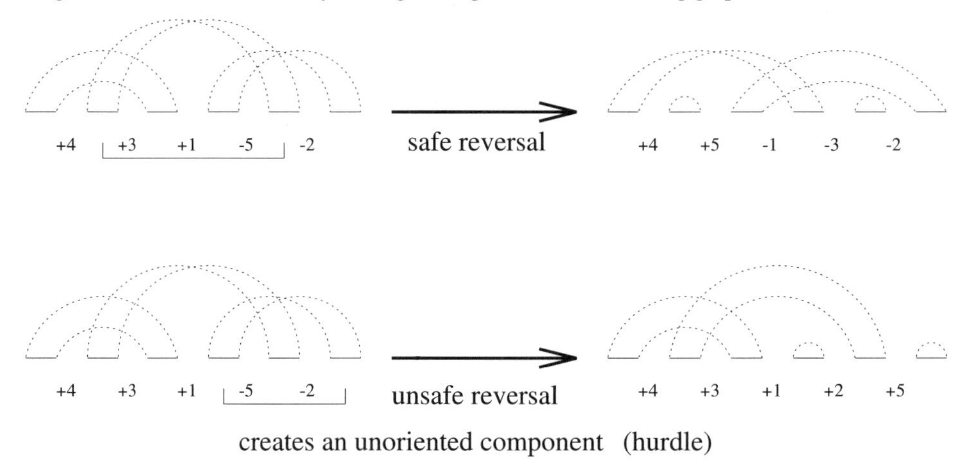

Figure 10.14: Examples of safe and unsafe reversals.

Lemma 10.14 *Let ρ and σ be reversals acting on two interleaving oriented cycles C and C', respectively, in $G(\pi)$. If C' belongs to an unoriented component $K_1(\rho)$ in $H_{\pi\rho}$, then*

- *every two vertices outside $K_1(\rho)$ that are adjacent in $H_{\pi\rho}$ are also adjacent in $H_{\pi\sigma}$, and*
- *the orientation of vertices outside $K_1(\rho)$ does not change in $H_{\pi\sigma}$ as compared to $H_{\pi\rho}$.*

Proof Let D and E be two vertices outside $K_1(\rho)$ connected by an edge in $H_{\pi\rho}$. If one of these vertices, say D, does not belong to $V(C)$ in H_π, then lemma 10.13

implies (i) (C', D) is not an edge in H_π and (ii) (D, E) is an edge in H_π. Therefore, by lemma 10.12, reversal σ preserves the edge (D, E) in $H_{\pi\sigma}$. If both vertices D and E belong to $V(C)$, then lemma 10.12 implies that (D, E) is not an edge in H_π. Since vertex C' and vertices D, E are in different components of $H_{\pi\rho}$, lemma 10.13 implies that (C', D) and (C', E) are edges in H_π. Therefore, by lemma 10.12, (D, E) is an edge in $H_{\pi\sigma}$. In both cases, σ preserves the edge (D, E) in $H_{\pi\sigma}$ and the first part of the lemma holds.

Lemma 10.13 implies that for every vertex D outside $K_1(\rho)$, $D \in V(C)$ if and only if $D \in V(C')$. This observation and lemma 10.12 imply that the orientation of vertices outside $K_1(\rho)$ does not change in $H_{\pi\sigma}$ as compared to $H_{\pi\rho}$. ■

Lemma 10.15 *Every unoriented component in the interleaving graph (of a simple permutation) contains at least two vertices.*

Proof By lemma 10.9, every gray edge in $G(\pi)$ has an interleaving gray edge. Therefore every unoriented (short) cycle in $G(\pi)$ has an interleaving cycle. ■

Theorem 10.9 *For every oriented component K in H_π, there exists a (safe) reversal $\rho \in \mathcal{R}(K)$ such that all components $K_1(\rho), K_2(\rho), \ldots$ are oriented in $H_{\pi\rho}$.*

Proof Assume that a reversal $\rho \in \mathcal{R}(K)$ "breaks" K into a number of connected components $K_1(\rho), K_2(\rho), \ldots$ in $H_{\pi\rho}$ and that the first m of these components are unoriented. Denote the overall number of vertices in these unoriented components as $index(\rho) = \sum_{i=1}^{m} |K_i(\rho)|$, where $|K_i(\rho)|$ is the number of vertices in $K_i(\rho)$. Let ρ be a reversal such that

$$index(\rho) = \min_{\sigma \in \mathcal{R}(K)} index(\sigma).$$

This reversal acts on a cycle C and breaks K into a number of components. If all these components are oriented (i.e., $index(\rho) = 0$) the theorem holds. Otherwise, $index(\rho) > 0$, and let $K_1(\rho), \ldots, K_m(\rho)$ $(m \geq 1)$ be unoriented components in $H_{\pi\rho}$. Below we find another reversal $\sigma \in \mathcal{R}(K)$ with $index(\sigma) < index(\rho)$, a contradiction.

Let V_1 be the set of vertices of the component $K_1(\rho)$ in $H_{\pi\rho}$. Note that $K_1(\rho)$ contains at least one vertex from $V(C)$, and consider the (non-empty) set $V = V_1 \cap V(C)$ of vertices from component $K_1(\rho)$ adjacent to C in H_π. Since $K_1(\rho)$ is an unoriented component in $\pi\rho$, all cycles from V are oriented in π and all cycles from $V_1 \setminus V$ are unoriented in π (lemma 10.12). Let C' be an (oriented) cycle in V, and let σ be the reversal acting on C' in $G(\pi)$. Lemma 10.14 implies that for $i \geq 2$, all edges of the component $K_i(\rho)$ in $H_{\pi\rho}$ are preserved in $H_{\pi\sigma}$, and the orientation of vertices in $K_i(\rho)$ does not change in $H_{\pi\sigma}$ as compared to $H_{\pi\rho}$. Therefore, all unoriented components $K_{m+1}(\rho), K_{m+2}(\rho), \ldots$ of $\pi\rho$ "survive" in $\pi\sigma$, and

$$index(\sigma) \leq index(\rho).$$

Below we prove that there exists a reversal σ acting on a cycle from V such that $index(\sigma) < index(\rho)$, a contradiction.

If $V_1 \neq V(C)$, then there exists an edge between an (oriented) cycle $C' \in V$ and an (unoriented) cycle $C'' \in V_1 \setminus V$ in $G(\pi)$. Lemma 10.12 implies that a reversal σ acting on C' in π orients the cycle C'' in $G(\pi)$. This observation and lemma 10.14 imply that σ reduces $index(\sigma)$ by at least 1 as compared to $index(\rho)$, a contradiction.

If $V_1 = V(C)$ (all cycles of K_1 interleave with C), then there exist at least two vertices in $V(C)$ (lemma 10.15). Moreover, there exist (oriented) cycles $C', C'' \in V_1$ such that (C', C'') are not interleaving in π (otherwise, lemma 10.12 would imply that $K_1(\rho)$ has no edges, a contradiction to the connectivity of $K_1(\rho)$). Define σ as a reversal acting on C'. Lemma 10.12 implies that σ preserves the orientation of C'', thus reducing $index(\sigma)$ by at least 1 as compared to $index(\rho)$, a contradiction.

The above discussion implies that there exists a reversal $\rho \in \mathcal{R}(K)$ such that $index(\rho) = 0$, i.e., ρ does not create new unoriented components. Then $\Delta c(\pi, \rho) = 1$ and $\Delta h(\pi, \rho) = 0$, implying that ρ is safe. ■

10.9 Clearing the Hurdles

If π has an oriented component, then theorem 10.9 implies that there exists a safe reversal in π. In this section we search for a safe reversal in the absence of any oriented component. Let $\prec$ be a partial order on a set P. We say that x is *covered* by y in P if $x \prec y$ and there is no element $z \in P$ for which $x \prec z \prec y$. The *cover graph* Ω of $\prec$ is an (undirected) graph with vertex set P and edge set $\{(x, y) : x, y \in P \text{ and } x \text{ is covered by } y\}$.

Let $\mathcal{U}_\pi$ be the set of unoriented components in H_π, and let $[U_{min}, U_{max}]$ be the interval between the leftmost and rightmost positions in an unoriented component $U \in \mathcal{U}_\pi$. Define $\overline{U}_{min} = \min_{U \in \mathcal{U}_\pi} U_{min}$ and $\overline{U}_{max} = \max_{U \in \mathcal{U}_\pi} U_{max}$, and let $[\overline{U}_{min}, \overline{U}_{max}]$ be the interval between the leftmost and rightmost positions among all the unoriented components of π. Let $\overline{U}$ be an (*artificial*) component associated with the interval $[\overline{U}_{min}, \overline{U}_{max}]$.

Define $\overline{\mathcal{U}}_\pi$ as the set of $|\mathcal{U}_\pi| + 1$ elements consisting of $|\mathcal{U}_\pi|$ elements $\{U : U \in \mathcal{U}_\pi\}$ combined with an additional element $\overline{U}$. Let $\prec \equiv \prec_\pi$ be the *containment* partial order on $\overline{\mathcal{U}}_\pi$ defined by the rule $U \prec W$ if and only if $[U_{min}, U_{max}] \subset$

$[W_{min}, W_{max}]$ for $U, W \in \overline{\mathcal{U}}_\pi$. If there exists the *greatest* unoriented component U in π (i.e., $[U_{min}, U_{max}] = [\overline{U}_{min}, \overline{U}_{max}]$), we assume that there exist two elements ("real" component U and "artificial" component $\overline{U}$) corresponding to the greatest interval and that $U \prec_\pi \overline{U}$. Let Ω_π be the *tree* representing the cover graph of the partial order $\prec_\pi$ on $\overline{\mathcal{U}}_\pi$ (Figure 10.15a). Every vertex in Ω_π except $\overline{U}$ is associated with an unoriented component in $\mathcal{U}_\pi$. In the case in which π has the greatest hurdle, we assume that the leaf $\overline{U}$ is associated with this greatest hurdle (i.e., in this case there are *two vertices* corresponding to the greatest hurdle, leaf $\overline{U}$, and its neighbor, the greatest hurdle $U \in \mathcal{U}_\pi$). Every leaf in Ω_π corresponding to a minimal element in $\prec_\pi$ is a hurdle. If $\overline{U}$ is a leaf in Ω_π, it is not necessarily a hurdle (for example, $\overline{U}$ is a leaf in Ω_π but is not a hurdle for the permutation π shown in Figure 10.11a). Therefore, the number of leaves in Ω_π coincides with the number of hurdles $h(\pi)$ except when [‡]

- there exists only one unoriented component in π (in this case Ω_π consists of two copies of this component and has two leaves, while $h(\pi) = 1$), or
- there exists the greatest element in $\mathcal{U}_\pi$ that is not a hurdle; i.e., this element separates other hurdles (in this case, the number of leaves equals $h(\pi) + 1$).

Every hurdle can be transformed into an oriented component by a reversal on an arbitrary cycle in this hurdle (Figure 10.10). Such an operation "cuts off" a leaf in the cover graph, as described in the following lemma.

Lemma 10.16 *(Hurdle cutting) Every reversal ρ on a cycle in a hurdle K cuts off the leaf K from the cover graph of π, i.e., $\Omega_{\pi\rho} = \Omega_\pi \setminus K$.*

Proof If ρ acts on an unoriented cycle of a component K in π, then K remains "unbroken" in $\pi\rho$. Also, lemma 10.9 implies that every reversal on an (unoriented) cycle of an (unoriented) component K orients at least one cycle in K. Therefore, ρ transforms K into an oriented component in $\pi\rho$ and deletes the leaf K from the cover graph. ■

Reversals cutting hurdles are not always safe. A hurdle $K \in \mathcal{U}_\pi$ *protects* a non-hurdle $U \in \mathcal{U}_\pi$ if deleting K from $\mathcal{U}_\pi$ transforms U from a non-hurdle into a hurdle (i.e., U is a hurdle in $\mathcal{U}_\pi \setminus K$). A hurdle in π is a *superhurdle* if it protects a non-hurdle $U \in \mathcal{U}_\pi$ and a *simple hurdle* otherwise. Components M, N, and U in Figure 10.15a are simple hurdles, while component L is a superhurdle (deleting L transforms non-hurdle K into a hurdle). In Figure 10.16a all three hurdles are superhurdles, while in Figure 10.16b there are two superhurdles and

[‡]Although the addition of an "artificial" component $\overline{U}$ might look a bit strange and unnecessary, we will find below that such an addition greatly facilitates the analysis of technical details.

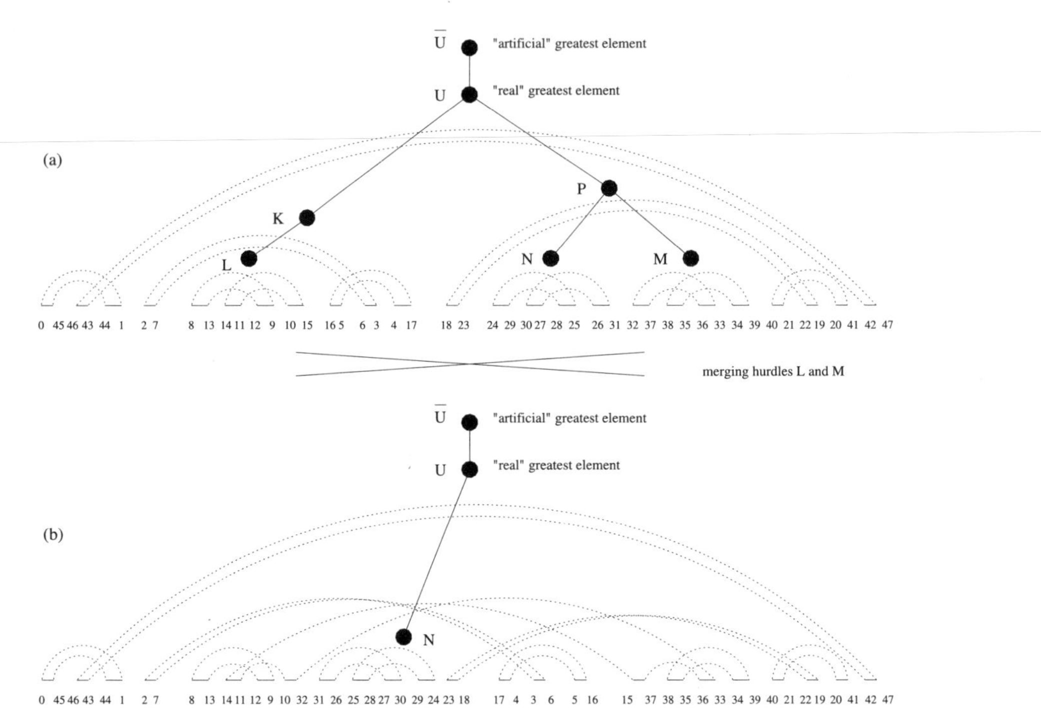

Figure 10.15: (a) A cover graph Ω_π of a permutation π with "real" unoriented components K, L, M, N, P, and U and an "artificial" component $\overline{U}$; (b) A reversal ρ merging hurdles L and M in π transforms unoriented components L, K, P, and M into an oriented component that "disappears" from $\Omega_{\pi\rho}$. This reversal transforms unoriented cycles $(32, 33, 36, 37, 32)$ and $(10, 11, 14, 15, 10)$ in π into an oriented cycle $(15, 14, 11, 10, 32, 33, 36, 37, 15)$ in $\pi\rho$. $LCA(L, M) = \overline{LCA}(L, M) = U$ and $PATH(A, F) = \{L, K, U, P, M\}$.

one simple hurdle (note that the cover graphs in Figure 10.16a and Figure 10.16b are the same!). The following lemma immediately follows from the definition of a simple hurdle.

Lemma 10.17 *A reversal acting on a cycle of a simple hurdle is safe.*

Proof Lemma 10.16 implies that for every reversal ρ acting on a cycle of a simple hurdle, $b(\pi) = b(\pi\rho)$, $c(\pi) = c(\pi\rho)$, and $h(\pi\rho) = h(\pi) - 1$, implying that ρ is safe. ■

Unfortunately, a reversal acting on a cycle of a superhurdle is unsafe, since it transforms a non-hurdle into a hurdle, implying $\Delta(c - h) = 0$. Below we define a new operation (hurdles merging) allowing one to search for safe reversals even in the absence of simple hurdles.

If L and M are two hurdles in π, define $PATH(L, M)$ as the set of (unoriented) components on the (unique) path from leaf L to leaf M in the cover graph Ω_π. If *both* L and M are *minimal* elements in $\prec$,

define $LCA(L, M)$ as an (unoriented) component that is the *least common ancestor* of L and M, and define $\overline{LCA}(L, M)$ as the *least common ancestor* of L and M *which does not separate* L and M. If L corresponds to the *greatest* hurdle U, there are two elements U and $\overline{U}$ in $\mathcal{U}_\pi$ corresponding to the same (greatest) interval $[U_{min}, U_{max}] = [\overline{U}_{min}, \overline{U}_{max}]$. In this case, define $LCA(L, M) = \overline{LCA}(L, M) = U$. Let $G(V, E)$ be a graph, $w \in V$ and $W \subset V$. A *contraction of* W *into* w *in* G is defined as a new graph with vertex set $V \setminus (W \setminus w)$ and edge set $\{(p(x), p(y)) : (x, y) \in E\}$, where $p(v) = w$ if $v \in W$, and $p(v) = v$ otherwise. Note that if $w \in W$, then a contraction reduces the number of vertices in G by $|W| - 1$, while if $w \notin W$, the number of vertices is reduced by $|W|$.

Let L and M be two hurdles in π, and let Ω_π be the cover graph of π. We define $\Omega_\pi(L, M)$ as the graph obtained from Ω_π by the contraction of $PATH(L, M)$ into $\overline{LCA}(L, M)$ (loops in the graph $\Omega_\pi(L, M)$ are ignored). Note that when

$$LCA(L, M) = \overline{LCA}(L, M),$$

$\Omega_\pi(L, M)$ corresponds to deleting the elements of $PATH(L, M) \setminus LCA(L, M)$ from the partial order $\prec_\pi$, while when

$$LCA(L, M) \neq \overline{LCA}(L, M),$$

$\Omega_\pi(L, M)$ corresponds to deleting the entire set $PATH(L, M)$ from $\prec_\pi$.

Lemma 10.18 *(Hurdles merging) Let π be a permutation with cover graph Ω_π, and let ρ be a reversal acting on black edges of (different) hurdles L and M in π. Then ρ acts on Ω_π as the contraction of $PATH(L, M)$ into $\overline{LCA}(L, M)$, i.e., $\Omega_{\pi\rho} = \Omega_\pi(L, M)$.*

Proof The reversal ρ acts on black edges of the cycles $C_1 \in L$ and $C_2 \in M$ in $G(\pi)$ and transforms C_1 and C_2 into an oriented cycle C in $G(\pi\rho)$ (Figure 10.15). It is easy to verify that every cycle interleaving with C_1 or C_2 in $G(\pi)$ interleaves with C in $G(\pi\rho)$. This implies that ρ transforms hurdles L and M in π into parts of an oriented component in $\pi\rho$, and, therefore L and M "disappear" from $\Omega_{\pi\rho}$.

Moreover, every (unoriented) component from $PATH(L, M) \setminus \overline{LCA}(L, M)$ has at least one cycle interleaving with C in $G(\pi\rho)$. This implies that every such component in π becomes a part of an oriented component in $\pi\rho$, and therefore "disappears" from $\Omega_{\pi\rho}$. Every component from $\mathcal{U}_\pi \setminus PATH(L, M)$ remains unoriented in $\pi\rho$. Component $LCA(L, M)$ remains unoriented if and only if $LCA(L, M) = \overline{LCA}(L, M)$. Every component that is covered by a vertex from $PATH(L, M)$ in $\prec_\pi$ will be covered by $\overline{LCA}(L, M)$ in $\prec_{\pi\rho}$. ■

We write $U < W$ for hurdles U and W if the rightmost position of U is smaller than the rightmost position of W, i.e., $U_{max} < W_{max}$. Order the hurdles of π in increasing order of their rightmost positions

$$U(1) < \ldots < U(l) \equiv L < \ldots < U(m) \equiv M < \ldots < U(h(\pi))$$

and define the sets of hurdles

$$BETWEEN(L, M) = \{U(i) : l < i < m\}$$

and

$$OUTSIDE(L, M) = \{U(i) : i \notin [l, m]\}.$$

Lemma 10.19 *Denote ρ be a reversal merging hurdles L and M in π. If both sets of hurdles $BETWEEN(L, M)$ and $OUTSIDE(L, M)$ are non-empty, then ρ is safe.*

Proof $U' \in BETWEEN(L, M)$ and $U'' \in OUTSIDE(L, M)$. Lemma 10.18 implies that the reversal ρ deletes the hurdles L and M from Ω_π. There is also a danger that ρ may add a new hurdle K in $\pi\rho$ by transforming K from a non-hurdle in π into a hurdle in $\pi\rho$. If this is the case, K does not separate L and M in π (otherwise, by lemma 10.18, K would be deleted from $\pi\rho$). Without loss of generality, we assume that $L < U' < M$.

If K is a *minimal* hurdle in $\pi\rho$, then either $L \prec_\pi K$ *or* $M \prec_\pi K$ (otherwise K would be a hurdle in π). Since K does not separate L and M in π, $L \prec_\pi K$ *and* $M \prec_\pi K$. Since U' is sandwiched between L and M, $U' \prec_\pi K$. Thus, $U' \prec_{\pi\rho} K$, a contradiction to the minimality of K in $\pi\rho$.

If K is the *greatest* hurdle in $\pi\rho$, then either $L, M \not\prec_\pi K$ or $L, M \prec_\pi K$ (if it were the case that $L \not\prec_\pi K$ and $M \prec_\pi K$, according to lemma 10.18, K would be deleted from $\pi\rho$). If $L, M \not\prec_\pi K$, then $L < U' \prec_\pi K < M$, i.e., K is sandwiched between L and M. Therefore U'' lies outside K in π and $U'' \not\prec_{\pi\rho} K$, a contradiction. If $L, M \prec_\pi K$ then, since K is a non-hurdle in π, K separates L, M from another hurdle N. Therefore K separates U' from N. Since both N and U' "survive" in $\pi\rho$, K separates N and U' in $\pi\rho$, a contradiction.

Therefore, ρ deletes hurdles L and M from Ω_π and does not add a new hurdle in $\pi\rho$, thus implying that $\Delta h = -2$. Since $b(\pi\rho) = b(\pi)$ and $c(\pi\rho) = c(\pi) - 1$, $\Delta(b - c + h) = -1$ and the reversal ρ is safe. ■

Lemma 10.20 *If $h(\pi) > 3$, then there exists a safe reversal merging two hurdles in π.*

Proof Order $h(\pi)$ hurdles of π in increasing order of their rightmost positions and let L and M be the first and $(1 + \frac{h(\pi)}{2})$-th hurdles in this order. Since $h(\pi) > 3$, both sets $BETWEEN(L, M)$ and $OUTSIDE(L, M)$ are non-empty, and by lemma 10.19, the reversal ρ merging L and M is safe. ■

Lemma 10.21 *If $h(\pi) = 2$, then there exists a safe reversal merging two hurdles in π. If $h(\pi) = 1$, then there exists a safe reversal cutting the only hurdle in π.*

Proof If $h(\pi) = 2$, then Ω_π either is a path graph or contains the greatest component separating two hurdles in π. In both cases, merging the hurdles in π is a safe reversal (lemma 10.18). If $h(\pi) = 1$, then lemma 10.16 provides a safe reversal cutting the only hurdle in π. ■

The previous lemmas show that hurdles merging provides a way to find safe reversals even in the absence of simple hurdles. On a negative note, hurdles merging does not provide a way to transform a superhurdle into a simple hurdle.

Lemma 10.22 *Let ρ be a reversal in π merging two hurdles L and M. Then every superhurdle in π (different from L and M) remains a superhurdle in $\pi\rho$.*

Proof Let U be a superhurdle in π (different from L and M) protecting a non-hurdle U'. Clearly, if U' is a minimal hurdle in $\mathcal{U}_\pi \setminus U$, then U remains a superhurdle in $\pi\rho$. If U' is the greatest hurdle in $\mathcal{U}_\pi \setminus U$, then U' does not separate any hurdles in $\mathcal{U}_\pi \setminus U$. Therefore U' does not belong to $PATH(L, M)$ and hence "survives" in $\pi\rho$ (lemma 10.18). This implies that U' remains protected by U in $\pi\rho$. ■

10.10 Duality Theorem for Reversal Distance

Lemmas 10.20 and 10.21 imply that unless Ω_π is a homeomorph of the *3-star* (a graph with three edges incident on the same vertex), there exists a safe reversal in π. On the other hand, if at least one hurdle in π is simple, then lemma 10.17 implies that there exists a safe reversal in π. Therefore, the only case in which a safe reversal might not exist is when Ω_π is a homeomorph of the 3-star with three superhurdles, called a *3-fortress* (Figure 10.16b).

Lemma 10.23 *If ρ is a reversal destroying a 3-fortress π (i.e., $\pi\rho$ is not a 3-fortress) then ρ is unsafe.*

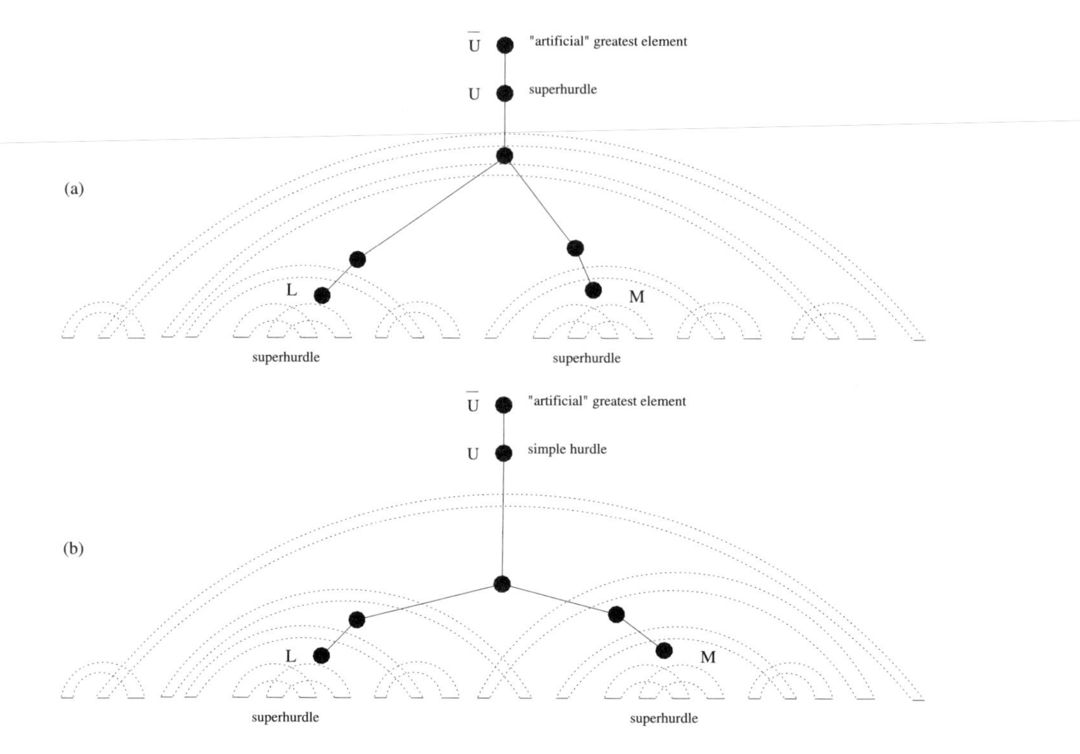

Figure 10.16: The permutation in (a) is a 3-fortress, while the permutation in (b), with the same cover graph, is not a fortress (hurdle U is not a superhurdle).

Proof Every reversal on a permutation π can reduce $h(\pi)$ by at most 2 and the *only* operation that can reduce the number of hurdles by 2 is merging of hurdles. On the other hand, lemma 10.18 implies that merging of hurdles in a 3-fortress can reduce $h(\pi)$ by at most 1. Therefore, $\Delta h \geq -1$. Note that for every reversal that does not act on edges of the *same* cycle, $\Delta c = -1$, and therefore, every reversal that does not act on edges of the same cycle in a 3-fortress is unsafe.

If ρ acts on a cycle in an unoriented component of a 3-fortress, then it does not reduce the number of hurdles. Since $\Delta c = 0$ for a reversal on an unoriented cycle, ρ is unsafe.

If ρ acts on a cycle in an oriented component of a 3-fortress, then it does not destroy any unoriented components in π and does not reduce the number of hurdles. If ρ increases the number of hurdles, then $\Delta h \geq 1$ and $\Delta c \leq 1$ imply that ρ is unsafe. If the number of hurdles in $\pi\rho$ remains the same, then every superhurdle in π remains a superhurdle in $\pi\rho$, thus implying that $\pi\rho$ is a 3-fortress, a contradiction.
■

Lemma 10.24 *If π is a 3-fortress, then $d(\pi) = n + 1 - c(\pi) + h(\pi) + 1$.*

Proof Lemma 10.23 implies that every sorting of 3-fortress contains at least one unsafe reversal. Therefore $d(\pi) \geq b(\pi) - c(\pi) + h(\pi) + 1$.

If π has oriented cycles, all oriented components in π can be destroyed by safe paddings (theorem 10.7) and safe reversals in oriented components (theorem 10.9) without affecting unoriented components.

If π is a 3-fortress without oriented cycles, then an (unsafe) reversal ρ merging arbitrary hurdles in π leads to a permutation $\pi\rho$ with two hurdles (lemma 10.18). Once again, oriented cycles appearing in $\pi\rho$ after such merging can be destroyed by safe paddings and safe reversals in oriented components (theorems 10.7 and 10.9), leading to a permutation σ with $h(\sigma) = 2$. Theorems 10.7 and 10.9 and lemma 10.21 imply that σ can be sorted by safe paddings and safe reversals. Hence, there exists a generalized sorting of π such that all paddings and all reversals but one in this sorting are safe. Therefore, this generalized sorting contains $n+1-c(\pi)+h(\pi)+1$ reversals. Lemma 10.11 implies that the generalized sorting of π mimics an optimal (genuine) sorting of π by $d(\pi) = n + 1 - c(\pi) + h(\pi) + 1$ reversals. ■

In the following, we try to avoid creating 3-fortresses in the course of sorting by reversals. If we are successful in this task, the permutation π can be sorted in $n + 1 - c(\pi) + h(\pi)$ reversals. Otherwise, we show how to sort π in $n + 1 - c(\pi) + h(\pi) + 1$ reversals and prove that such permutations cannot be sorted with a smaller number of reversals. A permutation π is called a *fortress* if it has an odd number of hurdles and all of these hurdles are superhurdles.

Lemma 10.25 *If ρ is a reversal destroying a fortress π with $h(\pi)$ superhurdles (i.e., $\pi\rho$ is not a fortress with $h(\pi)$ superhurdles), then either ρ is unsafe or $\pi\rho$ is a fortress with $h(\pi) - 2$ superhurdles.*

Proof Every reversal acting on a permutation can reduce the number of hurdles by at most two, and the *only* operation that can reduce the number of hurdles by two is a merging of hurdles. Arguments similar to the proof of lemma 10.23 demonstrate that if ρ does not merge hurdles, then ρ is unsafe. If a safe reversal ρ does merge (super)hurdles L and M in π, then lemma 10.18 implies that this reversal reduces the number of hurdles by two, and, if $h(\pi) > 3$, does not create new hurdles. Also, lemma 10.22 implies that every superhurdle in π except L and M remains a superhurdle in $\pi\rho$, thus implying that $\pi\rho$ is a fortress with $h(\pi)-2$ superhurdles. ■

Lemma 10.26 *If π is a fortress, then $d(\pi) \geq n + 1 - c(\pi) + h(\pi) + 1$.*

Proof Lemma 10.25 implies that every sorting of π either contains an unsafe reversal or gradually decreases the number of superhurdles in π by transforming a fortress with h (super)hurdles into a fortress with $h - 2$ (super)hurdles. Therefore, if a sorting of π uses only safe reversals, then it will eventually lead to a 3-fortress. Therefore, by lemma 10.23, every sorting of a fortress contains at least one unsafe reversal, and hence, $d(\pi) \geq n + 1 - c(\pi) + h(\pi) + 1$. ■

Finally, we formulate the duality theorem for sorting signed permutations by reversals:

Theorem 10.10 *(Hannenhalli and Pevzner, 1995 [154]) For every permutation π,*

$$d(\pi) = \begin{cases} n + 1 - c(\pi) + h(\pi) + 1, & \text{if } \pi \text{ is a fortress} \\ n + 1 - c(\pi) + h(\pi), & \text{otherwise.} \end{cases}$$

Proof If π has an even number of hurdles, then safe paddings (theorem 10.7), safe reversals in oriented components (theorem 10.9), and safe hurdles merging (lemmas 10.20 and 10.21) lead to a generalized sorting of π by $n+1-c(\pi)+h(\pi)$ reversals.

If π has an odd number of hurdles, at least one of which is simple, then there exists a safe reversal cutting this simple hurdle (lemma 10.17). This safe reversal leads to a permutation with an even number of hurdles. Therefore, similar to the previous case, there exists a generalized sorting of π using only safe paddings and $n + 1 - c(\pi) + h(\pi)$ safe reversals.

Therefore, if π is not a fortress, there exists a generalized sorting of π by $n+1-c(\pi) + h(\pi)$ reversals. Lemma 10.11 implies that this generalized sorting mimics optimal (genuine) sorting of π.

If π is a fortress there exists a sequence of safe paddings (theorem 10.7), safe reversals in oriented components (theorem 10.9), and safe hurdle mergings (lemma 10.20) leading to a 3-fortress that can be sorted by a series of reversals having at most one unsafe reversal. Therefore, there exists a generalized sorting of π using $n+1-c(\pi)+h(\pi)+1$ reversals. Lemma 10.26 implies that this generalized sorting mimics optimal (genuine) sorting of π with $d(\pi) = n+1-c(\pi)+h(\pi)+1$ reversals. ■

This theorem explains the mystery of the astonishing performance of approximation algorithms for sorting signed permutations by reversals. A simple explanation for this performance is that the bound $d(\pi) \geq n + 1 - c(\pi)$ is extremely tight, since $h(\pi)$ is small for "random" permutations.

10.11 Algorithm for Sorting by Reversals

Lemmas 10.11, 10.20, 10.17, and 10.21 and theorems 10.7, 10.9, and 10.10 motivate the algorithm $Reversal\_Sort$, which optimally sorts signed permutations.

Reversal_Sort(π)
1. **while** π is not sorted
2.   **if** π has a long cycle
3.   select a safe (g, b)-padding ρ of π (theorem 10.7)
4.   **else if** π has an oriented component
5.   select a safe reversal ρ in this component (theorem 10.9)
6.   **else if** π has an even number of hurdles
7.   select a safe reversal ρ merging two hurdles in π (lemmas 10.20 and 10.21)
8.   **else if** π has at least one simple hurdle
9.   select a safe reversal ρ cutting this hurdle in π (lemmas 10.17 and 10.21)
10.   **else if** π is a fortress with more than three superhurdles
11.   select a safe reversal ρ merging two (super)hurdles in π (lemma 10.20)
12.   **else**   /* π is a 3-fortress */
13.   select an (un)safe reversal ρ merging two arbitrary (super)hurdles in π
14. $\pi \leftarrow \pi \cdot \rho$
15. **endwhile**
16. mimic (genuine) sorting of π by the computed generalized sorting of π (lemma 10.11)

Theorem 10.11 *$Reversal\_Sort(\pi)$ optimally sorts a permutation π of order n in $O(n^4)$ time.*

Proof Theorem 10.10 implies that $Reversal\_Sort$ provides a generalized sorting of π by a series of reversals and paddings containing $d(\pi)$ reversals. Lemma 10.11 implies that this generalized sorting mimics an optimal (genuine) sorting of π by $d(\pi)$ reversals.

Note that every iteration of the **while** loop in $Reversal\_Sort$ reduces the quantity $complexity(\pi) + 3d(\pi)$ by at least 1, thus implying that the number of iterations of $Reversal\_Sort$ is bounded by $4n$. The most "expensive" iteration is a search for a safe reversal in an oriented component. Since for simple permutations it can be implemented in $O(n^3)$ time, the overall running time of $Reversal\_Sort$ is $O(n^4)$. ■

Below we describe a simpler version of $Reversal\_Sort$ that does not use paddings and runs in $O(n^5)$ time. Define

$$f(\pi) = \begin{cases} 1, & \text{if } \pi \text{ is a fortress} \\ 0, & \text{otherwise.} \end{cases}$$

A reversal ρ is *valid* if $\Delta(c - h - f) = 1$. The proofs of theorem 10.6 and lemma 10.26 imply that $\Delta(c - h - f) \geq -1$. This observation and theorem 10.10 imply the following:

Theorem 10.12 *For every permutation π, there exists a valid reversal in π. Every sequence of valid reversals sorting π is optimal.*

Theorem 10.12 motivates the simple version of $Reversal\_Sort$, which is very fast in practice:

Reversal_Sort_Simple(π)
1. **while** π is not sorted
2. select a valid reversal ρ in π (theorem 10.12)
3. $\pi \leftarrow \pi \cdot \rho$
4. **endwhile**

10.12 Transforming Men into Mice

Analysis of rearrangements in multichromosomal genomes makes use of the duality theorem for unichromosomal genomes (Hannenhalli and Pevzner, 1995 [154]) and two additional ideas called chromosome *flipping* and *capping*. In studies of genome rearrangements in multichromosomal genomes, a *chromosome* is defined as a *sequence* of genes, while a *genome* is defined as a *set* of chromosomes. Let $\Pi = \{\pi(1), \ldots, \pi(N)\}$ be a genome consisting of N chromosomes, and let $\pi(i) = \pi(i)_1 \ldots \pi(i)_{n_i}$, n_i be the number of genes in the i-th chromosome. Every chromosome π can be viewed either from "left to right" (i.e., as $\pi = \pi_1 \ldots \pi_n$) or from "right to left" (i.e., as $-\pi = -\pi_n \ldots - \pi_1$), leading to two equivalent representations of the same chromosome. From this perspective, a 3-chromosomal genome $\{\pi(1), \pi(2), \pi(3)\}$ is equivalent to $\{\pi(1), -\pi(2), \pi(3)\}$ or $\{-\pi(1), \pi(2), -\pi(3)\}$, i.e., the *directions* of chromosomes are irrelevant. The four most common elementary rearrangement events in multichromosomal genomes are *reversals*, *translocations*, *fusions*, and *fissions*.

We distinguish between *internal* reversals, which do not involve the ends of the chromosomes (i.e., the reversals $\rho(\pi, i, j)$ of an n-gene chromosome π with $1 < i < j < n$), and *prefix* reversals, involving ends of the chromosomes (i.e., either $i = 1$ or $j = n$). A translocation is *internal* if it is neither a fusion nor a fission.

For a chromosome $\pi = \pi_1 \ldots \pi_n$, the numbers $+\pi_1$ and $-\pi_n$ are called *tails* of π. Note that changing the direction of a chromosome does not change the set of its tails. Tails in an N-chromosomal genome Π comprise the set $\mathcal{T}(\Pi)$ of $2N$ tails. In this section we consider *co-tailed* genomes Π and Γ with $\mathcal{T}(\Pi) = \mathcal{T}(\Gamma)$. For co-tailed genomes, internal reversals and translocations are sufficient for genomic sorting— i.e., prefix reversals, fusions, and fissions can be ignored (the validity of this assumption will become clear later). For chromosomes $\pi = \pi_1 \ldots \pi_n$ and $\sigma = \sigma_1 \ldots \sigma_m$, denote the fusion $\pi_1 \ldots \pi_n \sigma_1 \ldots \sigma_m$ as $\pi + \sigma$ and the fusion $(\pi_1 \ldots \pi_n - \sigma_m \ldots - \sigma_1)$ as $\pi - \sigma$. Given an ordering of chromosomes $(\pi(1), \ldots, \pi(N))$ in a genome Π and a *flip* vector $s = (s(1), \ldots, s(N))$ with $s(i) \in \{-1, +1\}$, one can form a *concatenate* of Π as a permutation $\Pi(s) = s(1)\pi(1) + \ldots + s(N)\pi(N)$ on $\sum_{i=1}^{N} n_i$ elements. Depending on the choice of a flip vector, there exist 2^N concatenates of Π for each of $N!$ orderings of chromosomes in Π. If the order of chromosomes in a genome Π is fixed we call Π an *ordered* genome.

In this section we assume without loss of generality, that $\Gamma = (\gamma_1, \ldots, \gamma_N)$ is an (ordered) genome and that $\gamma = \gamma_1 + \ldots + \gamma_N$ is the identity permutation. We denote $d(\Pi) \equiv d(\Pi, \Gamma)$ and call the problem of a genomic sorting of Π into Γ simply a *sorting of a genome* Π.

We use the following idea to analyze co-tailed genomes. Given a concatenate π of a genome Π, one can optimally sort π by reversals. Every reversal in this sorting corresponds to a reversal or a translocation in a (not necessarily optimal) sorting of the genome Π. For example, a translocation $\rho(\pi, \sigma, i, j)$ acting on chromosomes $\pi = \pi_1 \ldots \pi_n$ and $\sigma = \sigma_1 \ldots \sigma_m$ (Figure10.17 can be alternatively viewed as a reversal $\rho(\pi - \sigma, i, n + (m - j + 1))$ acting on $\pi - \sigma$ (and vice versa). Define an *optimal concatenate* of Π as a concatenate π with minimum reversal distance $d(\pi)$ among all concatenates of Π. Below we prove that sorting of an optimal concatenate of Π mimics an optimal sorting of a genome Π. This approach reduces the problem of sorting Π to the problem of finding an optimal concatenate of Π.

In the following, by the number of cycles, we mean the number of cycles of length greater than 2, i.e., $c(\pi) - a(\pi)$, where $a(\pi)$ is the number of adjacencies (every adjacency is a cycle of length 2). Since $b(\pi) = n + 1 - a(\pi)$, the Hannenhalli-Pevzner theorem can be reformulated as $d(\pi) = b(\pi) - c(\pi) + h(\pi) + f(\pi)$, where $f(\pi) = 1$ if π is a fortress and $f(\pi) = 0$ otherwise.

Let π be a concatenate of $\Pi = (\pi(1), \ldots, \pi(N))$. Every tail of $\pi(i)$ corresponds to two vertices of the breakpoint graph $G(\pi)$, exactly one of which is a boundary (either leftmost or rightmost) vertex among the vertices of the chromosome $\pi(i)$ in the concatenate π. We extend the term *tail* to denote such vertices in $G(\pi)$. An edge in a breakpoint graph $G(\pi)$ of a concatenate π is *interchromosomal* if it connects vertices in different chromosomes of Π, and *intrachromosomal* otherwise. A component of π is *interchromosomal* if it contains an interchromosomal edge, and *intrachromosomal* otherwise.

Every interchromosomal black edge in $G(\pi)$ connects two tails. Let $b_{tail}(\Pi)$ (notice that $b_{tail}(\Pi) = N - 1$) be the number of interchromosomal black edges in $G(\pi)$. Note that for co-tailed genomes, tails in $G(\Pi)$ are adjacent only to tails, and hence a cycle containing a tail contains only tails. Let $c_{tail}(\Pi)$ be the number of cycles of $G(\pi)$ containing tails. Define $b(\Pi) = b(\pi) - b_{tail}(\pi)$ (notice that

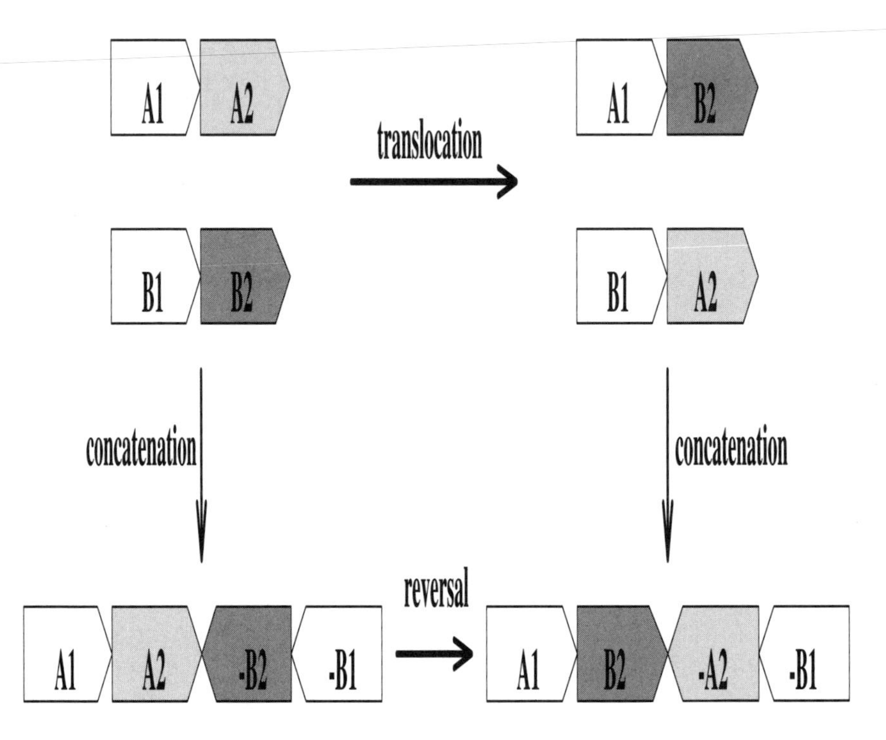

Figure 10.17: Translocations can be mimicked by reversals in a concatenated genome.

$b(\Pi) = n - N$ and $b(\pi) = n - 1$) and $c(\Pi) = c(\pi) - c_{tail}(\pi)$.

Consider the set of intrachromosomal unoriented components $\mathcal{IU}_\pi$ in π. Hurdles, superhurdles, and fortresses for the set $\mathcal{IU}_\pi$ are called *knots*, *superknots,* and *fortresses-of-knots* respectively. Let $k(\Pi)$ be the number of knots in a concatenate π of Π. Define $f(\Pi) = 1$ if π is a fortress-of-knots, and $f(\Pi) = 0$ otherwise. Clearly, $b(\Pi)$, $c(\Pi)$, $k(\Pi)$, and $f(\Pi)$ do not depend on the choice of a concatenate π.

Lemma 10.27 *For co-tailed genomes* Π *and* Γ*,* $d(\Pi) \geq b(\Pi) - c(\Pi) + k(\Pi) + f(\Pi)$.

Proof A more involved version of the proof of lemma 10.26. ■

Concatenates $\Pi(s)$ and $\Pi(s')$ of an (ordered) genome Π are *i-twins* if the directions of all chromosomes except the i-th one in $\Pi(s)$ and $\Pi(s')$ coincide, i.e.,

$s(i) = -s'(i)$ and $s(j) = s'(j)$ for $j \neq i$. A chromosome $\pi(i)$ is *properly flipped* in $\Pi(s)$ if all interchromosomal edges originating in this chromosome belong to oriented components in $\Pi(s)$. A concatenate π is *properly flipped* if every chromosome in π is properly flipped. The following lemma proves the existence of a properly flipped concatenate.

Lemma 10.28 *If a chromosome $\pi(i)$ is not properly flipped in $\pi = \Pi(s)$, then it is properly flipped in the i-twin π' of π. Moreover, every properly flipped chromosome in π remains properly flipped in π'.*

Proof Let g be an interchromosomal gray edge in π originating in the chromosome $\pi(i)$ and belonging to an unoriented component in π. Note that the orientation of any interchromosomal gray edge originating at $\pi(i)$ is different in π as compared to π' (i.e., a non-oriented edge in π becomes oriented in π', and vice versa). Since all edges interleaving with g in π are unoriented, every interchromosomal edge originating at $\pi(i)$ and interleaving with g in π is oriented in π'.

All interchromosomal edges originating in $\pi(i)$ that are not interleaving with g in π interleave with g in π'. Since g is oriented in π', all such edges belong to an oriented component containing g in π'. Therefore, $\pi(i)$ is properly flipped in π'.

Let $\pi(j)$ be a properly flipped chromosome in π. If $\pi(j)$ is not properly flipped in π', then there exists an interchromosomal unoriented component U having an interchromosomal gray edge originating at $\pi(j)$ in π'. If U does not have an edge originating at $\pi(i)$ in π', then U is an unoriented component in π, implying that $\pi(j)$ was not properly flipped in π, a contradiction. If U has an (unoriented) gray edge h originating at $\pi(i)$, then clearly, this edge does not interleave with g in π'. Therefore, h interleaves with g in π and h is oriented in π, thus implying that g belonged to an oriented component in π, a contradiction. ■

Lemma 10.28 implies the existence of a properly flipped concatenate $\pi = \Pi(s)$ with $h(\pi) = k(\Pi)$ and $f(\pi) = f(\Pi)$. Below we show that there exists a sorting of π by $b(\pi) - c(\pi) + h(\pi) + f(\pi)$ reversals that mimics a sorting of Π by $b(\Pi) - c(\Pi) + k(\Pi) + f(\Pi)$ internal reversals and translocations.

Theorem 10.13 *(Hannenhalli and Pevzner, 1995 [153]) For co-tailed genomes Π and Γ, $d(\Pi, \Gamma) \equiv d(\Pi) = b(\Pi) - c(\Pi) + k(\Pi) + f(\Pi)$.*

Proof Assume the contrary, and let Π be a genome with a minimum value of $b(\Pi) - c(\Pi) + h(\Pi) + f(\Pi)$ among the genomes for which the theorem fails. Let π be a properly flipped concatenate of Π with a minimal value of $b_{tail}(\pi) - c_{tail}(\pi)$ among all properly flipped concatenates of Π.

If $b_{tail}(\pi) = c_{tail}(\pi)$ (i.e., every interchromosomal black edge is involved in a cycle of length 2), then there exists an optimal sorting of π by $b(\pi) - c(\pi) + k(\pi) +$

$f(\pi)$ reversals that act on intrachromosomal black edges (Hannenhalli and Pevzner, 1995 [154]). Every such reversal ρ can be mimicked as an internal reversal or an internal translocation on Π, thus leading to a sorting of Π by $b(\pi) - c(\pi) + k(\pi) + f(\pi)$ internal reversals/translocations. Since π is a properly flipped concatenate, $b(\pi) = b(\Pi) + b_{tail}(\pi)$, $c(\pi) = c(\Pi) + c_{tail}(\pi)$, $h(\pi) = k(\Pi)$, and $f(\pi) = f(\Pi)$. Therefore, optimal sorting of π mimics an optimal sorting of Π by $b(\Pi) - c(\Pi) + k(\Pi) + f(\Pi)$ internal reversals/translocations.

If $b_{tail}(\pi) > c_{tail}(\pi)$, then there exists an interchromosomal black edge involved in a cycle of length greater than 2, and this edge belongs to an oriented component in π (since every interchromosomal black edge belongs to an oriented component in a properly flipped concatenate). Hannenhalli and Pevzner, 1995 [154] proved that if there exists an oriented component in π, then there exists a reversal ρ in π acting on the black edges of an oriented cycle in this component such that $c(\pi\rho) = c(\pi) + 1$. Moreover, this reversal does not create new unoriented components in $\pi\rho$, and $h(\pi\rho) = h(\pi)$ and $f(\pi\rho) = f(\pi)$. Note that every cycle containing tails of chromosomes belongs to an oriented component in π and consists entirely of edges between tails. Therefore, ρ acts either on two intrachromosomal black edges or on two interchromosomal black edges belonging to some oriented cycle of this component.

A reversal ρ acting on two interchromosomal black edges can be viewed as a transformation of a concatenate π of

$$\Pi = (\pi(1), \ldots, \pi(i-1), \pi(i), \ldots, \pi(j), \pi(j+1), \ldots, \pi(N))$$

into a concatenate $\pi\rho' = \Pi'(s')$, where Π' is a new ordering

$$(\pi(1), \ldots, \pi(i-1), \pi(j), \ldots, \pi(i), \pi(j+1), \ldots, \pi(N))$$

of the chromosomes and

$$s' = (s(1), \ldots, s(i-1), -s(j), \ldots, -s(i), s(j+1), \ldots, s(N)).$$

Therefore, $b_{tail}(\pi\rho) - c_{tail}(\pi\rho) = b_{tail}(\pi) - (c_{tail}(\pi) + 1)$ and $\pi\rho$ is a properly flipped concatenate of Π, a contradiction to the minimality of $b_{tail}(\pi) - c_{tail}(\pi)$.

If reversal ρ acts on two intrachromosomal black edges, then $\pi\rho$ is a properly flipped concatenate of $\Pi\rho$, implying that

$$\begin{gathered} b(\Pi) - c(\Pi) + k(\Pi) + f(\Pi) = (b(\pi) - b_{tail}(\pi)) - (c(\pi) - c_{tail}(\pi)) + h(\pi) + f(\pi) = \\ (b(\pi\rho) - b_{tail}(\pi\rho)) - (c(\pi\rho) - 1 - c_{tail}(\pi\rho)) + h(\pi\rho) + f(\pi\rho) = \\ b(\Pi\rho) - c(\Pi\rho) + h(\Pi\rho) + f(\Pi\rho) + 1 \,. \end{gathered}$$

Since $b(\Pi) - c(\Pi) + k(\Pi) + f(\Pi) > b(\Pi\rho) - c(\Pi\rho) + h(\Pi\rho) + f(\Pi\rho)$, the theorem holds for the genome $\Pi\rho$. Therefore, $d(\Pi) \leq d(\Pi\rho) + 1 = b(\Pi) - c(\Pi) + k(\Pi) + f(\Pi)$. ■

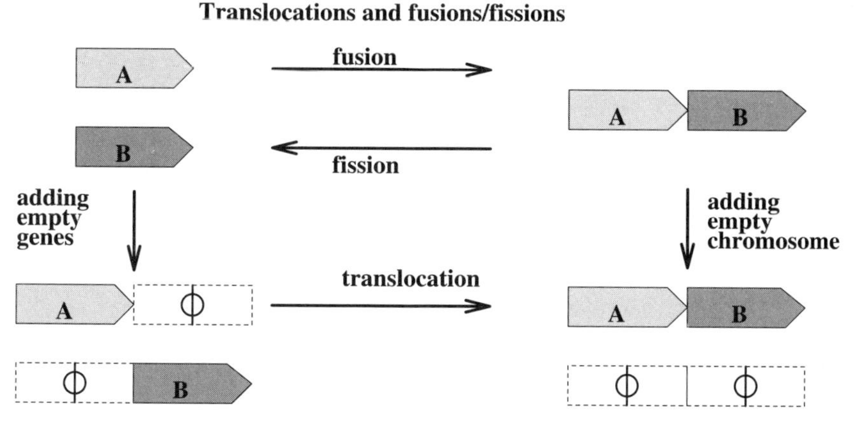

Figure 10.18: Fusions/fissions can be mimicked by translocations by introducing empty chromosomes.

10.13 Capping Chromosomes

We now turn to the general case in which genomes Π and Γ might have different sets of tails and different number of chromosomes. Below we describe an algorithm for computing $d(\Pi, \Gamma)$ that is polynomial in the number of genes but exponential in the number of chromosomes. This algorithm leads to the (truly) polynomial-time algorithm that is described in the following sections.

Let Π and Γ be two genomes with M and N chromosomes. Without loss of generality, assume that $M \leq N$ and extend Π by $N - M$ empty chromosomes (Figure 10.18). As a result, $\Pi = \{\pi(1), \ldots, \pi(N)\}$ and $\Gamma = \{\gamma(1), \ldots, \gamma(N)\}$ contain the same number of chromosomes. Let $\{cap_0, \ldots, cap_{2N-1}\}$ be a set of $2N$ distinct positive integers (called *caps*) that are different from the genes of Π (or equivalently, Γ). Let $\hat{\Pi} = \{\hat{\pi}(1), \ldots, \hat{\pi}(N)\}$ be a genome obtained from Π by adding caps to the ends of each chromosome, i.e.,

$$\hat{\pi}(i) = cap_{2(i-1)}, \pi(i)_1, \ldots, \pi(i)_{n_i}, cap_{2(i-1)+1}.$$

Note that every reversal/translocation in Π corresponds to an *internal* reversal/translocation in $\hat{\Pi}$. If this translocation is a fission, we assume that there are enough empty chromosomes in Π (the validity of this assumption will become clear later).

Every sorting of Π into Γ induces a sorting of $\hat{\Pi}$ into a genome

$$\hat{\Gamma} = \{\hat{\gamma}(1), \ldots, \hat{\gamma}(N)\}$$

(called a *capping* of Γ), where

$$\hat{\gamma}(i) = ((-1)^j cap_j, \hat{\gamma}(i)_1, \ldots, \hat{\gamma}(i)_{m_i}, (-1)^{k+1} cap_k)$$

for $0 \leq j, k \leq 2N-1$. Genomes $\hat{\Pi}$ and $\hat{\Gamma}$ are co-tailed, since $\mathcal{T}(\hat{\Pi}) = \mathcal{T}(\hat{\Gamma}) = \bigcup_{i=0}^{2N-1}(-1)^i cap_i$. There exist $(2N)!$ different cappings of Γ, each capping defined by the distribution of $2N$ caps of $\hat{\Pi}$ in $\hat{\Gamma}$. Denote the set of $(2N)!$ cappings of Γ as $\mathbf{\Gamma}$. The following lemma leads to an algorithm for computing genomic distance that is polynomial in the number of genes but exponential in the number of chromosomes N.

Lemma 10.29 $d(\Pi, \Gamma) = \min_{\hat{\Gamma} \in \mathbf{\Gamma}} b(\hat{\Pi}, \hat{\Gamma}) - c(\hat{\Pi}, \hat{\Gamma}) + k(\hat{\Pi}, \hat{\Gamma}) + f(\hat{\Pi}, \hat{\Gamma})$.

Proof Follows from theorem 10.13 and the observation that every sorting of $\hat{\Pi}$ into a genome $\hat{\Gamma} \in \mathbf{\Gamma}$ by internal reversals/translocations induces a sorting of Π into Γ. ∎

Let $\hat{\pi}$ and $\hat{\gamma}$ be arbitrary concatenates of (ordered) cappings $\hat{\Pi}$ and $\hat{\Gamma}$. Let $G(\hat{\Pi}, \hat{\Gamma})$ be a graph obtained from $G(\hat{\pi}, \hat{\gamma})$ by deleting all tails (vertices of $G(\hat{\pi}, \hat{\gamma})$) of genome $\hat{\Pi}$ (or equivalently, $\hat{\Gamma}$) from $G(\hat{\pi}, \hat{\gamma})$. Different cappings $\hat{\Gamma}$ correspond to different graphs $G(\hat{\Pi}, \hat{\Gamma})$. Graph $G(\hat{\Pi}, \hat{\Gamma})$ has $2N$ vertices corresponding to caps; gray edges incident on these vertices completely define the capping $\hat{\Gamma}$. Therefore, deleting these $2N$ gray edges transforms $G(\hat{\Pi}, \hat{\Gamma})$ into a graph $G(\Pi, \Gamma)$ that does not depend on capping $\hat{\Gamma}$ (Figure 10.19a, b, c, and d).

Graph $G(\Pi, \Gamma)$ contains $2N$ vertices of degree 1 corresponding to $2N$ caps of Π (called Π-*caps*) and $2N$ vertices of degree 1 corresponding to $2N$ tails of Γ (called Γ-*tails*). Therefore, $G(\Pi, \Gamma)$ is a collection of cycles and $2N$ paths, each path starting and ending with a black edge. A path is a $\Pi\Pi$-*path* ($\Gamma\Gamma$-*path*) if it starts and ends with Π-caps (Γ-tails), and a $\Pi\Gamma$-*path* if it starts with a Π-cap and ends with a Γ-tail. A vertex in $G(\Pi, \Gamma)$ is a $\Pi\Gamma$-*vertex* if it is a Π-cap in a $\Pi\Gamma$-path, and a $\Pi\Pi$-*vertex* if it is a Π-cap in a $\Pi\Pi$-path. $\Gamma\Pi$- and $\Gamma\Gamma$-vertices are defined similarly (see Figure 10.19d).

Every capping $\hat{\Gamma}$ corresponds to adding $2N$ gray edges to the graph $G(\Pi, \Gamma)$, each edge joining a Π-cap with a Γ-tail. These edges transform $G(\Pi, \Gamma)$ into the graph $G(\hat{\Pi}, \hat{\Gamma})$ corresponding to a capping $\hat{\Gamma}$ (Figure 10.19e).

Define $b(\Pi, \Gamma)$ as the number of black edges in $G(\Pi, \Gamma)$, and define $c(\Pi, \Gamma)$ as the overall number of cycles and paths in $G(\Pi, \Gamma)$. The parameter $b(\Pi, \Gamma) - b(\hat{\Pi}, \hat{\Gamma})$ does not depend on capping $\hat{\Gamma}$. Clearly, $c(\hat{\Pi}, \hat{\Gamma}) \leq c(\Pi, \Gamma)$, with $c(\hat{\Pi}, \hat{\Gamma}) = c(\Pi, \Gamma)$ if every path in $G(\Pi, \Gamma)$ is "closed" by a gray edge in $G(\hat{\Pi}, \hat{\Gamma})$. The observation that every cycle in $G(\hat{\Pi}, \hat{\Gamma})$ containing a $\Pi\Pi$-path contains at least one more path leads to the inequality $c(\hat{\Pi}, \hat{\Gamma}) \leq c(\Pi, \Gamma) - p(\Pi, \Gamma)$, where $p(\Pi, \Gamma)$ is the number of $\Pi\Pi$-paths (or equivalently, $\Gamma\Gamma$-paths) in $G(\Pi, \Gamma)$.

We define the notions of interleaving cycles/paths, oriented and unoriented components, etc. in the graph $G(\Pi, \Gamma)$ in the usual way by making no distinction between cycles and paths in $G(\Pi, \Gamma)$. We say that a vertex π_j is *inside* a component U of π if $j \in [\bar{U}_{min}, \bar{U}_{max}]$. An intrachromosomal component for genomes Π and Γ is called a *real* component if it has neither a Π-cap nor a Γ-tail inside.

For genomes Π and Γ, define $\mathcal{RU}(\Pi, \Gamma)$ as the set of real components and define $\mathcal{IU}(\Pi, \Gamma)$ as the set of intrachromosomal components (as defined by the graph $G(\Pi, \Gamma)$). Clearly $\mathcal{RU}(\Pi, \Gamma) \subseteq \mathcal{IU}(\Pi, \Gamma)$. Hurdles, superhurdles, and fortresses for the set $\mathcal{RU}(\Pi, \Gamma)$ are called *real-knots*, *super-real-knots*, and *fortresses-of-real-knots*. Let RK be the set of real-knots (i.e., hurdles for the set $\mathcal{RU}(\Pi, \Gamma)$), and let K be the set of knots (i.e., hurdles for the set $\mathcal{IU}(\Pi, \Gamma)$). A knot from the set $K \setminus RK$ is a *semi-knot* if it does not contain a $\Pi\Pi$- or $\Gamma\Gamma$-vertex inside. Clearly, every semi-knot contains a $\Pi\Gamma$-path (otherwise, it would be a real-knot). Denote the number of real-knots and semi-knots for genomes Π and Γ as $r(\Pi, \Gamma)$ and $s(\Pi, \Gamma)$, respectively. Clearly $k(\hat{\Pi}, \hat{\Gamma}) \geq r(\Pi, \Gamma)$, implying that

$$b(\hat{\Pi}, \hat{\Gamma}) - c(\hat{\Pi}, \hat{\Gamma}) + k(\hat{\Pi}, \hat{\Gamma}) \leq b(\Pi, \Gamma) - c(\Pi, \Gamma) + p(\Pi, \Gamma) + r(\Pi, \Gamma).$$

However, this bound is not tight, since it assumes that there exists a capping $\hat{\Gamma}$ that simultaneously maximizes $c(\hat{\Pi}, \hat{\Gamma})$ and minimizes $k(\hat{\Pi}, \hat{\Gamma})$. Taking $s(\Pi, \Gamma)$ into account leads to a better bound for genomic distance that is at most 1 rearrangement apart from the genomic distance.

10.14 Caps and Tails

Genomes Π and Γ are *correlated* if all the real-knots in $G(\Pi, \Gamma)$ are located on the same chromosome and *non-correlated* otherwise. In this section we restrict our analysis to non-correlated genomes (it turns out that the analysis of correlated genomes involves some additional technical difficulties) and prove a tight bound for $d(\Pi, \Gamma)$ (this bound leads to a rather complicated potential function used in the proof of the duality theorem):

$$b(\Pi, \Gamma) - c(\Pi, \Gamma) + p(\Pi, \Gamma) + r(\Pi, \Gamma) + \lceil \frac{s(\Pi, \Gamma)}{2} \rceil \leq d(\Pi, \Gamma) \leq$$

$$b(\Pi, \Gamma) - c(\Pi, \Gamma) + p(\Pi, \Gamma) + r(\Pi, \Gamma) + \lceil \frac{s(\Pi, \Gamma)}{2} \rceil + 1$$

The following lemmas suggest a way to connect some paths in $G(\Pi, \Gamma)$ by oriented edges.

Lemma 10.30 *For every $\Pi\Pi$-path and $\Gamma\Gamma$-path in $G(\Pi, \Gamma)$, there exists either an interchromosomal or an oriented gray edge that joins these paths into a $\Pi\Gamma$-path.*

Lemma 10.31 *For every two unoriented $\Pi\Gamma$-paths located on the same chromosome, there exists an oriented gray edge that joins these paths into a $\Pi\Gamma$-path.*

In a search for an optimal capping, we first ignore the term $f(\hat{\Pi}, \hat{\Gamma})$ in lemma 10.29 and find a capping whose genomic distance $d(\hat{\Pi}, \hat{\Gamma})$ is within 1 from the optimal. The following theorem suggests a way to find such an "almost optimal" capping $\hat{\Gamma}$.

Theorem 10.14 $\min_{\hat{\Gamma} \in \mathbf{\Gamma}} b(\hat{\Pi}, \hat{\Gamma}) - c(\hat{\Pi}, \hat{\Gamma}) + k(\hat{\Pi}, \hat{\Gamma}) = b(\Pi, \Gamma) - c(\Pi, \Gamma) + p(\Pi, \Gamma) + r(\Pi, \Gamma) + \lceil \frac{s(\Pi, \Gamma)}{2} \rceil$.

Proof Every capping $\hat{\Gamma}$ defines a transformation of $G(\Pi, \Gamma)$ into $G(\hat{\Pi}, \hat{\Gamma})$ by consecutively adding $2N$ gray edges to $G(\Pi, \Gamma)$: $G(\Pi, \Gamma) = G_0 \stackrel{g_1}{\to} G_1 \stackrel{g_2}{\to} \ldots \stackrel{g_{2N}}{\to} G_{2N} = G(\hat{\Pi}, \hat{\Gamma})$. For a graph G_i, the parameters $b_i = b(G_i)$, $c_i = c(G_i)$, $p_i = p(G_i)$, $r_i = r(G_i)$, and $s_i = s(G_i)$ are defined in the same way as for the graph $G_0 = G(\Pi, \Gamma)$. For a parameter ϕ, define $\Delta\phi_i$ as $\phi_i - \phi_{i-1}$, i.e., $\Delta c_i = c_i - c_{i-1}$, etc. Denote $\Delta_i = (c_i - p_i - r_i - \lceil \frac{s_i}{2} \rceil) - (c_{i-1} - p_{i-1} - r_{i-1} - \lceil \frac{s_{i-1}}{2} \rceil)$. Below we prove that $\Delta_i \le 0$ for $1 \le i \le 2N$, i.e., adding a gray edge does not increase the parameter $c(\Pi, \Gamma) - p(\Pi, \Gamma) - r(\Pi, \Gamma) - \lceil \frac{s(\Pi, \Gamma)}{2} \rceil$. For a fixed i, we ignore index i, i.e., denote $\Delta = \Delta_i$, etc.

Depending on the edge g_i, the following cases are possible (the analysis below assumes that Π and Γ are non-correlated):

Case 1: edge g_i "closes" a $\Pi\Gamma$-path (i.e., g_i connects a $\Pi\Gamma$-vertex with a $\Gamma\Pi$-vertex within the same $\Pi\Gamma$-path). If this vertex is the only $\Pi\Gamma$-vertex in a semi-knot, then $\Delta c = 0, \Delta p = 0, \Delta r = 1$, and $\Delta s = -1$ (note that this might not be true for correlated genomes). Otherwise $\Delta c = 0, \Delta p = 0, \Delta r = 0$, and $\Delta s = 0$. In both cases, $\Delta \le 0$.

Case 2: edge g_i connects a $\Pi\Gamma$-vertex with a $\Gamma\Pi$-vertex in a different $\Pi\Gamma$-path. This edge "destroys" at most two semi-knots, and $\Delta c = -1, \Delta p = 0, \Delta r = 0$, and $\Delta s \ge -2$. Therefore $\Delta \le 0$.

Case 3: edge g_i connects a $\Pi\Gamma$-vertex with a $\Gamma\Gamma$-vertex (or a $\Gamma\Pi$-vertex with a $\Pi\Pi$-vertex). This edge "destroys" at most one semi-knot, and $\Delta c = -1, \Delta p = 0, \Delta r = 0$, and $\Delta s \ge -2$. This implies $\Delta \le 0$.

Case 4. edge g_i connects a $\Pi\Pi$-vertex with a $\Gamma\Gamma$-vertex. This edge cannot destroy any semi-knots, and $\Delta c = -1, \Delta p = -1, \Delta r = 0$, and $\Delta s \ge 0$. This implies $\Delta \le 0$.

Note that $b_{2N} = b(\hat{\Pi}, \hat{\Gamma}) = b(\Pi, \Gamma) = b_0$, $c_{2N} = c(\hat{\Pi}, \hat{\Gamma})$, $p_{2N} = 0$, $s_{2N} = 0$, and $r_{2N} = k(\hat{\Pi}, \hat{\Gamma})$. Therefore, $b(\hat{\Pi}, \hat{\Gamma}) - c(\hat{\Pi}, \hat{\Gamma}) + k(\hat{\Pi}, \hat{\Gamma}) = b_{2N} - c_{2N} + p_{2N} + r_{2N} + \lceil \frac{s_{2N}}{2} \rceil \ge b_0 - c_0 + p_0 + r_0 + \lceil \frac{s_0}{2} \rceil = b(\Pi, \Gamma) - c(\Pi, \Gamma) + p(\Pi, \Gamma) + r(\Pi, \Gamma) + \lceil \frac{s(\Pi, \Gamma)}{2} \rceil$.

We now prove that there exists a capping $\hat{\Gamma}$ such that

$$b(\hat{\Pi},\hat{\Gamma})-c(\hat{\Pi},\hat{\Gamma})+k(\hat{\Pi},\hat{\Gamma}) = b(\Pi,\Gamma)-c(\Pi,\Gamma)+p(\Pi,\Gamma)+r(\Pi,\Gamma)+\lceil\frac{s(\Pi,\Gamma)}{2}\rceil$$

by constructing a sequence of $2N$ gray edges $g_1,\ldots,g_{2N}$ connecting Π-caps with Γ-tails in $G(\Pi,\Gamma)$ such that $\Delta_i = 0$ for all $1 \le i \le 2N$.

Assume that the first $i-1$ such edges are already found, and let G_{i-1} be the result of adding these $i-1$ edges to $G(\Pi,\Gamma)$. If G_{i-1} has a $\Pi\Pi$-path, then it has a $\Gamma\Gamma$-path as well, and by lemma 10.30, there exists an interchromosomal or oriented gray edge joining these paths into an oriented $\Pi\Gamma$-path. Clearly $\Delta c = -1$, $\Delta p = -1$, $\Delta r = 0$, and $\Delta s = 0$ for this edge, implying $\Delta = 0$.

If G_{i-1} has at least two semi-knots (i.e., $s_{i-1} > 1$), let v_1 and v_2 be a $\Pi\Gamma$- and a $\Gamma\Pi$-vertex in different semi-knots. If v_1 and v_2 are in different chromosomes of Π, then the gray edge $g_i = (v_1, v_2)$ "destroys" both semi-knots. Therefore $\Delta c = -1, \Delta p = 0, \Delta r = 0, \Delta s = -2$, and $\Delta = 0$. If v_1 and v_2 belong to the same chromosome, then by lemma 10.31 there exists an oriented gray edge joining these paths into an oriented $\Pi\Gamma$-path. This gray edge destroys two semi-knots. Therefore, $\Delta = 0$ in this case also.

If G_{i-1} has the only semi-knot, let P_1 be a $\Pi\Gamma$-path in this semi-knot. If it is the only $\Pi\Gamma$-path in the semi-knot, then for an edge g_i "closing" this path, $\Delta c = 0, \Delta p = 0, \Delta r = 1$, and $\Delta s = -1$, implying that $\Delta = 0$. Otherwise, $\Delta c = 0, \Delta p = 0, \Delta r = 0$, and $\Delta s = 0$, implying that $\Delta = 0$.

If G_{i-1} has neither a $\Pi\Pi$-path nor a semi-knot, then let g_i be an edge closing an arbitrary $\Pi\Gamma$-path in G_{i-1}. Since g_i does not belong to a semi-knot, $\Delta c = 0, \Delta p = 0, \Delta r = 0, \Delta s = 0$, and $\Delta = 0$. Therefore, the constructed sequence of edges $g_1,\ldots,g_{2N}$ transforms $G(\Pi,\Gamma)$ into $G(\hat{\Pi},\hat{\Gamma})$ such that $b(\hat{\Pi},\hat{\Gamma}) - c(\hat{\Pi},\hat{\Gamma}) + k(\hat{\Pi},\hat{\Gamma}) = b(\Pi,\Gamma) - c(\Pi,\Gamma) + p(\Pi,\Gamma) + r(\Pi,\Gamma) + \lceil\frac{s(\Pi,\Gamma)}{2}\rceil$. ■

Since $0 \le f(\Pi,\Gamma) \le 1$, lemma 10.29 and theorem 10.14 imply that $b(\Pi,\Gamma) - c(\Pi,\Gamma) + p(\Pi,\Gamma) + r(\Pi,\Gamma) + \lceil\frac{s(\Pi,\Gamma)}{2}\rceil$ is within one rearrangement from the genomic distance $d(\Pi,\Gamma)$ for non-correlated genomes. In the following section we close the gap between $b(\Pi,\Gamma) - c(\Pi,\Gamma) + p(\Pi,\Gamma) + r(\Pi,\Gamma) + \lceil\frac{s(\Pi,\Gamma)}{2}\rceil$ and $d(\Pi,\Gamma)$ for arbitrary genomes.

10.15 Duality Theorem for Genomic Distance

The major difficulty in closing the gap between $b(\Pi,\Gamma) - c(\Pi,\Gamma) + p(\Pi,\Gamma) + r(\Pi,\Gamma) + \lceil\frac{s(\Pi,\Gamma)}{2}\rceil$ and $d(\Pi,\Gamma)$ is "uncovering" remaining "obstacles" in the duality theorem. It turns out that the duality theorem involves seven (!) parameters, making it very hard to explain an intuition behind it. Theorem 10.14 provides such an intuition for the first five parameters. Two more parameters are defined below.

A component in $G(\Pi,\Gamma)$ containing a $\Pi\Gamma$-path is *simple* if it is not a semi-knot.

Lemma 10.32 *There exists an optimal capping $\hat{\Gamma}$ that closes all $\Pi\Gamma$-paths in simple components.*

Let $\overline{G}$ be a graph obtained from $G(\Pi,\Gamma)$ by closing all $\Pi\Gamma$-paths in simple components. Without confusion we can use the terms *real-knots, super-real-knots*, and *fortress-of-real-knots in* $\overline{G}$ and define $rr(\Pi,\Gamma)$ as the number of real-knots in $\overline{G}$. Note that $rr(\Pi,\Gamma)$ does not necessarily coincide with $r(\Pi,\Gamma)$.

Correlated genomes Π and Γ form a *weak-fortress-of-real-knots* if (i) they have an odd number of real-knots in $\overline{G}$, (ii) one of the real-knots is the greatest real-knot in $\overline{G}$, (iii) every real-knot but the greatest one is a super-real-knot in $\overline{G}$, and (iv) $s(\Pi,\Gamma)>0$. Notice that a weak-fortress-of-real-knots can be transformed into a fortress-of-real-knots by closing $\Pi\Gamma$-paths contained in one of the semi-knots. Define two more parameters as follows:

$$fr(\Pi,\Gamma)=\begin{cases}1, & \text{if } \Pi \text{ and } \Gamma \text{ form a fortress-of-real-knots or a weak-fortress-of-real-knots in } \overline{G}\\ 0, & \text{otherwise}\end{cases}$$

$$gr(\Pi,\Gamma)=\begin{cases}1, & \text{if there exists the greatest real-knot in } \overline{G} \text{ and } s(\Pi,\Gamma)>0\\ 0, & \text{otherwise}\end{cases}$$

The following theorem proves that *Genomic_Sort* (Figure 10.20) solves the genomic sorting problem. The running time of *Genomic_Sort* (dominated by the running time of sorting signed permutations by reversals) is $O(n^4)$, where n is the overall number of genes.

Theorem 10.15 *(Hannenhalli and Pevzner, 1995[153])*

$$d(\Pi,\Gamma)=$$

$$b(\Pi,\Gamma)-c(\Pi,\Gamma)+p(\Pi,\Gamma)+rr(\Pi,\Gamma)+\left\lceil\frac{s(\Pi,\Gamma)-gr(\Pi,\Gamma)+fr(\Pi,\Gamma)}{2}\right\rceil.$$

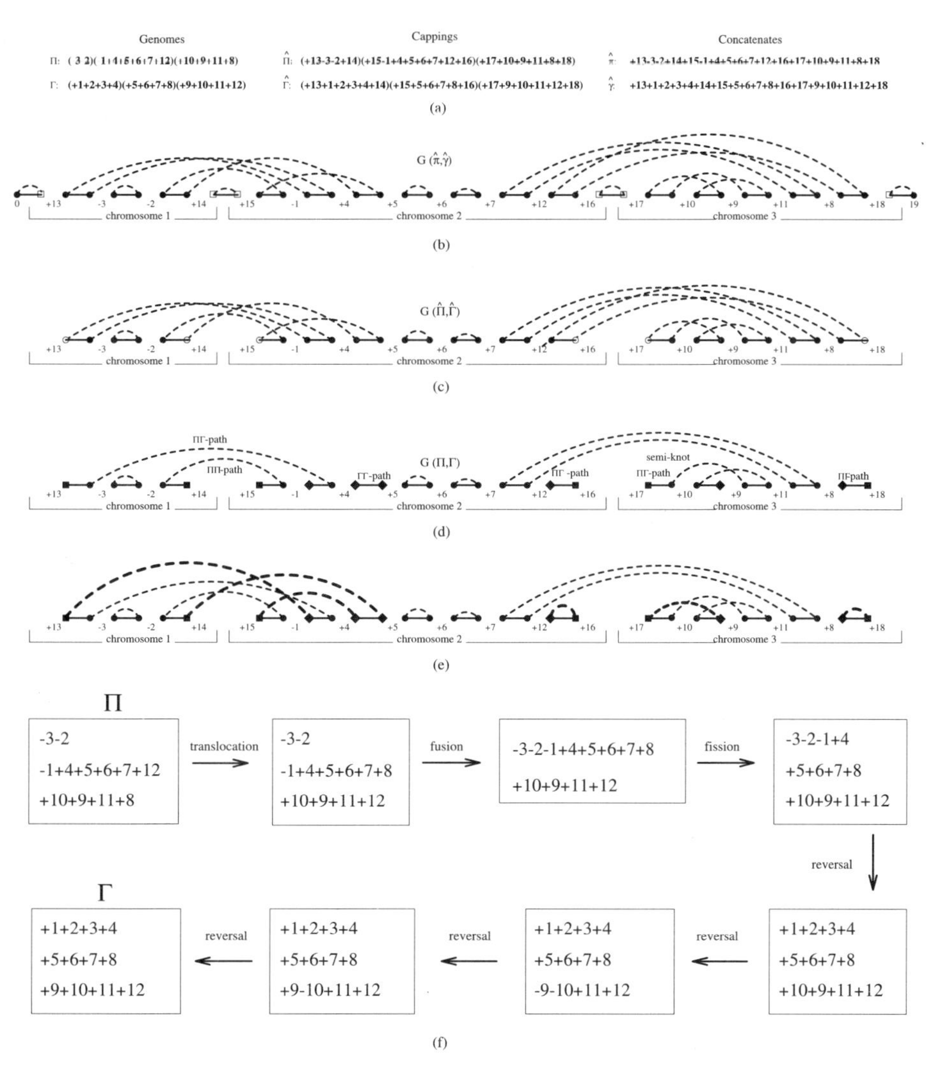

Figure 10.19: (a) Genomes Π and Γ, cappings $\hat{\Pi}$ and $\hat{\Gamma}$, and concatenates $\hat{\pi}$ and $\hat{\gamma}$. (b) Graph $G(\hat{\pi}, \hat{\gamma})$. Tails are shown as white boxes. (c) Graph $G(\hat{\Pi}, \hat{\Gamma})$ is obtained from $G(\hat{\pi}, \hat{\gamma})$ by deleting the tails. Caps are shown as white circles. (d) Graph $G(\Pi, \Gamma)$ with four cycles and six paths ($c(\Pi, \Gamma) = 10$). Π-caps are shown as boxes, while Γ-tails are shown by diamonds. For genomes Π and Γ, $b(\Pi, \Gamma) = 15$, $r(\Pi, \Gamma) = 0$, $p(\Pi, \Gamma) = 1$, $s(\Pi, \Gamma) = 1$, and $gr(\Pi, \Gamma) = fr(\Pi, \Gamma) = 0$. Therefore, $d(\Pi, \Gamma) = 15 - 10 + 1 + 0 + \lceil \frac{1-0+0}{2} \rceil = 7$. (e) Graph $G(\hat{\Pi}, \hat{\Gamma})$ corresponding to an optimal capping of $\hat{\Gamma} = (+13+1+2+3+4-15)(-14+5+6+7+8+18)(+17+9+10+11+12+16)$. Added gray edges are shown by thick dashed lines. (f) Optimal sorting of Π into Γ with seven rearrangements.

```
    Genomic_Sort (Π, Γ)
1.  Construct the graph G = G(Π, Γ)
2.  Close all ΠΓ-paths in simple components of G(Π, Γ) (lemma 10.32)
3.  Close all but one ΠΓ-path in components having more than one ΠΓ-path inside them
4.  while G contains a path
5.      if there exists a ΠΠ-path in G
6.          find an interchromosomal or an oriented edge g joining this ΠΠ-path with a ΓΓ-path
                                                                      (lemma 10.30)
7.      elseif G has more than 2 semi-knots
8.          find an interchromosomal or an oriented edge g joining ΠΓ-paths
                                                  in any two semi-knots (lemma 10.31)
9.      elseif G has 2 semi-knots
10.         if G has the greatest real-knot
11.             find an edge g closing the ΠΓ-path in one of these semi-knots
12.         else
13.             find an interchromosomal or an oriented edge g joining ΠΓ-paths
                                                  in these semi-knots (lemma 10.31)
14.     elseif G has 1 semi-knot
15.         find edge g closing the ΠΓ-path in this semi-knot
16.     else
17.         find edge g closing arbitrary ΠΓ-path
18.     add edge g to the graph G, i.e., G ← G + {g}
19. find a capping Γ̂ defined by the graph G = G(Π̂, Γ̂)
20. sort genome Π̂ into Γ̂ (theorem 10.14)
21. sorting of Π̂ into Γ̂ mimics sorting of Π into Γ
```

Figure 10.20: Algorithm *Genomic_Sort*.

10.16 Genome Duplications

Doubling of the entire genome is a common and lethal accident of reproduction. However, if this doubling can be resolved in the organism and fixed as a normal state in a population, it represents the duplication of the *entire* genome. Such an event may even lead to evolutionary advantages, since a double genome has two copies of each gene that can evolve independently. Since genes may develop novel functions, genome duplication may lead to rapid evolutionary progress. There is evidence that the vertebrate genome underwent duplications two hundred million years ago (Ohno et al., 1968 [256]), with more recent duplications in some vertebrate lines (Postlethwait, 1998 [278]). Comparative genetic maps of plant genomes also reveal multiple duplications (Paterson et al., 1996 [260]).

Yeast sequencing revealed evidence for an ancient doubling of the yeast genome a hundred million years ago (Wolfe and Shields, 1997 [369]). Originally, the duplicated genome contained two identical copies of each chromosome, but through inversions, translocations, fusions and fissions these two copies got disrupted. The solution to thc problem of reconstructing of the gene order in the ancient yeast genome prior to doubling was proposed by El-Mabrouk et al., 1999 [96].

A rearranged duplicated genome contains two copies of each gene. The *genome duplication* problem is to calculate the minimum number of translocations required to transform a rearranged duplicated genome into some *perfect duplicated genome* with an even number of chromosomes that contains two identical copies of each chromosome. For example, a rearranged duplicated genome $\{abc, def, aef, dbc\}$ consisting of four chromosomes can be transformed into a perfect duplicated genome $\{abc, def, abc, def\}$ by a single translocation of chromosomes aef and dbc. El-Mabrouk et al., 1999 [96] proposed a polynomial algorithm for the *genome duplication* problem in the case when the rearrangement operations are translocations only. The algorithm uses the Hannenhalli, 1995 [151] duality theorem regarding the translocation distance between multichromosomal genomes. The problem of devising a more adequate genome duplication analysis with both translocations and reversals remains unsolved.

10.17 Some Other Problems and Approaches

10.17.1 Genome rearrangements and phylogenetic studies

Sankoff et al., 1992 [304] pioneered the use of rearrangement distance for molecular evolution studies. A generalization of the genomic distance problem for multiple genomes corresponds to the following:

Multiple Genomic Distance Problem Given a set of permutations $\pi^1, \ldots, \pi^k$, find a permutation σ such that $\sum_{i=1,k} d(\pi^i, \sigma)$ is minimal (d is the distance between genomes π^i and σ).

In the case in which $d(\pi, \sigma)$ is a reversal distance between π and σ, the Multiple Genomic Distance Problem has been shown to be NP-hard (Caprara, 1999 [56]). Similarly to evolutionary tree multiple alignment, there exists a generalization of the Multiple Genomic Distance Problem for the case when a phylogenetic tree is not known in advance (Figure 10.21). Since Multiple Genomic Distance is difficult in the case of reversal distance, most genomic molecular evolution studies are based on *breakpoint distance*. The breakpoint distance $d(\pi, \sigma)$ between permutations π and σ is defined as the number of breakpoints in $\pi\sigma^{-1}$. Although Multiple Genomic Distance in this formulation is also NP-hard, Sankoff and Blanchette, 1998 [301] suggested practical heuristics for this problem.

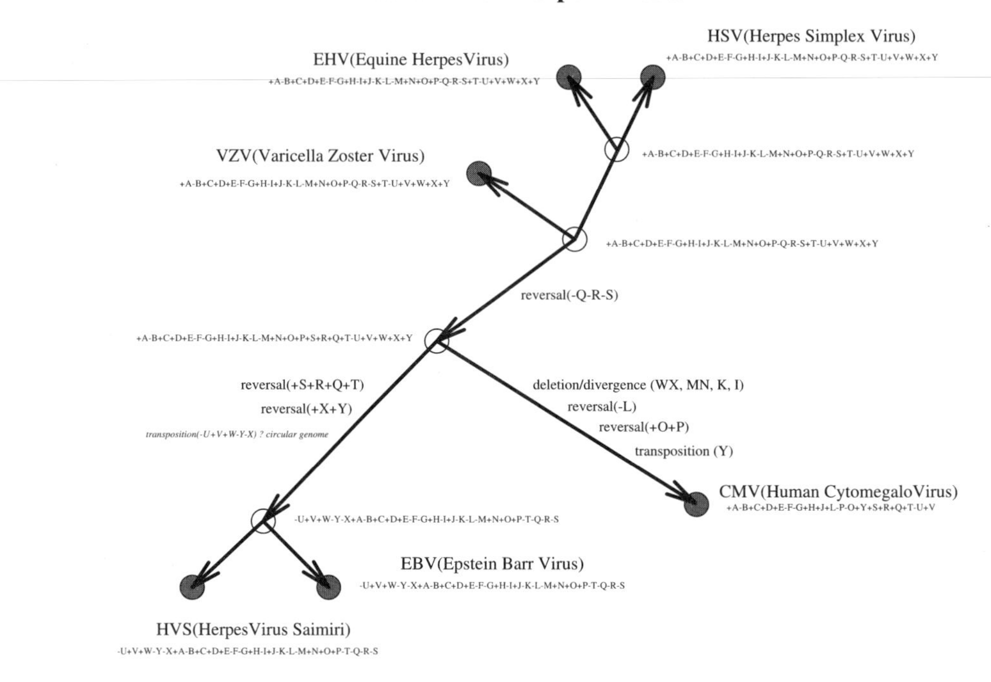

Figure 10.21: Putative scenario of herpes virus evolution.

10.17.2 Fast algorithm for sorting by reversals

Berman and Hannenhalli, 1996 [33] and Kaplan et al., 1997 [185] devised fast algorithms for sorting signed permutations by reversals. The Kaplan et al., 1997 [185] quadratic algorithm bypasses the equivalent transformations step of the Hannenhalli-Pevzner algorithm and explores the properties of the interleaving graph of *gray edges* (rather than the interleaving graph of cycles). This leads to a more elegant, simple proof of theorem 10.10.

Chapter 11

Computational Proteomics

11.1 Introduction

In a few seconds, a mass-spectrometer is capable of breaking a peptide into fragments and measuring their masses (the spectrum of the peptide). The peptide sequencing problem is to derive the sequence of a peptide given its spectrum. For an ideal fragmentation process (each peptide is cleaved between every two consecutive amino acids) and an ideal mass-spectrometer, the peptide sequencing problem is simple. In practice, the fragmentation processes are far from ideal, thus making *de novo* peptide sequencing difficult.

Database search is an alternative to *de novo* peptide sequencing, and mass-spectrometry is very successful in identification of proteins already present in genome databases (Patterson and Aebersold, 1995 [261]). Database search in mass-spectrometry (Mann and Wilm, 1994 [230], Eng et al., 1994 [97], Taylor and Johnson, 1997 [335], Fenyo et al., 1998 [101]) relies on the ability to "look the answer up in the back of the book" when studying genomes of extensively sequenced organisms. An experimental spectrum can be compared with theoretical spectra for each peptide in a database, and the peptide from the database with the best fit usually provides the sequence of the experimental peptide. In particular, Eng et al., 1994 [97] identified proteins from the class II MHC complex, while Clauser et al., 1999 [72] identified proteins related to the effects of preeclampsia. However, in light of the dynamic nature of samples introduced to a mass spectrometer and potential multiple mutations and modifications, the reliability of database search methods that rely on precise or almost precise matches may be called into question. *De novo* algorithms that attempt to interpret tandem mass spectra in the absence of a database are invaluable for identification of unknown proteins, but they are most useful when working with high-quality spectra.

Since proteins are parts of complex systems of cellular signalling and metabolic regulation, they are subject to an almost uncountable number of biological modifications (such as phosphorylation and glycosylation) and genetic variations (Gooley and Packer, 1997 [134], Krishna and Wold, 1998 [207]). For example, at least 1,000 kinases exist in the human genome, indicating that phosphorylation is a common mechanism for signal transmission and enzyme activation. Almost all protein sequences are post-translationally modified, and as many as 200 types of modifications of amino acid residues are known. Since currently post-translational modifications cannot be inferred from DNA sequences, finding them will remain an open problem even after the human genome is completed. This also raises a challenging computational problem for the post-genomic era: given a very large collection of spectra representing the human proteome, find out which of 200 types of modifications are present in each human gene.

Starting from the classical Biemann and Scoble, 1987 [35] paper, there have been a few success stories in identifying modified proteins by mass-spectrometry. The computational analysis of modified peptides was pioneered by Mann and Wilm, 1994 [230] and Yates et al., 1995 [374], [373]. The problem is particularly important since mass-spectrometry techniques sometimes introduce chemical modifications to native peptides and make these peptides "invisible" to database search programs. Mann and Wilm, 1994[230] used a combination of a partial *de novo* algorithm and database search in their *Peptide Sequence Tag* approach. A Peptide Sequence Tag is a short, clearly identifiable substring of a peptide that is used to reduce the search to the peptides containing this tag. Yates et al., 1995 [374] suggested an exhaustive search approach that (implicitly) generates a virtual database of all modified peptides for a small set of modifications and matches the spectrum against this virtual database. It leads to a large combinatorial problem, even for a small set of modification types. Another limitation is that extremely bulky modifications such as glycosylation disrupt the fragmentation pattern and would not be amenable to analysis by this method.

Mutation-tolerant database search in mass-spectrometry can be formulated as follows: given an experimental spectrum, find the peptide that best matches the spectrum among the peptides that are at most k mutations apart from a database peptide. This problem is far from simple since very similar peptides may have very different spectra. Pevzner et al., 2000 [270] introduced a notion of spectral similarity that led to an algorithm that identifies related spectra even if the corresponding peptides have multiple modifications or mutations. The algorithm reveals potential peptide modifications without an exhaustive search and therefore does not require generating a virtual database of modified peptides.

Although database search is very useful, a biologist who attempts to clone a *new* gene based on mass spectrometry data needs *de novo* rather than database matching algorithms. However, until recently, sequencing by mass-spectrometry was not widely practiced and had a limited impact on the discovery of new proteins. There are precious few examples of cloning of a gene on the basis of mass-spectrometry-derived sequence information alone (Lingner et al., 1997 [223]).

The recent progress in *de novo* peptide sequencing, combined with automated mass spectrometry data acquisition, may open a door to "proteome sequencing." Long stretches of protein sequences could be assembled following the generation and sequencing of overlapping sets of peptides from treatment of protein mixtures with proteolytic enzymes of differing specificity. Complete protein sequence determination has already been demonstrated with such a strategy on a single protein (Hopper et al., 1989 [166]).

11.2 The Peptide Sequencing Problem

Let A be the set of amino acids with molecular masses $m(a)$, $a \in A$. A *peptide* $P = p_1, \ldots, p_n$ is a sequence of amino acids, and the (parent) mass of peptide P is $m(P) = \sum m(p_i)$. A *partial peptide* P' is a substring $p_i \ldots p_j$ of P of mass $\sum_{i \leq t \leq j} m(p_t)$.

Peptide fragmentation in a *tandem mass-spectrometer* can be characterized by a set of numbers $\Delta = \{\delta_1, \ldots, \delta_k\}$ representing *ion-types*. A δ-*ion* of a partial peptide $P' \subset P$ is a modification of P' that has mass $m(P') - \delta$. For tandem mass-spectrometry, the *theoretical* spectrum of peptide P can be calculated by subtracting all possible ion-types $\delta_1, \ldots, \delta_k$ from the masses of all partial peptides of P (every partial peptide generates k masses in the theoretical spectrum). An (experimental) spectrum $S = \{s_1, \ldots, s_m\}$ is a set of masses of (fragment) ions. The *match* between spectrum S and peptide P is the number of masses that the experimental and theoretical spectra have in common (shared peaks count). Dancik et al., 1999 [79] addressed the following

Peptide Sequencing Problem Given spectrum S, the set of ion-types Δ, and the mass m, find a peptide of mass m with the maximal match to spectrum S.

Denote partial *N-terminal* peptide $p_1, \ldots, p_i$ as P_i and partial *C-terminal* peptide $p_{i+1}, \ldots, p_n$ as P_i^-, $i = 1, \ldots, n$. In practice, a spectrum obtained by tandem mass-spectrometry (MS/MS) consists mainly of some of the δ-ions of partial N-terminal and C-terminal peptides. To reflect this, a theoretical MS/MS spectrum consists only of ions of N-terminal and C-terminal peptides (Figure 11.1). For example, the most frequent N-terminal ions are usually b-ions (b_i corresponds to P_i with $\delta = -1$) and the most frequent C-terminal ions are usually y-ions (y_i corresponds to P_i^- with $\delta = 19$). Other frequent N-terminal ions for an ion-trap mass-spectrometer (a, b–H_2O, and b–NH_3) are shown in Figure 11.2. Also, instead of the shared peaks count, the existing database search and *de novo* algorithms use more sophisticated objective functions (such as the weighted shared peaks count).

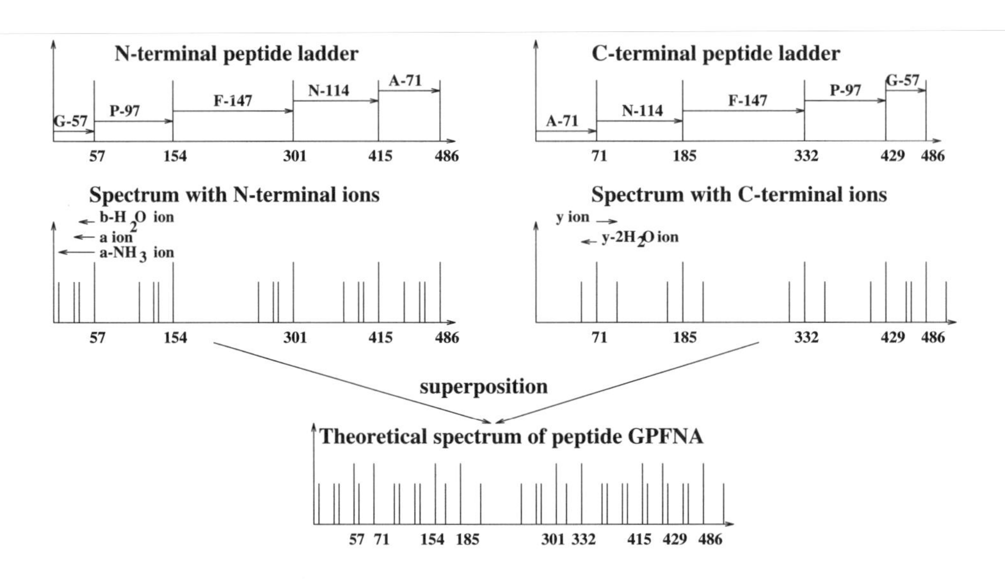

Figure 11.1: Theoretical MS/MS spectrum of peptide GPFNA with parent mass $57 + 97 + 147 + 114 + 71 = 486$.

11.3 Spectrum Graphs

The development of peptide sequencing algorithms have followed either exhaustive search or spectrum graph paradigms. The former approach (Sakurai et al., 1984 [294]) involves the generation of all amino acid sequences and corresponding theoretical spectra. The goal is to find a sequence with the best match between the experimental and theoretical spectra. Since the number of sequences grows exponentially with the length of the peptide, different pruning techniques have been designed to limit the combinatorial explosion in global methods. Prefix pruning (Hamm et al., 1986 [149], Johnson and Biemann, 1989 [182], Zidarov et al., 1990 [378], Yates et al., 1991 [375]) restricts the computational space to sequences whose prefixes match the experimental spectrum well. The difficulty with the prefix approach is that pruning frequently discards the correct sequence if its prefixes are poorly represented in the spectrum. Another problem is that the spectrum information is used only *after* the potential peptide sequences are generated.

Spectrum graph approaches tend to be more efficient because they use spectral information *before* any candidate sequence is evaluated. In this approach, the

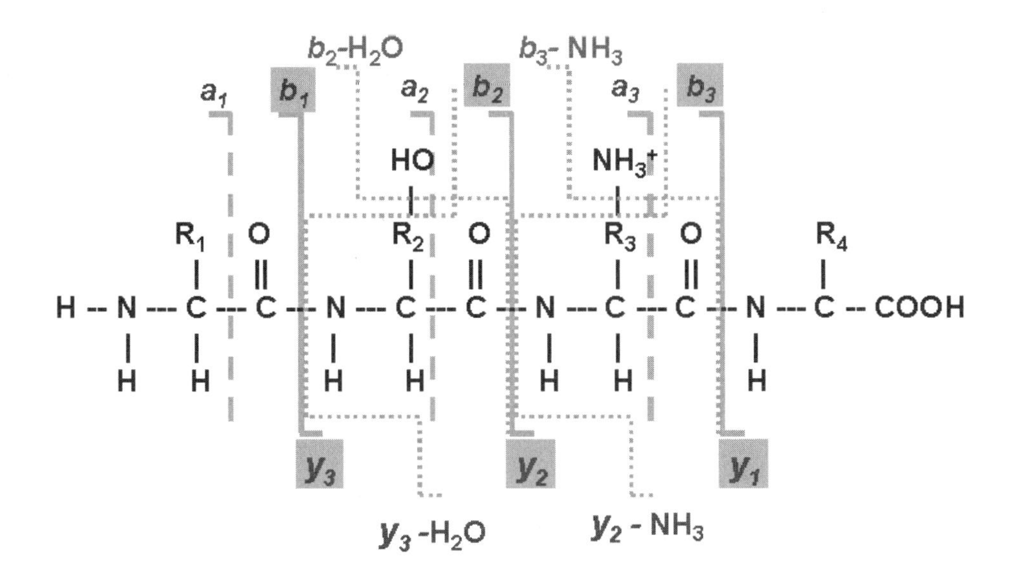

Figure 11.2: Typical fragmentation patterns in tandem mass-spectrometry.

peaks in a spectrum are transformed into a *spectrum graph* (Bartels, 1990 [26], Fernández-de-Cossío et al., 1995 [103], Taylor and Johnson, 1997 [335], Dancik et al., 1999 [79]). The peaks in the spectrum serve as vertices in the spectrum graph, while the edges of the graph correspond to linking vertices differing by the mass of an amino acid. Each peak in an experimental spectrum is transformed into several vertices in a spectrum graph; each vertex represents a possible fragment ion-type assignment for the peak. The Peptide Sequencing Problem is thus cast as finding the longest path in the resulting directed acyclic graph. Since efficient algorithms for finding the longest paths in directed acyclic graphs are known (Cormen et al., 1989 [75]), such approaches have the potential to efficiently prune the set of all peptides to the set of high-scoring paths in the spectrum graph.

The spectrum graph approach is illustrated in Figure 11.3. Since "meaningful" peaks in the spectrum are generated from a peptide ladder (masses of partial peptides) by δ-shifts, one might think that reverse δ-shifts will reconstruct the ideal spectrum, thus leading to peptide sequencing. Figure 11.3 illustrates that this is not true and that a more careful analysis (based on the notion of a spectrum graph) of reverse δ-shifts is required.

Assume, for the sake of simplicity, that an MS/MS spectrum $S = \{s_1, \ldots, s_m\}$ consists mainly of N-terminal ions, and transform it into a spectrum graph $G_\Delta(S)$

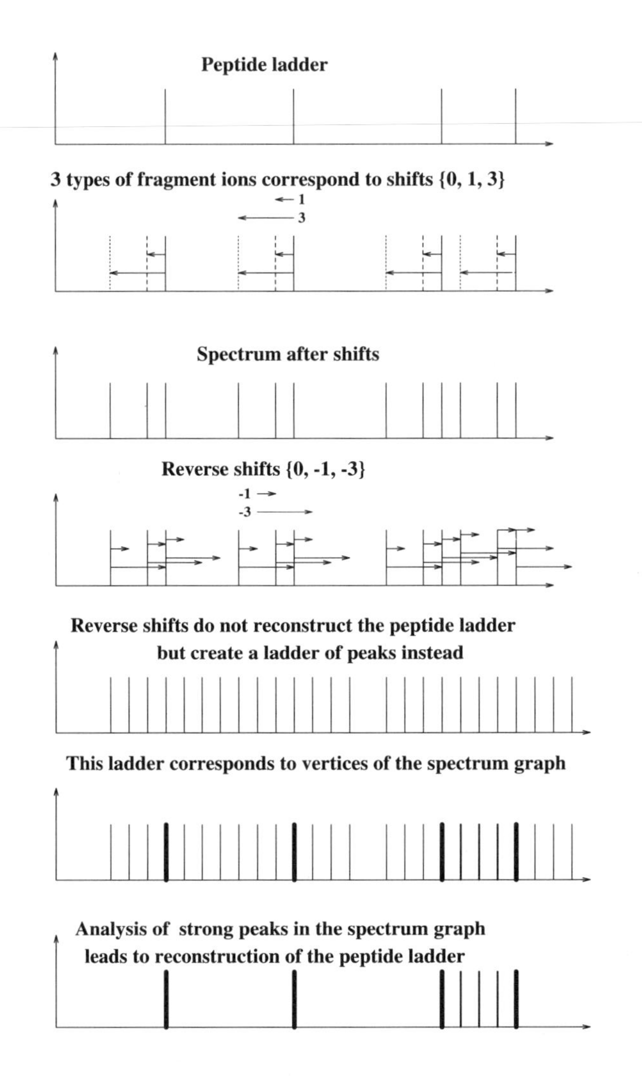

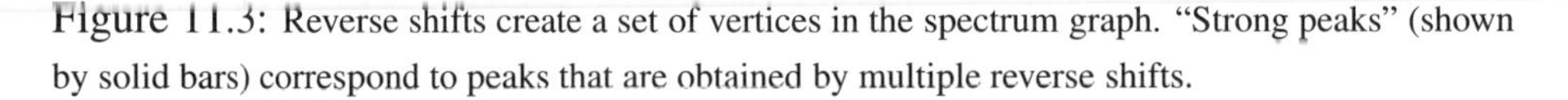

Figure 11.3: Reverse shifts create a set of vertices in the spectrum graph. “Strong peaks” (shown by solid bars) correspond to peaks that are obtained by multiple reverse shifts.

(Bartels, 1990 [26]). Vertices of the graph are integers $s_i + \delta_j$ representing potential masses of partial peptides. Every peak of spectrum $s \in S$ generates k vertices $V(s) = \{s + \delta_1, \ldots, s + \delta_k\}$. The set of vertices of a spectrum graph, then, is $\{s_{initial}\} \cup V(s_1) \cup \cdots \cup V(s_m) \cup \{s_{final}\}$, where $s_{initial} = 0$ and $s_{final} = m(P)$.

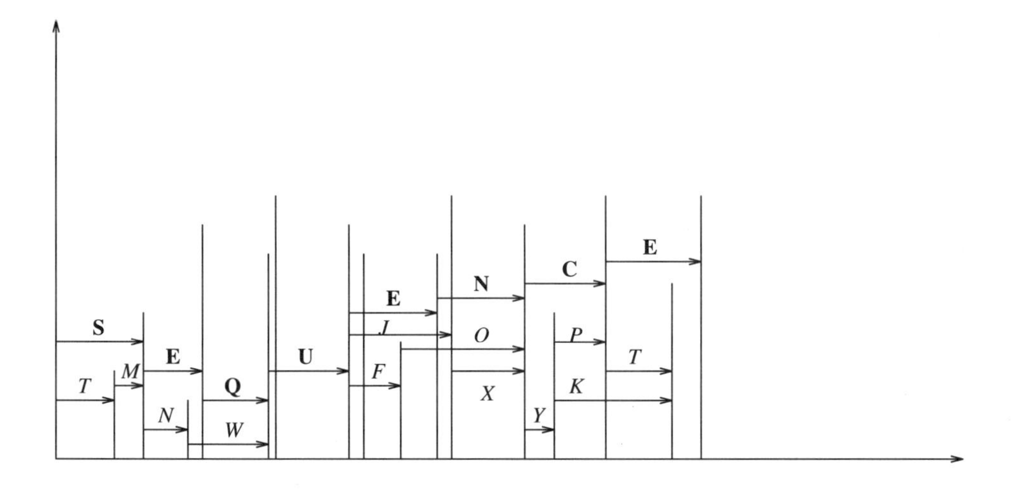

Figure 11.4: Multiple paths in a spectrum graph.

Two vertices u and v are connected by a directed edge from u to v if $v - u$ is the mass of some amino acid and the edge is *labeled* by this amino acid. If we look at vertices as potential N-terminal peptides, the edge from u to v implies that the sequence at v may be obtained by extending the sequence at u by one amino acid (Figures 11.4 and 11.5).

A spectrum S of a peptide P is called *complete* if S contains at least one ion-type corresponding to P_i for every $1 \leq i \leq n$. The use of a spectrum graph is based on the observation that for a complete spectrum there exists a path of length n from $s_{initial}$ to s_{final} in $G_\Delta(S)$ that is labeled by P. This observation casts the tandem mass-spectrometry Peptide Sequencing Problem as one of finding the correct path in the set of all paths.

Unfortunately, experimental spectra are frequently incomplete. Another problem is that MS/MS experiments performed with the same peptide but a different type of mass-spectrometers will produce significantly different spectra. Different ionization methods have a dramatic impact on the propensities for producing particular fragment ion-types. Therefore, every algorithm for peptide sequencing should be adjusted for a particular type of mass-spectrometer. To address this problem, Dancik et al., 1999 [79] described an algorithm for an *automatic* learning of ion-types from a sample of experimental spectra of known sequences. They introduced the *offset frequency function*, which evaluates the ion-type tendencies for particular mass-spectrometers and leads to an instrument-independent peptide sequencing algorithm.

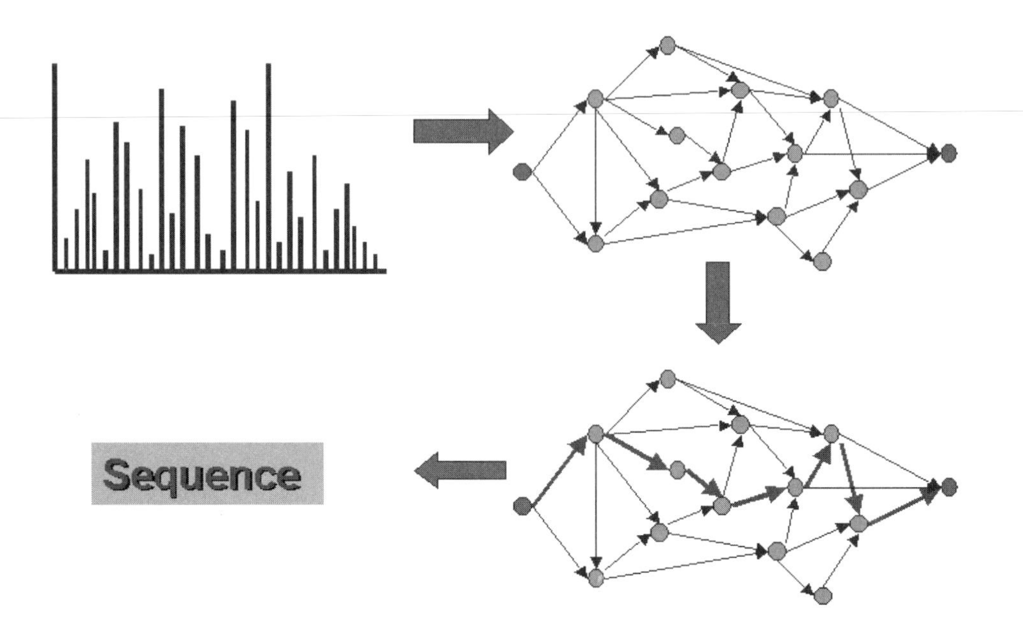

Figure 11.5: Noise in a spectrum generates many "false" edges and vertices in the spectrum graph and disguises edges corresponding to the real peptide sequence. Sequence reconstruction corresponds to finding an optimal path in the spectrum graph.

11.4 Learning Ion-Types

If the ion-types $\Delta = \{\delta_1, \ldots, \delta_k\}$ produced by a given mass-spectrometer are not known, the spectrum cannot be interpreted. Below we show how to learn the set Δ and ion propensities from a sample of experimental spectra of known sequences *without* any prior knowledge of the fragmentation patterns.

Let $S = \{s_1, \ldots, s_m\}$ be a spectrum corresponding to the peptide P. A partial peptide P_i and a peak s_j have an offset $x_{ij} = m(P_i) - s_j$; we can treat x_{ij} as a random variable. Since the probability of offsets corresponding to "real" fragment ions is much greater than the probability of random offsets, the peaks in the empirical distribution of the offsets reveal fragment ions. The statistics of offsets over all ions and all partial peptides provides a reliable learning algorithm for ion-types (Dancik et al., 1999 [79]).

Given spectrum S, offset x, and precision ε, let $H(x, S)$ be the number of pairs (P_i, s_j), $i = 1, \ldots, n-1$, $j = 1, \ldots, m$ that have offset $m(P_i) - s_j$ within distance ε from x. The *offset frequency function* is defined as $H(x) = \sum_S H(x, S)$,

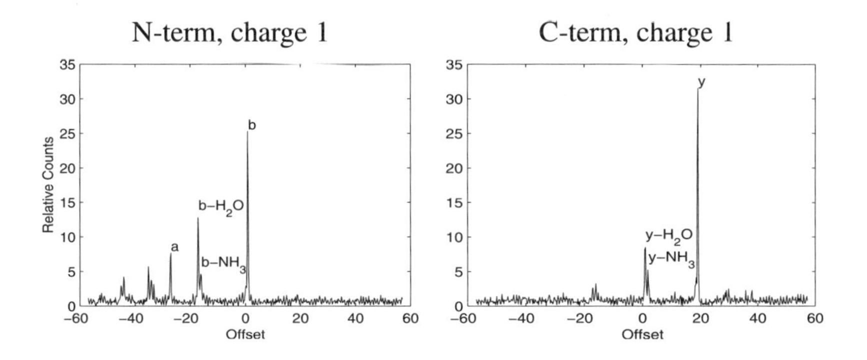

Figure 11.6: Offset frequency function for N-terminal (left) and C-terminal (right) peptides. Horizontal axes represent offsets between peaks in spectra and masses of partial peptide molecules. Vertical axes represent normalized offset counts, with 1 being the average count.

where the sum is taken over all spectra from the learning sample (Figure11.6). To learn about C-terminal ions, we do the same for pairs (P_i^-, s_j). Offsets $\Delta = \{\delta_1, \ldots, \delta_k\}$ corresponding to peaks of $H(x)$ represent the ion-types produced by a given mass-spectrometer.

Peaks in a spectrum differ in *intensity*, and one has to set a threshold for distinguishing the signal from noise in a spectrum prior to transforming it into a spectrum graph. Low thresholds lead to excessive growth of the spectrum graph, while high thresholds lead to fragmentation of the spectrum graph. The offset frequency function allows one to set up the intensity thresholds in a rigorous way (Dancik et al., 1999 [79]).

11.5 Scoring Paths in Spectrum Graphs

The goal of scoring is to quantify how well a candidate peptide "explains" a spectrum and to choose the peptide that explains the spectrum the best. If $p(P, S)$ is the probability that a peptide P produces spectrum S, then the goal is to find a peptide P maximizing $p(P, S)$ for a given spectrum S. Below we describe a probabilistic model, evaluate $p(P, S)$, and derive a scoring schema for paths in the spectrum graph. The longest path in the weighted spectrum graph corresponds to the peptide P that "explains" spectrum S the best.

In a probabilistic approach, tandem mass-spectrometry is characterized by a set of ion-types $\Delta = \{\delta_1, \ldots, \delta_k\}$ and their probabilities $\{p(\delta_1), \ldots, p(\delta_k)\}$ such that δ_i-ions of a partial peptide are produced independently with probabilities $p(\delta_i)$. A mass-spectrometer also produces a "random" (chemical or electronic) noise that in any position may generate a peak with probability q_R. Therefore, a peak at a position corresponding to a δ_i-ion is generated with probability $q_i = p(\delta_i) + (1 - p(\delta_i))q_R$, which can be estimated from the observed empirical distributions. A partial peptide may theoretically have up to k corresponding peaks in the spectrum. It has all k peaks with probability $\prod_{i=1}^{k} q_i$, and it has no peaks with probability $\prod_{i=1}^{k}(1-q_i)$. The probabilistic model defines the probability $p(P, S)$ that a peptide P produces a spectrum S. Below we describe how to compute $p(P, S)$ and derive scoring that leads to finding a peptide maximizing $p(P, S)$ for a given spectrum P.

Suppose that a candidate partial peptide P_i produces ions $\delta_1, \ldots, \delta_l$ ("present" ions) and does not produce ions $\delta_{l+1}, \ldots, \delta_k$ ("missing" ions) in the spectrum S. The l "present" ions will result in a vertex in the spectrum graph corresponding to P_i. How should we score this vertex? A naive approach would be to use a "premium for explained ions" approach, suggesting that the score for this vertex should be proportional to $q_1 \cdots q_l$ or maybe $\frac{q_1}{q_R} \cdots \frac{q_l}{q_R}$ to normalize the probabilities against the noise. However, such an approach has a serious deficiency, and significant improvement results from a different, "premium for present ions, penalty for missing ions" approach. The (probability) score of the vertex is then given by

$$\frac{q_1}{q_R} \cdots \frac{q_l}{q_R} \frac{(1 - q_{l+1})}{(1 - q_R)} \cdots \frac{(1 - q_k)}{(1 - q_R)}.$$

We explain the role of this principle for the resolution of a simple alternative between dipeptide GG and amino acid N of the same mass. In the absence of a penalty for missing ions, GG is selected over N in the presence of *any* peak supporting the position of the first G (even a very weak one corresponding to random noise). Such a rule leads to many wrong GG-abundant interpretations and indicates that a better rule is to vote for GG if it is supported by major ion-types with sufficient intensities, which is automatically enforced by "premium for present ions, penalty for missing ions" scoring.

For the sake of simplicity, we assume that all partial peptides are equally likely and ignore the intensities of peaks. We discretize the space of all masses in the interval from 0 to the parent mass $m(P) = M$, denote $T = \{0, \ldots, M\}$, and represent the spectrum as an M-mer vector $S = \{s_1, \ldots, s_M\}$ such that s_t is the indicator of the presence or absence of peaks at position t ($s_t = 1$ if there is a peak at position t and $s_t = 0$ otherwise). For a given peptide P and position t, s_t is a 0-1 random variable with probability distribution $p(P, s_t)$.

Let $T_i = \{t_{i1}, \ldots, t_{ik}\}$ be the set of positions that represent Δ-ions of a partial peptide P_i where $\Delta = \{\delta_1, \ldots, \delta_k\}$. Let $R = T \setminus \bigcup_i T_i$ be the set of positions that are not associated with any partial peptides. The probability distribution $p(P, s_t)$ depends on whether $t \in T_i$ or $t \in R$. For a position $t = t_{ij} \in T_i$, the probability

$p(P, s_t)$ is given by

$$p(P, s_t) = \begin{cases} q_j\,, & \text{if } s_t = 1 \text{ (a peak at position } t) \\ 1 - q_j\,, & \text{otherwise.} \end{cases} \tag{11.1}$$

Similarly, for $t \in R$, the probability $p(P, s_t)$ is given by

$$p_R(P, s_t) = \begin{cases} q_R\,, & \text{if } s_t = 1 \text{ (random noise at position } t) \\ 1 - q_R\,, & \text{otherwise.} \end{cases} \tag{11.2}$$

The overall probability of "noisy" peaks in the spectrum is $\prod_{t \in R} p_R(P, s_t)$.

Let $p(P_i, S) = \prod_{t \in T_i} p(P, s_t)$ be the probability that a peptide P_i produces a given spectrum at positions from the set T_i (all other positions are ignored). For the sake of simplicity, assume that each peak of the spectrum belongs only to one set T_i and that all positions are independent. Then

$$p(P, S) = \prod_{t=1}^{M} p(P, s_t) = \left(\prod_{i=1}^{n} p(P_i, S) \right) \prod_{t \in R} p_R(P, s_t).$$

For a given spectrum S, the value $\prod_{t \in T} p_R(P, s_t)$ does not depend on P, and the maximization of $p(P, S)$ is the same as the maximization of

$$\frac{p(P, S)}{p_R(S)} = \frac{\prod_{i=1}^{n} \prod_{j=1}^{k} p(P, s_{t_{ij}}) \prod_{t \in R} p_R(P, s_t)}{\prod_{t \in T} p_R(P, s_t)} = \prod_{i=1}^{n} \prod_{j=1}^{k} \frac{p(P, s_{t_{ij}})}{p_R(P, s_{t_{ij}})}$$

where $p_R(S) = \prod_{t \in T} p_R(P, s_t)$.

In logarithmic scale, the above formula implies the additive "premium for present ions, penalty for missing ions" scoring of vertices in the spectrum graph (Dancik et al., 1999 [79]).

11.6 Peptide Sequencing and Anti-Symmetric Paths

After the weighted spectrum graph is constructed, we cast the Peptide Sequencing Problem as the *longest path problem in a directed acyclic graph* which is solved by a fast linear time algorithm. Unfortunately, this simple algorithm does not quite work in practice. The problem is that every peak in the spectrum may be interpreted as either an N-terminal ion or a C-terminal ion. Therefore, every "real" vertex (corresponding to a mass m) has a "fake" *twin* vertex (corresponding to a mass

$m(P) - m$). Moreover, if the real vertex has a high score, then its fake twin also has a high score. The longest path in the spectrum graph then tends to include *both* the real vertex and its fake twin since they both have high scores. Such paths do not correspond to feasible protein reconstructions and should be avoided. However, the known longest path algorithms do not allow us to avoid such paths.

Therefore, the reduction of the Peptide Sequencing Problem to the longest path problem described earlier is inadequate. Below we formulate the *anti-symmetric longest path* problem, which adequately models the peptide sequence reconstruction.

Let G be a graph and let T be a set of *forbidden pairs* of vertices of G (twins). A path in G is called anti-symmetric if it contains at most one vertex from every forbidden pair. The *anti-symmetric longest path problem* is to find a longest anti-symmetric path in G with a set of forbidden pairs T. Unfortunately, the anti-symmetric longest path problem is NP-hard (Garey and Johnson, 1979 [119]), indicating that efficient algorithms for solving this problem are unlikely. However, this negative result does not imply that there is no hope of finding an efficient algorithm for tandem mass-spectrometry peptide sequencing, since this problem has a *special structure* of forbidden pairs.

Vertices in a spectrum graph are modeled by numbers that correspond to masses of potential partial peptides. Two forbidden pairs of vertices (x_1, y_1) and (x_2, y_2) are *non-interleaving* if the intervals (x_1, y_1) and (x_2, y_2) do not interleave. A graph G with a set of forbidden pairs is called *proper* if every two forbidden pairs of vertices are non-interleaving.

The tandem mass-spectrometry Peptide Sequencing Problem corresponds to the anti-symmetric longest path problem in a proper graph. Dancik et al., 1999 [79] proved that there exists an efficient algorithm for anti-symmetric longest path problem in a proper graph.

11.7 The Peptide Identification Problem

Pevzner et al., 2000 [270] studied the following

Peptide Identification Problem Given a database of peptides, spectrum S, set of ion-types Δ, and parameter k, find a peptide with the maximal match to spectrum S that is at most k mutations or modifications apart from a database entry.

The major difficulty in the Peptide Identification Problem comes from the fact that very similar peptides P_1 and P_2 may have very different spectra S_1 and S_2. Our goal is to define a notion of spectral similarity that correlates well with sequence similarity. In other words, if P_1 and P_2 are a few substitutions, insertions, deletions, or modifications apart, the spectral similarity between S_1 and S_2 should be high. The shared peaks count is, of course, an intuitive measure of spectral similarity. However, this measure diminishes very quickly as the number of mutations increases, thus leading to limitations in detecting similarities in an MS/MS database search. Moreover, there are many correlations between spectra of related peptides and only a small portion of them is captured by the "shared peaks" count. The PEDANTA algorithm (Pevzner et al., 2000 [270]) captures *all* correlations between related spectra for any k and handles the cases in which mutations in the peptide significantly change the fragmentation pattern. For example, replacing amino acids like H, K, R, and P may dramatically alter the fragmentation. Even in an extreme case—as when a single mutation changes the fragmentation pattern from "only b-ions" to "only y-ions"—PEDANTA still reveals the similarity between the corresponding spectra.

11.8 Spectral Convolution

Let S_1 and S_2 be two spectra. Define *spectral convolution* $S_2 \ominus S_1 = \{s_2 - s_1 : s_1 \in S_1, s_2 \in S_2\}$ and let $(S_2 \ominus S_1)(x)$ be the multiplicity of element x in this set. In other words, $(S_2 \ominus S_1)(x)$ is the number of pairs $s_1 \in S_1, s_2 \in S_2$ such that $s_2 - s_1 = x$. If $M(P)$ is the parent mass of peptide P with the spectrum S, then $S^R = M(P) - S$ is the *reversed spectrum* of S (every b-ion (y-ion) in S corresponds to a y-ion (b-ion) in S^R). The *reversed spectral convolution* $(S_2 \ominus S_1^R)(x)$ is the number of pairs $s_1 \in S_1, s_2 \in S_2$ such that $s_2 + s_1 - M(P) = x$.

To illustrate the idea of this approach, consider two copies P_1 and P_2 of the same peptide. The number of peaks in common between S_1 and S_2 (shared peaks count) is the value of $S_2 \ominus S_1$ at $x = 0$. Many MS/MS database search algorithms implicitly attempt to find a peptide P in the database that maximizes $S_2 \ominus S_1$ at $x = 0$, where S_2 is an experimental spectrum and S_1 is a theoretical spectrum of peptide P. However, if we start introducing k mutations in P_2 as compared to P_1, the value of $S_2 \ominus S_1$ at $x = 0$ quickly diminishes. As a result, the discriminating power of the shared peaks count falls significantly at $k = 1$ and almost disappears at $k > 1$.

The peaks in spectral convolution allow one to detect mutations and modifications without an exhaustive search. Let P_2 differ from P_1 by only mutation ($k = 1$) with amino acid difference $\delta = M(P_2) - M(P_1)$. In this case, $S_2 \ominus S_1$ is expected to have two approximately equal peaks at $x = 0$ and $x = \delta$. If the mutation ocurrs at position t in the peptide, then the peak at $x = 0$ corresponds to b_i-ions for $i < t$ and y_i-ions for $i \geq t$. The peak at $x = \delta$ corresponds to b_i-ions for $i \geq t$ and y_i-ions for $i < t$. A mutation in P_2 that changes $M(P_1)$ by δ also "mutates" the spectrum S_2 by shifting some peaks by δ. As a result, the number of shared peaks between S_1 and "mutated" S_2 may increase as compared to the number of shared peaks between S_1 and S_2. This increase is bounded by $(S_2 \ominus S_1)(\delta)$, and

$(S_2 \ominus S_1)(0) + (S_2 \ominus S_1)(\delta)$ is an upper bound on the number of shared peaks between S_1 and "mutated" S_2.

The other set of correlations between spectra of mutated peptides is captured by the reverse spectral convolution $S_2 \ominus S_1^R$, reflecting the pairings of N-terminal and C-terminal ions. $S_2 \ominus S_1^R$ is expected to have two peaks at the *same* positions 0 and δ.

Now assume that P_2 and P_1 are two substitutions apart, one with mass difference δ_1 and another with mass difference $\delta - \delta_1$. These mutations generate two new peaks in the spectral convolution at $x = \delta_1$ and at $x = \delta - \delta_1$. For uniform distribution of mutations in a random peptide, the ratio of the expected heights of the peaks at $0, \delta, \delta_1, \delta - \delta_1$ is $2 : 2 : 1 : 1$.

To increase the signal-to-noise ratio, we combine the peaks in spectral and reverse spectral convolution:

$$S = S_2 \ominus S_1 \; + \; S_2 \ominus S_1^R$$

Furthermore, we combine the peaks at 0 and δ (as well as at δ_1 and $\delta - \delta_1$) by introducing the *shift function*

$$F(x) = \frac{1}{2}(S(x) + S(\delta - x)).$$

Note that $F(x)$ is symmetric around the axis $x = \frac{\delta}{2}$ with $F(0) = F(\delta)$ and $F(\delta_1) = F(\delta - \delta_1)$. We are interested in the peaks of $F(x)$ for $x \geq \frac{\delta}{2}$.

Define $x_1 = \delta = M(P_2) - M(P_1)$ and $y_1 = F(\delta) = F(0)$. Let $y_2 = F(x_2), y_3 = F(x_3), \ldots, y_k = F(x_k)$ be the $k - 1$ largest peaks of $F(x)$ for $x \geq \delta/2$ and $x \neq \delta$. Define

$$SIM_k(S_1, S_2) = \sum_{i=1}^{k} y_i$$

as an estimate of the similarity between spectra S_1 and S_2 under the assumption that the corresponding peptides are k mutations apart. SIM_k is usually the overall height of k highest peaks of the shift function. For example, $SIM_1(S_1, S_2) = y_1$ is an upper bound for the number of shared peaks between S_1 and "mutated" S_2 if $k = 1$ mutation in P_2 is allowed.

Although spectral convolution helps to identify mutated peptides, it has a serious limitation which is described below.

Let

$$S = \{10, 20, 30, 40, 50, 60, 70, 80, 90, 100\}$$

be a spectrum of peptide P, and assume for simplicity that P produces only b-ions. Let

$$S' = \{10, 20, 30, 40, 50, 55, 65, 75, 85, 95\}$$

and

$$S'' = \{10, 15, 30, 35, 50, 55, 70, 75, 90, 95\}$$

be two theoretical spectra corresponding to peptides P' and P'' from the database. Which peptide (P' or P'') fits spectrum S the best? The shared peaks count does not allow one to answer this question, since both S' and S'' have five peaks in common with S. Moreover, the spectral convolution also does not answer this question, since both $S \ominus S'$ and $S \ominus S''$ (and corresponding shift functions) reveal strong peaks of the same height at 0 and 5. This suggests that both P' and P'' can be obtained from P by a single mutation with mass difference 5. However, a more careful analysis shows that although this mutation can be realized for P' by introducing a shift 5 after mass 50, it cannot be realized for P''. The major difference between S' and S'' is that the matching positions in S' come in clumps while the matching positions in S'' don't. Below we describe the spectral alignment approach, which addresses this problem.

11.9 Spectral Alignment

Let $A = \{a_1, \ldots, a_n\}$ be an ordered set of natural numbers $a_1 < a_2 \ldots < a_n$. A *shift* Δ_i transforms A into $\{a_1, \ldots a_{i-1}, a_i + \Delta_i, \ldots, a_n + \Delta_i\}$. We consider only the shifts that do not change the order of elements, i.e., the shifts with $\Delta_i \geq a_{i-1} - a_i$. Given sets $A = \{a_1, \ldots, a_n\}$ and $B = \{b_1, \ldots, b_m\}$, we want to find a series of k shifts of A that make A and B as similar as possible. The *k-similarity* $D(k)$ between sets A and B is defined as the maximum number of elements in common between these sets after k shifts. For example, a shift -5_6 transforms

$$S = \{10, 20, 30, 40, 50, 60, 70, 80, 90, 100\}$$

into

$$S' = \{10, 20, 30, 40, 50, 55, 65, 75, 85, 95\},$$

and therefore $D(1) = 10$ for these sets. The set

$$S'' = \{10, 15, 30, 35, 50, 55, 70, 75, 90, 95\}$$

has five elements in common with S (the same as S') but there is no shift transforming S into S'', and $D(1) = 6$. Below we describe a dynamic programming algorithm for computing $D(k)$.

Define a *spectral product* $A \otimes B$ as an $a_n \times b_m$ two-dimensional matrix with nm 1s corresponding to all pairs of indices (a_i, b_j) and remaining elements being

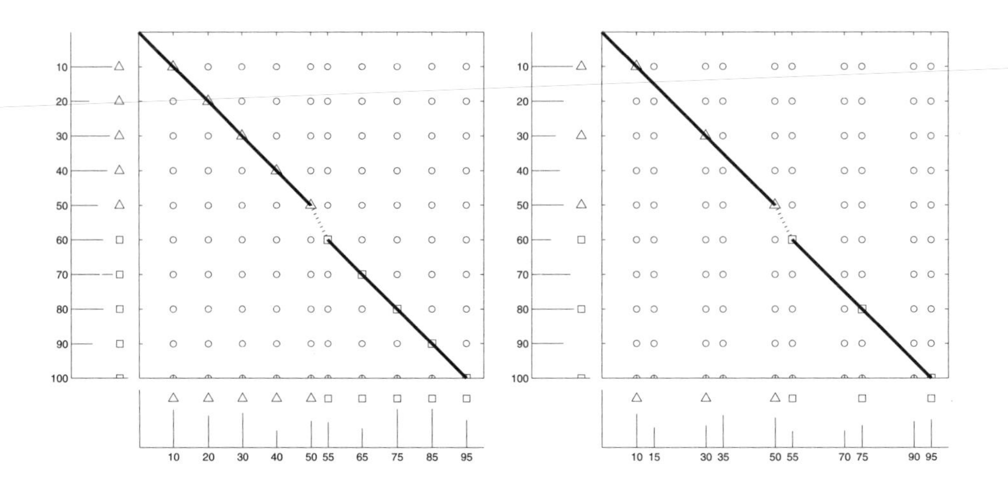

Figure 11.7: Spectrum S can be transformed into S' by a single mutation and $D(1) = 10$ (left matrix). Spectrum S cannot be transformed into S'' by a single mutation and $D(1) = 6$ (right matrix).

zeroes. The number of 1s at the main diagonal of this matrix describes the shared peaks count between spectra A and B, or in other words, 0-similarity between A and B. Figure 11.7 shows the spectral products $S \otimes S'$ and $S \otimes S''$ for the example from the previous section. In both cases the number of 1s on the main diagonal is the same, and $D(0) = 5$. The δ-shifted peaks count is the number of 1s on the diagonal $(i, i + \delta)$. The limitation of the shift function is that it considers diagonals separately without combining them into feasible mutation scenarios. *k-similarity* between spectra is defined as the maximum number of 1s on a path through the spectral matrix that uses at most $k + 1$ diagonals, and *k-optimal spectral alignment* is defined as a path using these $k + 1$ diagonals. For example, 1-similarity is defined by the maximum number of 1s on a path through this matrix that uses at most two diagonals. Figure 11.7 reveals that the notion of 1-similarity allows one to find out that S is closer to S' than to S'', since in the first case the 2-diagonal path covers 10 ones (left matrix), versus 6 in the second case (right matrix). Figure 11.8 illustrates that the spectral alignment allows one to detect more and more subtle similarities between spectra by increasing k. Below we describe a dynamic programming algorithm for spectral alignment.

Let A_i and B_j be the i-prefix of A and j-prefix of B, correspondingly. Define $D_{ij}(k)$ as the k-similarity between A_i and B_j such that the last elements of A_i and B_j are matched. In other words, $D_{ij}(k)$ is the maximum number of 1s on a path to (a_i, b_j) that uses at most $k+1$ diagonals. We say that (i', j') and (i, j) are *co-diagonal* if $a_i - a_{i'} = b_j - b_{j'}$ and that $(i', j') < (i, j)$ if $i' < i$ and $j' < j$. To take care of the initial conditions, we introduce a fictitious element $(0, 0)$ with $D_{0,0}(k) = 0$ and assume that $(0, 0)$ is co-diagonal with any other (i, j). The dynamic programming recurrency for $D_{ij}(k)$ is

$$D_{ij}(k) = \max_{(i',j')<(i,j)} \begin{cases} D_{i'j'}(k) + 1, & \text{if } (i', j') \text{ and } (i, j) \text{ are co-diagonal} \\ D_{i'j'}(k-1) + 1, & \text{otherwise.} \end{cases}$$

The k-similarity between A and B is given by $D(k) = \max_{ij} D_{ij}(k)$.

The described dynamic programming algorithm for spectral alignment is rather slow (running time $O(n^4 k)$ for n-element spectra), and below we describe an $O(n^2 k)$ algorithm for solving this problem. Define $diag(i, j)$ as the maximal co-diagonal pair of (i, j) such that $diag(i, j) < (i, j)$. In other words, $diag(i, j)$ is the position of the previous 1 on the same diagonal as (a_i, b_j) or $(0, 0)$ if such a position does not exist. Define

$$M_{ij}(k) = max_{(i',j') \leq (i,j)} D_{i'j'}(k).$$

Then the recurrency for $D_{ij}(k)$ can be re-written as

$$D_{ij}(k) = \max \begin{cases} D_{diag(i,j)}(k) + 1, \\ M_{i-1,j-1}(k-1) + 1. \end{cases}$$

The recurrency for $M_{ij}(k)$ is given by

$$M_{ij}(k) = \max \begin{cases} D_{ij}(k) \\ M_{i-1,j}(k) \\ M_{i,j-1}(k) \end{cases}$$

The described transformation of the dynamic programming graph is achieved by introducing horizontal and vertical edges that provide switching between diagonals (Figure 11.9). The score of a path is the number of 1s on this path, while k corresponds to the number of switches (number of used diagonals minus 1).

11.10 Aligning Peptides Against Spectra

The simple description above hides many details that make the spectral alignment problem difficult. A spectrum is usually a combination of an increasing (N-terminal ions) and a decreasing (C-terminal ions) number series. These series form

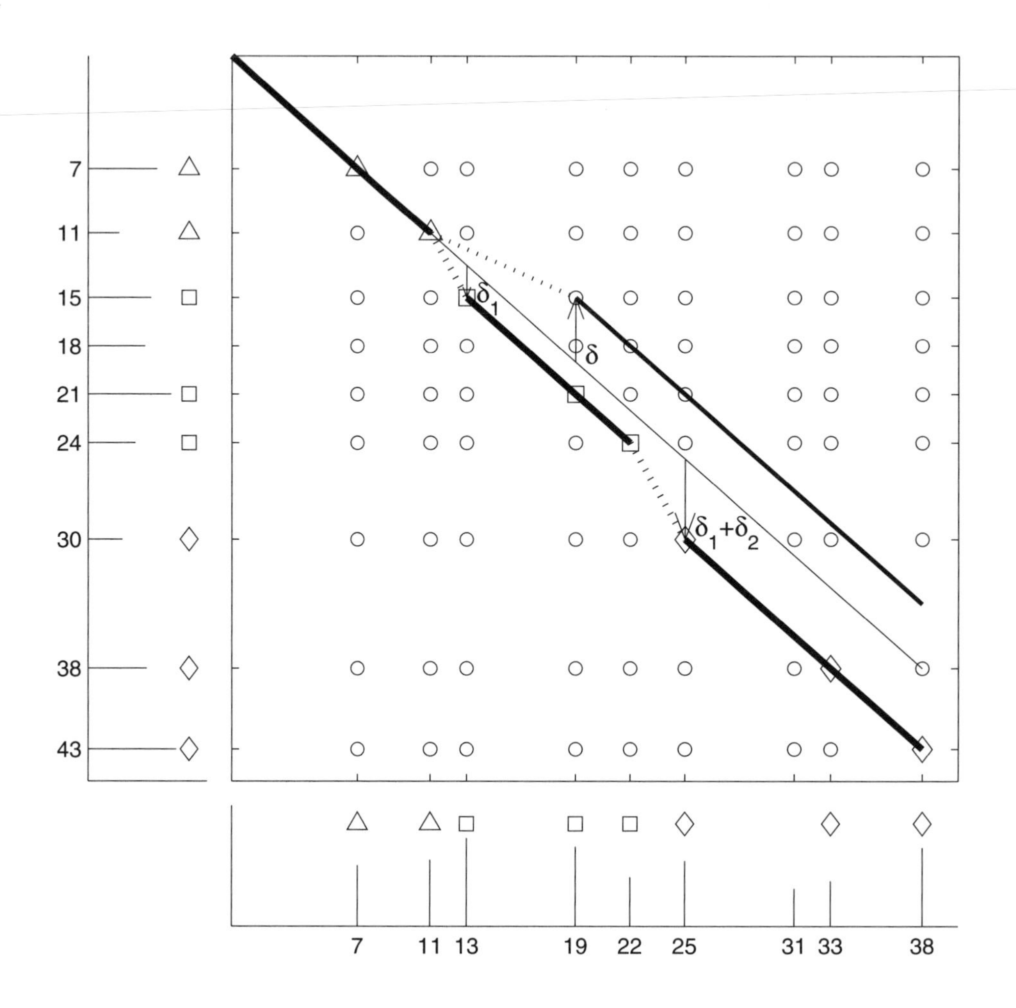

Figure 11.8: Aligning spectra. The shared peaks count reveals only $D(0) = 3$ matching peaks on the main diagonal, while spectral alignment reveals more hidden similarities between spectra ($D(1) = 5$ and $D(2) = 8$) and detects the corresponding mutations.

two diagonals in the spectral product $S \otimes S$, the main diagonal and the perpendicular diagonal, which corresponds to pairings of N-terminal and C-terminal ions. The described algorithm does not capture this detail and deals with the main diagonal only.

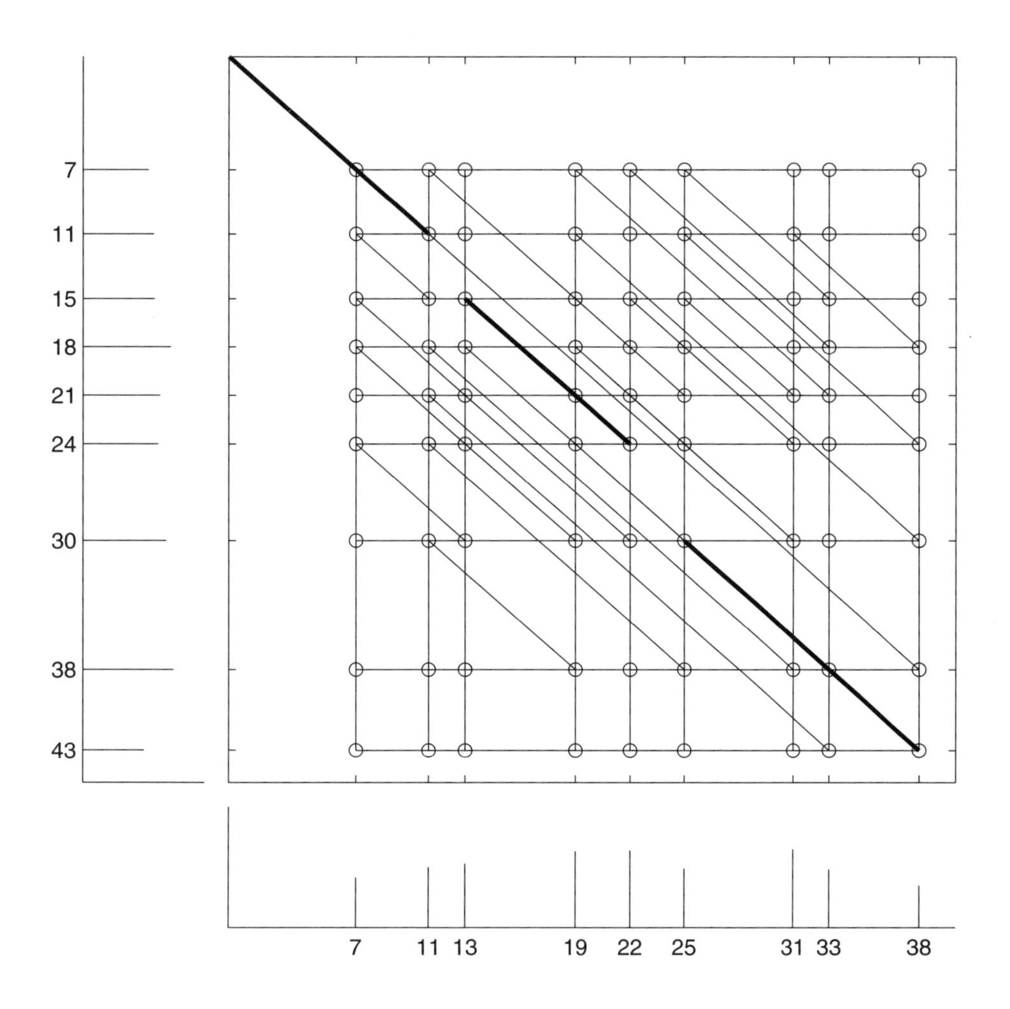

Figure 11.9: Modification of a dynamic programming graph leads to a fast spectral alignment algorithm.

To combine N-terminal and C-terminal series together, we work with $(S_1 \cup S_1^R) \otimes (S_2 \cup S_2^R)$, where S^R is the reversed spectrum of peptide P. This transformation creates a "b-version" for every y-ion and a "y-version" for every b-ion, thus increasing noise (since every noisy peak is propagated twice). Another and even more serious difficulty is that every 1 in the spectral product will have a reversed twin, and only one of these twins should be counted in the feasible spectral alignment. Ignoring this problem may lead to infeasible solutions that are sorted out in the anti-symmetric path approach (Dancik et al., 1999 [79]).

The described algorithm also does not capture all the relevant details in the case of the "sequence against the spectrum" comparison. In this case the horizontal and vertical arcs in the dynamic programming graph (Figure 11.9) are limited by the possible shifts reflecting mass differences between amino acids participating in the mutation. Let $P = p_1 \ldots p_n$ be a peptide that we compare with the spectrum $S = \{s_1, \ldots, s_m\}$. The d-prefix of spectrum S contains all peaks of S with $s_i \leq d$. We introduce a new variable $H_{i,d}(k)$ that describes the "best" transformation of the i-prefix of peptide P into the d-prefix of spectrum S with at most k substitutions in P_i. More precisely, $H_{i,d}(k)$ describes the number of 1s on the optimal path with k shifts between diagonals from $(0,0)$ to the position (i,d) of the properly defined "peptide versus spectrum" $P \otimes S$ matrix. For the sake of simplicity, assume that the theoretical spectrum of P contains only b-ions.

Let $H_{i,d}(k)$ be the "best" transformation of P_i into S_d with k substitutions (i.e., a transformation that uses the maximum number of 1s on a path with at most k shifts between diagonals). However, in this case, the jumps between diagonals are not arbitrary but are restricted by mass differences of mutated amino acids (or mass differences corresponding to chemical modifications). Below we describe the dynamic programming algorithm for the case of substitutions (deletions, insertions, and modifications lead to similar recurrencies). Define $x(d) = 1$ if $d \in S$ and $x(d) = 0$ otherwise. Then $H_{i,d}(k)$ is described by the following recurrency ($m(a)$ is the mass of amino acid a):

$$H_{i,d}(k) = \max \begin{cases} H_{i-1,d-m(p_i)}(k) + x(d) \\ \max_{a=1,20} H_{i,d-(m(a)-m(p_i))}(k-1) \end{cases}$$

11.11 Some Other Problems and Approaches

11.11.1 From proteomics to genomics

Mass-spectrometry is very successful for the identification of proteins whose genes are contained in sequence databases. However, *de novo* interpretation of tandem mass-spectra remained a complex and time-consuming problem and, as a result, mass-spectrometry has not yet had a significant impact for discovery of *new* genes. As recently as in 1995, Mann and Wilm (Mann and Wilm, 1995 [231]) remarked that they cannot find an example in the literature of a gene that was cloned on the basis of MS/MS-derived sequence information *alone*. This situation changed in the last 5 years, in particular, the reverse genetics studies of the catalytic subunit of telomerase (Lingner et al., 1997 [223]) required de novo sequencing of 14 peptides with further design of PCR primers for gene amplification.

11.11.2 Large-scale protein analysis

Complex protein mixtures can be separated by highly-resolving two-dimensional gel-electrophoresis. After separation, the identity of each "spot" (peptide) in 2-D gel is unknown and has to be identified by mass-spectrometry. This approach requires efficient methods for extracting resulting peptides from the gel and transferring them into mass-spectrometer.

Chapter 12

Problems

12.1 Introduction

Molecular biology has motivated many interesting combinatorial problems. A few years ago Pevzner and Waterman, 1995 [275] compiled a collection of 57 open problems in computational molecular biology. Just five years later a quarter of them have been solved. For this reason I don't explicitly say which of the problems below are open: they may be solved by the time you read this sentence.

12.2 Restriction Mapping

Suppose a DNA molecule is digested twice, by two restriction enzymes. The interval graph of resulting fragments is a *bipartite interval graph* (Waterman and Griggs, 1986 [361]).

Problem 12.1 *Design an efficient algorithm to recognize a bipartite interval graph.*

Problem 12.2 *Let v be a vertex of even degree in a balanced colored graph. Prove that $d(v)$ edges incident to v can be partitioned into $d(v)/2$ pairs such that edges in the same pair have different colors.*

Let $P = x_1 \ldots x_m$ be a path in an l-colored balanced graph $G(V, E)$. A color c is *critical* for P if (i) it is different from the color of the last edge (x_{m-1}, x_m) in P and (ii) it is the most frequent color among the edges of $E \setminus EP$ incident to x_m (EP denotes the edge set of path P). The edges of the set $E \setminus EP$ incident to x_m and having a critical color are called the *critical* edges. A path is called critical if it is obtained by choosing a critical edge at every step.

Problem 12.3 *Show how to use critical paths for constructing alternating Eulerian cycles.*

The following problem asks for an analog of the BEST theorem for bicolored graphs.

Problem 12.4 *Find the number of alternating Eulerian cycles in a bicolored Eulerian graph.*

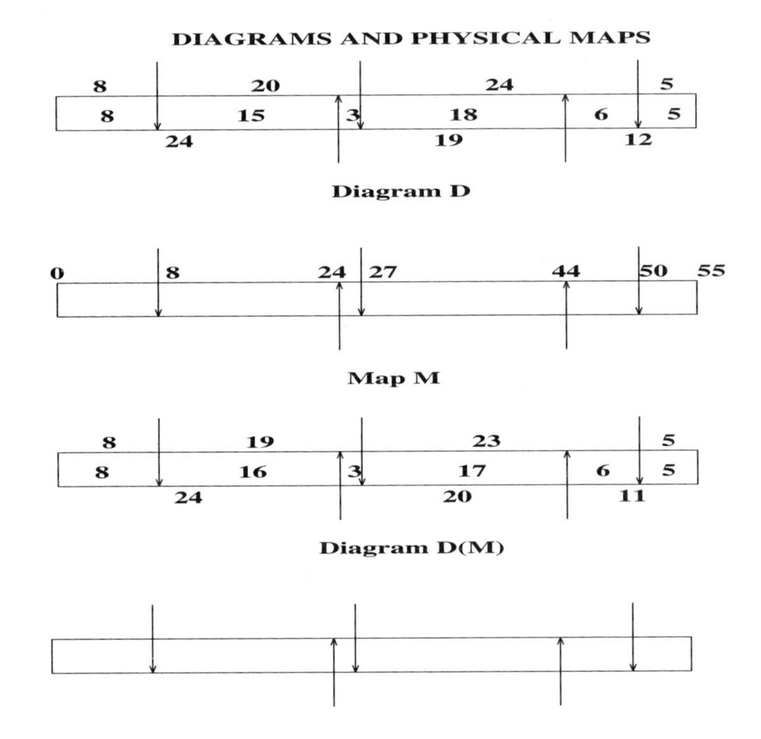

Figure 12.1: Diagrams and physical maps.

Most algorithms for the Double Digest Problem are based on generation and analysis of hypotheses about the order of restriction fragments in a physical map. Each such hypothesis corresponds to a *mapping diagram* D showing the order of sites and fragment lengths (Figure 12.1). Note that the coordinates of the sites are not shown in the mapping diagram. A physical map M provides information about both the order and the coordinates of the sites. Every physical map corresponds to a diagram $D(M)$ (Figure 12.1) with the lengths of the fragments corresponding to the distances between sites. The opposite is not true; not every diagram corresponds to a physical map. The question then arises of how to construct a physical map that best approximates the diagram. Let $D = (d_1, \ldots, d_n)$ and $M = (m_1, \ldots, m_n)$ be the lengths of all the fragments in diagrams D and

$D(M)$ given in the same order. The distance between diagram D and map M is

$$d(D, M) = \max_{i=1,n} |d_i - m_i|.$$

For example, in Figure 12.1, $D = (8, 20, 24, 5, 24, 19, 12, 8, 15, 3, 18, 6, 5)$, $M = (8, 19, 23, 5, 24, 20, 11, 8, 16, 3, 17, 6, 5)$, and $d(D, M) = 1$. The *diagram adjustment problem* is to find a map within a shortest distance from a diagram:

Problem 12.5 *Given a diagram D, find a map M minimizing $d(D, M)$.*

Problem 12.6 *Given two maps from the same equivalence class, find a shortest series of cassette transformations to transform one map into another.*

A generalization of the Double Digest Problem is to have three enzymes A, B, and C and to get experimental data about the lengths of the fragments in *single* digestions A, B, and C, *double* digestions AB, BC, and CA, and *triple* digestion ABC. Such an experiment leads to the *Multiple Digest Problem.*

Problem 12.7 *Find a physical map of three enzymes A, B, and C given provided six sets of experimental data (digestions A, B, C, AB, BC, CA, and ABC).*

Problem 12.8 *Characterize cassette transformations of multiple maps (three or more enzymes).*

The Rosenblatt and Seymour, 1982 [289] PDP algorithm is pseudo-polynomial.

Problem 12.9 *Does a polynomial algorithm exist for PDP?*

Skiena et al., 1990 [314] proved that the maximum number $H(n)$ of strongly homometric sets on n elements is bounded by $\frac{1}{2}n^{0.6309} \leq H(n) \leq \frac{1}{2}n^{2.5}$. The upper bound seems to be rather pessimistic.

Problem 12.10 *Derive tighter bounds for $H(n)$.*

Problem 12.11 *Prove that every 5-point set is reconstructible.*

Problem 12.12 *Design an efficient algorithm for the Probed Partial Digest Problem.*

Problem 12.13 *Derive upper and lower bounds for the maximum number of solutions for an n-site Probed Partial Digest Problem.*

The input to the optical mapping problem is a 0-1 $n \times m$ matrix $S = (s_{ij})$ where each row corresponds to a DNA molecule (straight or reversed), each column corresponds to a position in that molecule, and $s_{ij} = 1$ if there is a cut in position j of molecule i. The goal is to reverse the orientation of a subset of molecules (subset of rows in S) and to declare a subset of the t columns "real cut sites" so that the number of ones in cut site columns is maximized (Karp and Shamir, 1998 [190]).

A naive approach to this problem is to find t columns with a large proportion of ones and declare them potential cut sites. However, in this approach every real site will have a reversed twin. Let $w(i, j)$ be the number of molecules with both cut sites i and j present (in either direct or reverse orientation). In a different approach, a graph on vertices $\{1, \ldots, m\}$ is constructed and two vertices are connected by an edge (i, j) of weight $w(i, j)$.

Problem 12.14 *Establish a connection between optical mapping and the anti-symmetric longest path problem.*

12.3 Map Assembly

Problem 12.15 *Find the number of different interleavings of n clones.*

An interleaving of n clones can be specified by a sequence of integers $a_1 \ldots a_n$, where a_i is the number of clones that end before the clone i starts. For example, the interleaving of nine clones in Figure 3.2 corresponds to the sequence 000112267. Not every sequence of integers $a_1 \ldots a_n$ specifies a valid interleaving. Moreover, if a probe specifies a run $[i, j]$ of clones, this run implies the inequalities $a_j \leq i - 1$ (clones j and i overlap) and $a_{j+1} \geq i - 1$ (clones $i - 1$ and $j + 1$ do not overlap).

Problem 12.16 *Formulate the Shortest Covering String Problem with a given order of clones as a linear integer program, and solve it.*

Problem 12.17 *Let k be the maximal number of pairwise disjoint intervals in a collection of intervals $\mathcal{I}$ on a line. Prove that there exist k points on the line such that each interval in $\mathcal{I}$ contains at least one of these points.*

The intersection graphs corresponding to collections of arcs on a circle are called *circular-arc graphs*. If a collection of arcs on a circle does not cover some point x on the circle, then the *circular-arc graph* of this collection is an interval graph (cut the circle at x and straighten it out).

Problem 12.18 *Design an efficient algorithm to recognize circular-arc graphs.*

Problem 12.19 *If N random clones of length L are chosen from a genome of length G, the expected fraction of the genome represented in these clones is approximately $1 - e^c$, where $c = \frac{NL}{G}$ is the coverage.*

In *cosmid contig mapping* (Zhang et al., 1994 [376]), clone overlap information is generated from hybridization data. A set of clones is placed on a filter for colony hybridization, and the filter is probed with a clone that has been radioactively labeled. This process produces overlap information as to which probes overlap with other clones. If only a subset of clones are used as probes, overlap information is not available between clones that are not probes. A graph is a *probe interval graph* if its vertex set can be partitioned into subsets P (clones used as probes) and N (clones not used as probes), with an interval assigned to each vertex, such that two vertices are adjacent if and only if 'their corresponding intervals overlap and at least one of the vertices is in P (McMorris et al., 1998 [235]).

Problem 12.20 *Devise an algorithm to recognize probe interval graphs.*

Inner Product Mapping (Perlin and Chakravarti, 1993 [263]) is a clone mapping approach that probes a set of radiation hybrid clones twice, once with BACs and once with STSs, to obtain a map of BACs relative to STSs. Inner Product Mapping requires two sets of data: a hybrid screening matrix with STSs and a hybrid screening matrix with BACs.

Problem 12.21 *Given hybrid screening matrices with STSs and BACs, construct a map of BACs relative to STSs.*

Elements $\pi_i \pi_j \pi_k$ for $1 \le i < j < k \le n$ form an *ordered triple* in a permutation $\pi = \pi_1 \ldots \pi_n$. Let $\Phi(\pi)$ be a collection of all $\binom{n}{3}$ ordered triples for π. Radiation hybrid mapping motivates the following problem:

Problem 12.22 *Given an arbitrary set T of ordered triples of an n-element set, find a permutation π such that $T \subset \Phi(\pi)$.*

Elements $\pi_i \pi_j \pi_k$ form an *unordered triple* if either $1 \le i < j < k \le n$ or $1 \le k < j < i \le n$. Let $\Theta(\pi)$ be a collection of all unordered triples for π.

Problem 12.23 *Given an arbitrary set T of unordered triples of an n-element set, find a permutation π such that $T \subset \Theta(\phi)$.*

12.4 Sequencing

The simplest heuristic for the Shortest Superstring Problem is the GREEDY algorithm: repeatedly merge a pair of strings with maximum overlap until only one string remains. Tarhio and Ukkonen, 1988 [333] defined the *compression* of an SSP algorithm as the number of symbols saved by this algorithm compared to plainly concatenating all the strings.

Problem 12.24 *Prove that the GREEDY algorithm achieves at least $\frac{1}{2}$ the compression of an optimal superstring, i.e.,* $\frac{GREEDY\ compression}{optimal\ compression} \geq \frac{1}{2}$.

A performance guarantee with respect to compression does not imply a performance guarantee with respect to length. Since an example for which the approximation ratio of GREEDY is worse than 2 is unknown, Blum et al., 1994 [37] raised the following conjecture:

Problem 12.25 *Prove that GREEDY achieves a performance guarantee of 2.*

Let $\mathcal{S} = \{s_1, \ldots, s_n\}$ be a collection of linear strings and $\mathcal{C} = \{c_1, \ldots, c_m\}$ be a collection of circular strings. We say that $\mathcal{C}$ is a *circulation* of $\mathcal{S}$ if every s_i is contained in one of the circular strings c_j for $1 \leq j \leq m$. The length of circulation $|\mathcal{C}| = \sum_{j=1}^{m} |c_j|$ is the overall length of the strings from $\mathcal{C}$.

Problem 12.26 *Find the shortest circulation for a collection of linear strings.*

Let $P = \{s_1, \ldots, s_m\}$ be a set of *positive* strings and $N = \{t_1, \ldots, t_k\}$ be a set of *negative* strings. We assume that no negative string t_i is a substring of any positive string s_j. A *consistent superstring* for (P, N) is a string s such that each s_i is a substring of s and no t_i is a substring of s (Jiang and Li, 1994 [180]).

Problem 12.27 *Design an approximation algorithm for the shortest consistent superstring problem.*

Short fragments read by sequencing contain errors that lead to complications in fragment assembly. Introducing errors leads to *shortest k-approximate superstring problem* (Jiang and Li, 1996 [181]):

Problem 12.28 *Given a set S of strings, find a shortest string w such that each string x in S matches some substring of w with at most k errors.*

Suppose we are given a set $\mathcal{S}$ of n *random* strings of a fixed size l in an A-letter alphabet. If n is large (of the order of A^l), the length of the shortest common superstring $E(\mathcal{S})$ for the set $\mathcal{S}$ is of the order n. If n is small, $E(\mathcal{S})$ is of the order $n \cdot l$.

Problem 12.29 *Estimate $E(\mathcal{S})$ as a function of l, n, and A.*

Given strings s and t, $overlap(s,t)$ is the length of a maximal prefix of s that matches a suffix of t.

Problem 12.30 *Given a collection of i.i.d. strings $\{s_1, \ldots, s_n\}$ of fixed length, find the distribution of* $\max\{overlap(s_i, s_j) : 1 \leq i \neq j \leq n\}$.

Given a collection of reads $\mathcal{S} = \{s_1, \ldots, s_n\}$ from a DNA sequencing project and an integer l, the spectrum of $\mathcal{S}$ is a set $\mathcal{S}_l$ of all l-tuples from strings $s_1, \ldots, s_n$. Let Δ be an upper bound on the number of errors in each DNA read. One approach to the fragment assembly problem is to first correct the errors in each read and then assemble the correct reads into contigs. This motivates the following problem:

Problem 12.31 *Given $\mathcal{S}$, Δ, and l, introduce up to Δ corrections in each string in $\mathcal{S}$ in such a way that $|\mathcal{S}_l|$ is minimized.*

12.5 DNA Arrays

Problem 12.32 *Prove that the information-theoretic lower bound for the number of probes needed to unambiguously reconstruct an arbitrary string of length n is $\Omega(n)$.*

Problem 12.33 *Devise an algorithm for SBH sequence reconstruction by data with errors (false positive and false negative).*

Given two strings with the same l-tuple composition, the *distance* between them is the length of the *shortest* series of transpositions transforming one into the other.

Problem 12.34 *Devise an algorithm for computing or approximating the distance between two strings with the same l-tuple composition.*

Problem 12.35 *What is the largest distance between two n-letter strings with the same l-tuple composition?*

Continuous stacking hybridization assumes an additional hybridization of short probes that continuously extends duplexes formed by the target DNA fragment and the probes from the sequencing array. In this approach, additional hybridization with a short m-tuple on the array $C(k)$ provides information about some $(k+m)$-tuples contained in the sequence.

Problem 12.36 *Given a spectrum S that does not provide an unambiguous SBH reconstruction, determine a minimum number of continuous stacking hybridization experiments needed to unambiguously reconstruct the target fragment.*

Problem 12.37 *Reconstruct the sequence of a DNA fragment given a spectrum S and the results of additional continuous stacking hybridizations.*

A *reduced binary array* of order l is an array with memory $2 \cdot 2^l$ composed of all multiprobes of two kinds

$$\underbrace{\{W,S\},\{W,S\},\ldots,\{W,S\}}_{l} \text{ and } \underbrace{\{R,Y\},\{R,Y\},\ldots,\{R,Y\}}_{l}.$$

For example, for $l = 2$, the reduced binary array consists of 8 multiprobes: WW, WS, SW, SS, RR, RY, YR, and YY. Each multiprobe is a pool of four dinucleotides.

Problem 12.38 *Compute the branching probability of reduced binary arrays and compare it with the branching probability of uniform arrays.*

We call an array *k-bounded* if all probes in the array have a length of at most k.

Problem 12.39 *Given m and k, find a k-bounded array with m (multi)probes maximizing the resolving power.*

An easier version of the previous problem is the following:

Problem 12.40 *Do binary arrays $C_{bin}(k-1)$ provide asymptotically the best resolving power among all k-bounded arrays?*

Although binary arrays provide better resolving power than uniform arrays, an efficient algorithm for reconstruction of a DNA fragment from its spectrum over a binary array is still unknown.

Problem 12.41 *Does there exist a polynomial algorithm for SBH sequence reconstruction by binary arrays?*

The proof of theorem 5.6 considers the switch s_{i+1} at x, y and the switch s_i at z, u. The proof implicitly assumes that the sets of vertices $\{x, y\}$ and $\{z, u\}$ do not overlap.

Problem 12.42 *Adjust the proof for the case in which these sets do overlap.*

Let $\mathcal{L}$ be a set of strings. Consider a set E (*strings precedence data*) of all ordered pairs of different l-tuples such that they occur in some string from $\mathcal{L}$ in the given order but at arbitrary distances. Chetverin and Kramer, 1993 [66] suggested the *nested strand hybridization* approach to DNA arrays, which results in the following problem (Rubinov and Gelfand, 1995 [291]):

Problem 12.43 *Reconstruct $\mathcal{L}$ given strings precedence data.*

Two-dimensional Gray codes are optimal for minimizing the border length of uniform DNA arrays. However, for an *arbitrary* array, the problem of minimizing the overall border lengths of photolithographic masks remains unsolved.

Problem 12.44 *Find an arrangement of probes in an (arbitrary) array minimizing the overall border lengths of masks for photolithographic array design.*

12.6 Sequence Comparison

Fitting a sequence V into a sequence W is a problem of finding a substring W' of W that maximizes the score of alignment $s(V, W')$ among all substrings of W.

Problem 12.45 *Devise an efficient algorithm for the fitting problem.*

Problem 12.46 *Estimate the number of different alignments between two n-letter sequences.*

Problem 12.47 *Devise an algorithm to compute the number of distinct optimal alignments between a pair of strings.*

Problem 12.48 *For a pair of strings $v_1 \ldots v_n$ and $w_1 \ldots w_m$ show how to compute, for each (i, j), the value of the best alignment that aligns the character v_i with character w_j.*

Problem 12.49 *For a parameter k, compute the global alignment between two strings, subject to the constraint that the alignment contains at most k gaps (blocks of consecutive indels).*

The *k-difference alignment* problem is to find the best global alignment of strings V and W containing at most k mismatches, insertions, or deletions.

Problem 12.50 *Devise an $O(kn)$ k-difference global alignment algorithm for comparing two n-letter strings.*

Chao et al., 1992 [64] described an algorithm for aligning two sequences within a diagonal band that requires only $O(nw)$ computation time and $O(n)$ space, where n is the length of the sequences and w is the width of the band.

Problem 12.51 *Can an alignment within a diagonal band be implemented with $O(w)$ space?*

Myers and Miller, 1988 [246] studied the following:

Problem 12.52 *Develop a linear-space version of global sequence alignment with affine gap penalties.*

Huang and Miller, 1991 [169] studied the following:

Problem 12.53 *Develop a linear-space version of the local alignment algorithm.*

In the space-efficient approach to sequence alignment, the original problem of size $n \times m$ is reduced to two subproblems of sizes $i \times \frac{m}{2}$ and $(n-i) \times \frac{m}{2}$. In a fast parallel implementation of sequence alignment, it is desirable to have a *balanced partitioning* that breaks the original problem into sub-problems of equal sizes.

Problem 12.54 *Design a space-efficient alignment algorithm with balanced partitioning.*

The score of a local alignment is not normalized over the length of the matching region. As a result, a local alignment with score 100 and length 100 will be chosen over a local alignment with score 99 and length 10, although the latter one is probably more important biologically. To reflect the length of the local alignment in scoring, the score $s(I, J)$ of local alignment involving substrings I and J may be adjusted by dividing $s(I, J)$ by the total length of the aligned regions: $\frac{s(I,J)}{|I|+|J|}$. The *normalized local alignment* problem is to find substrings I and J that maximize $\frac{s(I,J)}{|I|+|J|}$ among all substrings I and J with $|I| + |J| \geq k$, where k is a threshold for the minimal overall length of I and J.

Problem 12.55 *Devise an algorithm for solving the normalized local alignment problem.*

A string X is called a *supersequence* of a string V if V is a subsequence of X.

Problem 12.56 *Given strings V and W, devise an algorithm to find the shortest supersequence for both V and W*

Let P be a pattern of length n, and let T be a text of length m. The *tandem repeat* problem is to find an interval in T that has the best global alignment with some tandem repeat of P. Let P^m be the concatenation of P with itself m times. The tandem repeat problem is equivalent to computing the local alignment between P^m and T, and the standard local alignment algorithm solves this problem in $O(nm^2)$ time.

Problem 12.57 *Find an approach that solves the tandem repeat problem in $O(nm)$ time.*

An alignment of circular strings is defined as an alignment of linear strings formed by cutting (linearizing) these circular strings at arbitrary positions.

Problem 12.58 *Find an optimal alignment (local and global) of circular strings.*

A local alignment between two different strings A and B finds a pair of substrings, one in A and the other in B, with maximum similarity. Suppose that we want to find a pair of (non-overlapping) substrings *within* string A with maximum similarity (*optimal inexact repeat problem*). Computing the local alignment between A and itself does not solve the problem, since the resulting alignment may correspond to overlapping substrings. This problem was studied by Miller (unpublished manuscript) and later by Kannan and Myers, 1996 [184] and Schmidt, 1998 [308].

Problem 12.59 *Devise an algorithm for the optimal inexact repeat problem.*

Schoniger and Waterman, 1992 [310] extended the range of edit operations in sequence alignment to include *non-overlapping* reversals in addition to insertions, deletions, and substitutions.

Problem 12.60 *Devise an efficient algorithm for sequence alignment with non-overlapping reversals.*

In the *chimeric alignment* problem (Komatsoulis and Waterman, 1997 [205]), a string V and a database of strings $\mathcal{W} = \{W_1, \dots W_N\}$ are given, and the problem is to find $\max_{1 \leq i \neq j \leq N} s(V, W_i \oplus W_j)$ where $W_i \oplus W_j$ is the concatenation of W_i and W_j.

Problem 12.61 *Devise an efficient algorithm for the chimeric alignment problem.*

Problem 12.62 *Show that in any permutation of n distinct integers, there is either an increasing subsequence of length at least $\sqrt{n}$ or a decreasing subsequence of length at least $\sqrt{n}$.*

A *Catalan* sequence is a permutation $x_1 \dots x_{2n}$ of n ones and n zeros such that for any prefix $x_1 \dots x_i$, the number of ones is at least as great as the number of zeros. The n-th *Catalan* number C_n is the number of such sequences.

Problem 12.63 *Prove the following:*

- *C_n is the number of standard Young tableaux with two rows of length n.*
- *C_n is the number of permutations $\pi \in S_n$ with a longest decreasing subsequence of length at most 2.*
- *C_n is the number of sequences of positive integers $1 \leq a_1 \leq a_2 \leq \ldots \leq a_n$ such that $a_i \leq i$ for all i.*

Problem 12.64 *Prove the recurrence $C_{n+1} = C_n C_o + C_{n-1} C_1 + \ldots + C_o C_n$.*

Problem 12.65 *Prove that the length of the longest decreasing subsequence of permutation π is the length of the first column of the Young tableau $P(\pi)$.*

A subsequence σ of permutation π is *k-increasing* if, as a set, it can be written as

$$\sigma = \sigma_1 \cup \sigma_2 \cup \ldots \cup \sigma_k$$

where any given σ_i is an increasing subsequence of π.

Problem 12.66 *Devise an algorithm to find longest k-increasing subsequences.*

Chang and Lampe [61] suggested an analog of the Sankoff-Mainville conjecture for the case of the edit distance $d(V, W)$ between n-letter i.i.d. strings V and W:

Problem 12.67 *For random i.i.d. n-letter strings in a k-letter alphabet,*

$$\frac{Expectation(d(V, W))}{n} = 1 - \frac{1}{\sqrt{k}} + o(\frac{1}{\sqrt{k}}).$$

Gusfield et al., 1994 [146] proved that the number of convex polygons in the parameter space decomposition for *global* alignment is bounded by $O(n^{2/3})$, where n is the length of the sequences. Fernandez-Baca et al., 1999 [102] studied the following:

Problem 12.68 *Generate a pair of sequences of length n that have an order of $\Omega(n^{2/3})$ regions in the decomposition of the parameter space.*

For a fixed-length alphabet, no examples of sequences with $\Omega(n^{2/3})$ regions in the parameter space decomposition are known.

Problem 12.69 *Improve the bound $O(n^{2/3})$ on the number of regions in space decomposition for global alignment in the case of a bounded alphabet.*

Problem 12.70 *Derive bounds for the expected number of regions in space decomposition for global alignment of two random sequences of length n.*

Problem 12.71 *Generalize bounds for the number of regions in space decomposition for the case of multiple alignment.*

Parameter space decomposition for local alignment usually contains more regions than parameter space decomposition for global alignment. Vingron and Waterman, 1994 [346] studied the links between the parametric sequence alignment and the phase transition. In this connection, it is interesting to study the parameter decomposition of logarithmic area.

Problem 12.72 *Derive bounds for the expected number of regions in space decomposition of logarithmic area for local alignment of two random sequences of length n.*

The Gusfield et al., 1994 [146] algorithm for parametric sequence alignment of two sequences runs in $O(R + E)$ time per region, where R is the number of regions in the parametric decomposition and E is the time needed to perform a single alignment. In the case of unweighted scoring schemes, $R = O(E)$, so the cost per region is $O(E)$. When one uses a weight matrix, little is known about R. Gusfield formulated the following:

Problem 12.73 *Estimate the number of regions in a convex decomposition in the case of weight matrices.*

Problem 12.74 *Devise a fast algorithm for space decomposition in the case of weight matrices.*

Since *energy* parameters for RNA folding are estimated with errors, it would be useful to study parametric RNA folding. For example, comparison of regions corresponding to cloverleafs for tRNA parameter space decomposition would provide an estimate of the accuracy of currently used RNA energy parameters.

Problem 12.75 *Develop an algorithm for parametric RNA folding.*

Let $S_n(\mu, \delta)$ be a random variable corresponding to the score (# matches – μ# mismatches – σ# indels) of the global alignment between two random i.i.d. strings of length n. Arratia and Waterman, 1994 [14] defined $a(\mu, \delta) = \lim_{n\to\infty} \frac{S_n(\mu,\delta)}{n}$ and demonstrated that $\{a = 0\} = \{(\mu, \delta) : a(\mu, \delta) = 0\}$ is a continuous phase transition curve.

Problem 12.76 *Characterize the curve $a(\mu, \delta) = 0$.*

12.7 Multiple Alignment

Problem 12.77 *Devise a space-efficient algorithm for multiple alignment.*

Problem 12.78 *Devise an algorithm that assembles multiple alignments from 3-way alignments.*

Problem 12.79 *Construct an example for which the Vingron-Argos matrix multiplication algorithm requires $\Omega(L)$ iterations, where L is the length of sequences.*

Jiang and Li, 1994 [180] formulated the following:

Problem 12.80 *Can shortest common supersequences (SCSs) and longest common supersequences (LCSs) on binary alphabets be approximated with a ratio better than 2?*

One could argue that the notion of NP-completeness is somewhat misleading for some computational biology problems, because it is insensitive to the limited parameter ranges that are often important in practice. For example, in many applications, we would be happy with efficient algorithms for multiple alignment with $k \leq 10$. What we currently have is the $O((2n)^k)$ dynamic programming algorithm. The NP-completeness of the multiple alignment problem tells us almost nothing about what to expect if we fix our attention on the range of $k \leq 10$. It could even be the case that there is a linear-time algorithm for every fixed value of k! For example, it would be entirely consistent with NP-completeness if the problem could be solved in time $O(2^k n)$.

The last decade has seen the development of algorithms that are particularly applicable to problems such as multiple alignment for fixed parameter ranges. We presently do not know whether the complexity obtained by dynamic programming is the "last word" on the complexity of the multiple alignment problem (Bodlaender et al., 1995 [38]). Mike Fellows formulated the following conjecture:

Problem 12.81 *The longest common subsequence problem for k sequences in a fixed-size alphabet can be solved in time $f(k)n^\alpha$ where α is independent of k.*

12.8 Finding Signals in DNA

Problem 12.82 *Describe a winning strategy for B in the best bet for simpletons.*

Problem 12.83 *Describe the best strategy for A in the best bet for simpletons (i.e., the strategy that minimizes losses).*

If a coin in the best bet for simpletons is biased (e.g., $p(0) > p(1)$), it makes sense for A to choose a word such as $0 \ldots 0$ to improve the odds.

Problem 12.84 *Study the best bet for simpletons with a biased coin. Does B still have an advantage over A in this case?*

Problem 12.85 *Derive the variance of the number of occurrences of a given word in the case of linear strings.*

Problem 12.86 *Devise an approximation algorithm for the consensus word problem.*

The Decoding Problem can be formulated as a longest path problem in a directed acyclic graph. This motivates a question about a space-efficient version of the Viterbi algorithm.

Problem 12.87 *Does there exist a linear-space algorithm for the decoding problem?*

12.9 Gene Prediction

The spliced alignment algorithm finds exons in genomic DNA by using a related protein as a template. What if a template is not a protein but other (uninterpreted) genomic DNA? In particular, can we use (unannotated) mouse genomic DNA to predict human genes?

Problem 12.88 *Generalize the spliced alignment algorithm for alignment of one genomic sequence against another.*

Problem 12.89 *Generalize the similarity-based approach to gene prediction for the case in which multiple similar proteins are available.*

Sze and Pevzner, 1997 [332] formulated the following:

Problem 12.90 *Modify the spliced alignment algorithm for finding suboptimal spliced alignments.*

The "Twenty Questions Game with a Liar" assumes that every answer in the game is false with probability p. Obviously, if $p = \frac{1}{2}$, the game is lost since the liar does not communicate any information to us.

Problem 12.91 *Design an efficient strategy for the "Twenty Questions Game with a Liar" that finds k unknown integers from the interval $[1, n]$ if the probability of a false answer is $p \neq \frac{1}{2}$.*

Problem 12.92 *Estimate the expected number of questions in the "Twenty Questions Game with a Liar" if the probability of a false answer is* $p \neq \frac{1}{2}$.

Problem 12.93 *Design experimental and computational protocols to find all alternatively spliced variants for a given genomic sequence.*

The observation that PCR-based queries can be used to test a potentially exponential number of hypotheses about splicing variants leads to a reformulation of the above problem.

Problem 12.94 *Given a graph* $G(V, E)$ *with a collection* $\mathcal{C}$ *of (unknown) paths, reconstruct* $\mathcal{C}$ *by asking the minimum number of queries of the form: "Does a collection* $\mathcal{C}$ *contain a path passing through vertices* v *and* w *from* G*?"*

Let S be a fixed set of probes, and let C be a DNA sequence. A fingerprint of C is a subset of probes from S that hybridize with C. Let G be a genomic sequence containing a gene represented by a cDNA clone C. Mironov and Pevzner, 1999 [241] studied the following fingerprint-based gene recognition problem:

Problem 12.95 *Given a genomic sequence* G *and the fingerprint of the corresponding cDNA clone* C*, predict a gene in* G *(i.e., predict all exons in* G*).*

12.10 Genome Rearrangements

Sorting by reversals corresponds to eliminating breakpoints. However, for some permutations (such as 563412), no reversal reduces the number of breakpoints. All three strips (maximal intervals without breakpoints) in 563412 are increasing.

Problem 12.96 *Prove that if an unsigned permutation has a decreasing strip, then there is a reversal that reduces the number of breakpoints by at least one.*

A 2-greedy algorithm for sorting π by reversals chooses reversals ρ and σ such that the number of breakpoints in $\pi \cdot \rho \cdot \sigma$ is minimal among all pairs of reversals.

Problem 12.97 *Prove that 2-greedy is a performance guarantee algorithm for sorting by reversals.*

In the case in which the sequence of genes contains duplications, sorting *permutations* by reversals is transformed into sorting *words* by reversals. For example, the shortest sequence of reversals to transform the word 43132143 into the word 42341314 involves two inversions: $4\mathbf{3132}143 \rightarrow 423\mathbf{13143} \rightarrow 42341314$.

Problem 12.98 *Devise a performance guarantee algorithm for sorting words by reversals.*

Kececioglu and Sankoff, 1994 [193] studied the bounds for the diameter $D(n)$ in the case of signed permutations and proved that $n-1 \leq D(n) \leq n$. They also conjectured the following:

Problem 12.99 *For signed circular permutations, $D(n) = n$ for sufficiently large n.*

Problem 12.100 *Characterize the set of "hard-to-sort" signed permutations on n elements such than $d(\pi) = D(n)$.*

Problem 12.101 *Improve the lower bound and derive an upper bound for the expected reversal distance.*

Problem 12.102 *Estimate the variance of reversal distance.*

Gates and Papadimitriou, 1979 [120] conjectured that a particular permutation on n elements requires at least $\frac{19}{16}n$ reversals to be sorted. Heydari and Sudborough, 1997 [161] disproved their conjecture by describing $\frac{18}{16}n + 2$ reversals sorting the Gates-Papadimitriou permutation.

Problem 12.103 *Find the prefix reversal diameter of the symmetric group.*

Genomes evolve not only by inversions but by *transpositions* as well. For a permutation π, a *transposition* $\rho(i,j,k)$ (defined for all $1 \leq i < j \leq n+1$ and all $1 \leq k \leq n+1$ such that $k \notin [i,j]$) "inserts" an interval $[i, j-1]$ of π between π_{k-1} and π_k. Thus $\rho(i,j,k)$ corresponds to the permutation

$$\begin{pmatrix} 1 \ \ldots \ i-1 & \boxed{\mathbf{i \ i{+}1 \ \ldots \ \ldots \ \ldots \ j{-}2 \ j{-}1}} & \boxed{\mathrm{j} \ \ldots \ \mathrm{k{-}1}} & k \ \ldots \ n \\ 1 \ \ldots \ i-1 & \boxed{\mathrm{j} \ \ldots \ \mathrm{k{-}1}} & \boxed{\mathbf{i \ i{+}1 \ \ldots \ \ldots \ \ldots \ j{-}2 \ j{-}1}} & k \ \ldots \ n \end{pmatrix}$$

Given permutations π and σ, the *transposition distance* is the length of the shortest series of transpositions $\rho_1, \rho_2, \ldots, \rho_t$ transforming π into $\pi \cdot \rho_1 \cdot \rho_2 \cdots \rho_t = \sigma$. *Sorting π by transpositions* is the problem of finding the transposition distance $d(\pi)$ between π and the identity permutation $\imath$. Bafna and Pevzner, 1998 [20] devised a 1.5 performance guarantee algorithm for sorting by transpositions and demonstrated that the transposition diameter $D_t(n)$ of the symmetric group is bounded by $\frac{n}{2} \leq D_t(n) \leq \frac{3n}{4}$.

Problem 12.104 *Find the transposition diameter of the symmetric group.*

The well-known variant of sorting by transpositions is sorting by transpositions $\rho(i, i+1, i+2)$ where the operation is an exchange of adjacent elements. A simple bubble-sort algorithm solves this problem for linear permutations. Solving the problem for circular permutations is more difficult.

Problem 12.105 *Design an algorithm for sorting circular permutations by exchanges of adjacent elements.*

Problem 12.106 *Every circular permutation can be sorted in* $2\lceil\frac{n-1}{2}\rceil\lfloor\frac{n-1}{2}\rfloor$ *exchanges of adjacent elements.*

We represent a circular permutation as elements $\pi_1 \ldots \pi_n$ equally spaced on a circle. Figure 12.2 presents circular permutations $\pi = \pi_1 \ldots \pi_n$ and $\sigma = \sigma_1 \ldots \sigma_n$ positioned on two concentric circles and n edges $e_1 \ldots e_n$ such that e_i joins element i in π with element i in σ. We call such a representation of π and σ an *embedding*, and we are interested in embeddings minimizing the number C of *crossing* edges. Edges in an embedding can be directed either clockwise or counterclockwise; notice that the overall number of crossing edges depends on the choice of directions. For example, the embedding in Figure 12.2a corresponds to $C = 2$, while the embedding in Figure 12.2b corresponds to $C = 3$. An n-mer *direction* vector $\mathbf{v} = v_1, \ldots v_n$ with $v_i \in \{+1, -1\}$ defines an embedding by directing an edge e_i clockwise if $v_i = +1$ and counterclockwise otherwise.

For convenience we choose the "twelve o'clock" vertex on a circle to represent a "starting" point of a circular permutation. Choosing an element r as a starting point of π defines a *rotation of* π. For the sake of simplicity, we assume that $\sigma = 1 \ldots n$ is the identity permutation and that the starting point of σ is 1.

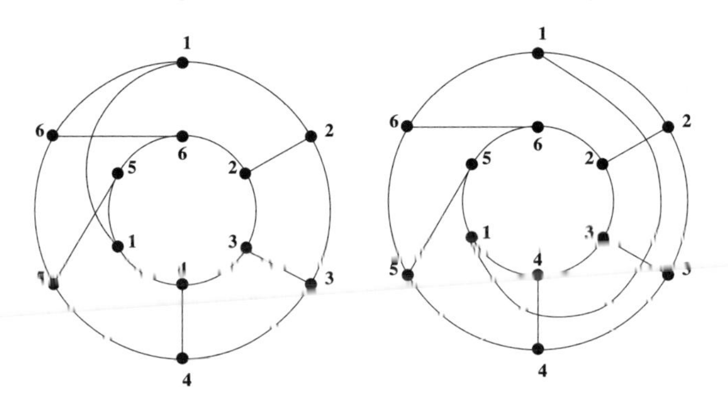

Figure 12.2: Crossing edges in embeddings.

Every rotation of π with r as a starting point and every vector $\mathbf{v}$ define an embedding with a number of *crossing* edges $C(r, \mathbf{v})$. Sankoff and Goldstein, 1989 [303] studied the following *optimal embedding* problem:

Problem 12.107 *Find*

$$\min_{r,\mathbf{v}} C(r, \mathbf{v}).$$

Let $d_{ij}(\pi)$ be the distance from element i to element j in permutation π counted clockwise. The *length* of a clockwise edge e_i is defined as $\hat{\imath} = (d_{1i}(\sigma) - d_{1i}(\pi))$ mod(n), while the length of a counterclockwise edge is defined as $\check{\imath} = n - \hat{\imath}$. For a rotation r, define a *canonical* direction vector $\mathbf{v}(r) = (v_1(r) \dots v_n(r))$ according to the rule that $v_i(r) = +1$ if clockwise edge e_i is shorter than counterclockwise edge e_i (i.e., if $\hat{\imath} < \check{\imath}$) and -1 otherwise.

Problem 12.108 *Prove*

$$\min_{r} C(r, \mathbf{v}(r)) \leq \lceil \frac{n-1}{2} \rceil \lfloor \frac{n-1}{2} \rfloor.$$

Problem 12.109 *Prove that for every $0 \leq r \leq n-1$, circular permutation π can be sorted in at most $C(r, \mathbf{v}(r))$ exchanges of adjacent elements.*

An $\mathcal{A}$-permutation on n elements is a permutation of $\{1, 2, \dots, n\}$ interspersed with letters from an alphabet $\mathcal{A}$. For example, 3aa1baa4a52b is an $\mathcal{A}$-permutation of 31452 with $\mathcal{A} = \{a, b\}$. A reversal of an $\mathcal{A}$-permutation is valid if its starting and ending elements coincide. An identity $\mathcal{A}$-permutation is a permutation in which $\{1, 2, \dots, n\}$ appear in order (with arbitrary assignment of elements from $\mathcal{A}$).

Problem 12.110 *Design a test deciding whether an $\mathcal{A}$-permutation can be sorted by valid reversals.*

Problem 12.111 *Design an algorithm for sorting $\mathcal{A}$-permutations by valid reversals.*

12.11 Computational Proteomics

Mass-spectrometry has become a source of new protein sequences, some of them previously unknown at the DNA level. This raises the problem of bridging gap between proteomics and genomics, i.e., problem of sequencing DNA based on information derived from large-scale MS/MS peptide sequencing. This problem is complicated since peptides sequenced by mass-spectrometry may come in short pieces with potential ambiguities (such as transposition of adjacent amino acids and wild cards).

Problem 12.112 *Given a set of peptides (with ambiguities) from a given protein, design experimental and computational protocols to find the genomic sequence corresponding to this protein.*

Problem 12.113 *Design an algorithm for searching peptides with ambiguities in protein databases.*

Consider a mixture of (unknown) proteins subject to complete digestion by a protease (e.g., trypsin). This results in a collection of peptides ranging in length from 10 to 20 amino acids, and the problem is to decide which peptides belong to the same proteins and to reconstruct the order of peptides in each of the proteins. The mass-spectrometer is capable of (partial) sequencing of all peptides in the mixture, but it is typically unknown which peptides come from the same protein and what the order is of peptides in the proteins. Protein sequencing of protein mixtures is a problem of assembling peptides into individual proteins.

Problem 12.114 *Devise experimental and computational protocols for sequencing protein mixtures by mass-spectrometry.*

The spectrum graph approach to *de novo* peptide sequencing does not take into account internal ions and multiple-charged ions.

Problem 12.115 *Devise a peptide sequencing algorithm taking into account internal and multiple-charged ions.*

Let $M(P)$ be the set of masses of all partial peptides of peptide P. Using digestion of P by different non-specific proteases, one can obtain a set of experimentally measured masses of partial peptides $M \subset M(P)$.

Problem 12.116 *Given a set of masses $M \subset M(P)$, reconstruct P.*

Accurate determination of the peptide parent mass is extremely important in *de novo* peptide sequencing. An error in parent mass leads to systematic errors in the masses of vertices for C-terminal ions, thus making peptide reconstruction difficult. In practice, the offsets between the real peptide masses (given by the sum of amino acids of a peptide) and experimentally observed parent masses are frequently so large that errors in peptide reconstruction become almost unavoidable.

Problem 12.117 *Given an MS/MS spectra (without parent mass), devise an algorithm that estimates the parent mass.*

Chapter 13

All You Need to Know about Molecular Biology

Well, not really, of course, see Lewin, 1999 [220] for an introduction.

DNA is a string in the four-letter alphabet of *nucleotides* A, T, G, and C. The entire DNA of a living organism is called its *genome*. Living organisms (such as humans) have trillions of cells, and each cell contains the same genome. DNA varies in length from a few million letters (bacteria) to a few billion letters (mammals). DNA forms a helix, but that is not really important for this book. What is more important is that DNA is usually double-stranded, with one strand being the *Watson-Crick complement* (T pairs with A and C pairs with G) of the other, like this:

$$\begin{array}{ccccccccc} A & T & G & C & T & C & A & G & G \\ | & | & | & | & | & | & | & | & | \\ T & A & C & G & A & G & T & C & C \end{array}$$

DNA makes the workhorses of the cell called *proteins*. Proteins are short strings in the *amino acid* 20-letter alphabet. The human genome makes roughly 100,000 proteins, with each protein a few hundred amino acids long. Bacteria make 500—1500 proteins, this is close to the lower bound for a living organism to survive. Proteins are made by fragments of DNA called *genes* that are roughly three times longer than the corresponding proteins. Why three? Because every three nucleotides in the DNA alphabet code one letter in the protein alphabet of amino acids. There are $4^3 = 64$ triplets (*codons*), and the question arises why nature needs so many combinations to code 20 amino acids. Well, genetic code (Figure 13.1) is redundant, not to mention that there exist *Stop* codons signaling the end of protein.

Biologists divide the world of organisms into *eukaryotes* (whose DNA is enclosed into a nucleus) and *prokaryotes*. A eukaryotic genome is usually not a single string (as in prokaryotes), but rather a set of strings called *chromosomes*. For our purposes, the major difference to remember between prokaryotes and eukaryotes is that in prokaryotes genes are continuous strings, while they are broken into pieces (called *exons*) in eukaryotes. Human genes may be broken into as many as 50 exons, separated by seemingly meaningless pieces called *introns*.

A gene broken into many pieces still has to produce the corresponding protein. To accomplish this, cells have to cut off the introns and concatenate all the exons together. This is done in *mRNA*, an intermediary molecule similar to short, single-stranded DNA, in a process called *transcription*. There are signals in DNA to start transcription that are called *promoters*. The protein-synthesizing machinery then *translates* codons in mRNA into a string of amino acids (protein). In the laboratory, mRNA can also be used as a template to make a complementary copy called *cDNA* that is identical to the original gene with cut-out introns.

Second position

First position	T	C	A	G
T	TTT **PHE**	TCT **SER**	TAT **TYR**	TGT **CYS**
T	TTC **PHE**	TCC **SER**	TAC **TYR**	TGC **CYS**
T	TTA **LEU**	TCA **SER**	TAA **Stop**	TGA **Stop**
T	TTG **LEU**	TCG **SER**	TAG **Stop**	TGG **TRP**
C	CTT **LEU**	CCT **PRO**	CAT **HIS**	CGT **ARG**
C	CTC **LEU**	CCC **PRO**	CAC **HIS**	CGC **ARG**
C	CTA **LEU**	CCA **PRO**	CAA **GLN**	CGA **ARG**
C	CTG **LEU**	CCG **PRO**	CAG **GLN**	CGG **ARG**
A	ATT **ILE**	ACT **THR**	AAT **ASN**	AGT **SER**
A	ATC **ILE**	ACC **THR**	AAC **ASN**	AGC **SER**
A	ATA **ILE**	ACA **THR**	AAA **LYS**	AGA **ARG**
A	ATG **MET**	ACG **THR**	AAG **LYS**	AGG **ARG**
G	GTT **VAL**	GCT **ALA**	GAT **ASP**	GGT **GLY**
G	GTC **VAL**	GCC **ALA**	GAC **ASP**	GGC **GLY**
G	GTA **VAL**	GCA **ALA**	GAA **GLU**	GGA **GLY**
G	GTG **VAL**	GCG **ALA**	GAG **GLU**	GGG **GLY**

Figure 13.1: Genetic code.

Over the years biologists have learned how to make many things with DNA. They have also learned how to copy DNA in large quantities for further study. One way to do this, *PCR* (polymerase chain reaction), is the Gutenberg printing press of DNA. PCR amplifies a short (100 to 500-nucleotide) DNA fragment and produces a large number of identical strings. To use PCR, one has to know a pair of short (20 to 30-letter) strings flanking the area of interest and design two

PCR primers, synthetic DNA fragments identical to these strings. Why do we need a large number of short identical DNA fragments? From a computer science perspective, having the same string in 10^{18} copies does not mean much; it does not increase the amount of information. It means a lot to biologists however, since most biological experiments require using a lot of strings. For example, PCR can be used to detect the existence of a certain DNA fragment in a DNA sample.

Another way to copy DNA is to *clone* it. In contrast to PCR, cloning does not require any prior information about flanking primers. However, in cloning, biologists do not have control over what fragment of DNA they amplify. The process usually starts with breaking DNA into small pieces. To study an individual piece, biologists obtain many identical copies of each piece by *cloning* the pieces. Cloning incorporates a fragment of DNA into a *cloning vector.* A cloning vector is a DNA molecule (usually originated from a virus or DNA of a higher organism) into which another DNA fragment can be inserted. In this operation, the cloning vector producing an does not lose its ability for self-replication. Vectors introduce foreign DNA into host cells (such as bacteria) where they can be reproduced in large quantities. The self-replication process creates a large number of copies of the fragment, thus enabling its structure to be investigated. A fragment reproduced in this way is called a *clone.* Biologists can make *clone libraries* consisting of thousands of clones (each representing a short, randomly chosen DNA fragment) from the same DNA molecule.

Restriction enzymes are molecular scissors that cut DNA at every occurrence of certain words. For example, the *Bam*HI restriction enzyme cuts DNA into *restriction fragments* at every occurrence of GGATCC. Proteins also can be cut into short fragments (called *peptides*) by another type of scissors, called *proteases.*

The process of joining two complementary DNA strands into a double-stranded molecule is called *hybridization.* Hybridization of a short *probe* complementary to a known DNA fragment can be used to detect the presence of this DNA fragment. A probe is a short, single-stranded, fluorescently labeled DNA fragment that is used to detect whether a complementary sequence is present in a DNA sample. Why do we need to fluorescently label the probe? If a probe hybridizes to a DNA fragment, then we can detect this using a spectroscopic detector.

Gel-electrophoresis is a technique that allows biologists to measure the size of DNA fragments without sequencing them. DNA is a negatively charged molecule that migrates toward a positive pole in the electric field. The speed of migration is a function of fragment size, and therefore, measurement of the migration distances allows biologists to estimate the sizes of DNA fragments.

Bibliography

[1] A.V. Aho and M.J. Corasick. Efficient string matching: an aid to bibliographic search. *Communication of ACM,* 18:333–340, 1975.

[2] D.J. Aldous and P. Diaconis. Hammersley's interacting particle process and longest increasing subsequences. *Probability Theory and Related Fields,* 103:199–213, 1995.

[3] F. Alizadeh, R.M. Karp, L.A. Newberg, and D.K. Weisser. Physical mapping of chromosomes: A combinatorial problem in molecular biology. *Algorithmica,* 13:52–76, 1995.

[4] F. Alizadeh, R.M. Karp, D.K. Weisser, and G. Zweig. Physical mapping of chromosomes using unique probes. *Journal of Computational Biology,* 2:159–184, 1995.

[5] S. Altschul, W. Gish, W. Miller, E. Myers, and J. Lipman. Basic local alignment search tool. *Journal of Molecular Biology,* 215:403–410, 1990.

[6] S.F. Altschul. Amino acid substitution matrices from an information theoretic perspective. *Journal of Molecular Biology,* 219:555–565, 1991.

[7] S.F. Altschul, T.L. Madden, A.A. Schaffer, J. Zhang, Z. Zhang, W. Miller, and D.J. Lipman. Gapped Blast and Psi-Blast: a new generation of protein database search programs. *Nucleic Acids Research,* 25:3389–3402, 1997.

[8] T.S. Anantharaman, B. Mishra, and D.C. Schwartz. Genomics via optical mapping. II: Ordered restriction maps. *Journal of Computational Biology,* 4:91–118, 1997.

[9] A. Apostolico. Improving the worst-case performance of the Hunt-Szymanski strategy for the longest common subsequence of two strings. *Information Processing Letters,* 23:63–69, 1986.

[10] A. Apostolico and F. Preparata. Data structures and algorithms for the string statistics problem. *Algorithmica*, 15:481–494, 1996.

[11] R. Arratia, E.S. Lander, S. Tavare, and M.S. Waterman. Genomic mapping by anchoring random clones: a mathematical analysis. *Genomics*, 11:806–827, 1991.

[12] R. Arratia, D. Martin, G. Reinert, and M.S. Waterman. Poisson process approximation for sequence repeats, and sequencing by hybridization. *Journal of Computational Biology*, 3:425–464, 1996.

[13] R. Arratia and M.S. Waterman. The Erdös-Rényi strong law for pattern matching with a given proportion of mismatches. *Annals of Probability*, 17:1152–1169, 1989.

[14] R. Arratia and M.S. Waterman. A phase transition for the score in matching random sequences allowing deletions. *Annals of Applied Probability*, 4:200–225, 1994.

[15] R. Baer and P. Brock. Natural sorting over permutation spaces. *Math. Comp.*, 22:385–410, 1968.

[16] R.A. Baeza-Yates and G.H. Gonnet. A new approach to text searching. In *Proceedings of the Twelfth Annual International ACM SIGIR Conference on Research and Development in Information Retrieval*, pages 168–175, Cambridge, Massachussets, 1989.

[17] R.A. Baeza-Yates and C.H. Perleberg. Fast and practical approximate string matching. In *Third Annual Symposium on Combinatorial Pattern Matching*, volume 644 of *Lecture Notes in Computer Science*, pages 185–192, Tucson, Arizona, April/May 1992. Springer-Verlag.

[18] V. Bafna, E.L. Lawler, and P.A. Pevzner. Approximation algorithms for multiple sequence alignment. *Theoretical Computer Science*, 182:233–244, 1997.

[19] V. Bafna and P.A. Pevzner. Genome rearrangements and sorting by reversals. *SIAM Journal on Computing*, 25:272–289, 1996.

[20] V. Bafna and P.A. Pevzner. Sorting by transpositions. *SIAM Journal on Discrete Mathematics*, 11:224–240, 1998.

[21] J. Baik, P.A. Deift, and K. Johansson. On the distribution of the length of the longest subsequence of random permutations. *Journal of the American Mathematical Society*, 12:1119–1178, 1999.

[22] W. Bains. Multan: a program to align multiple DNA sequences. *Nucleic Acids Research*, 14:159–177, 1986.

[23] W. Bains and G. Smith. A novel method for nucleic acid sequence determination. *Journal of Theoretical Biology*, 135:303–307, 1988.

[24] P. Baldi and S. Brunak. *Bioinformatics: The Machine Learning Approach*. The MIT Press, 1997.

[25] E. Barillot, B. Lacroix, and D. Cohen. Theoretical analysis of library screening using an N-dimensional pooling strategy. *Nucleic Acids Research*, 19:6241–6247, 1991.

[26] C. Bartels. Fast algorithm for peptide sequencing by mass spectroscopy. *Biomedical and Environmental Mass Spectrometry*, 19:363–368, 1990.

[27] G.J. Barton and M.J.E. Sternberg. A strategy for the rapid multiple alignment of protein sequences. *Journal of Molecular Biology*, 198:327–337, 1987.

[28] A. Baxevanis and B.F. Ouellette. *Bioinformatics: A Practical Guide to the Analysis of Genes and Proteins*. Wiley-Interscience, 1998.

[29] R. Bellman. *Dynamic Programming*. Princeton University Press, 1957.

[30] G. Benson. Sequence alignment with tandem duplication. In S. Istrail, P.A. Pevzner, and M.S. Waterman, editors, *Proceedings of the First Annual International Conference on Computational Molecular Biology (RECOMB-97)*, pages 27–36, Santa Fe, New Mexico, January 1997. ACM Press.

[31] G. Benson. An algorithm for finding tandem repeats of unspecified pattern size. In S. Istrail, P.A. Pevzner, and M.S. Waterman, editors, *Proceedings of the Second Annual International Conference on Computational Molecular Biology (RECOMB-98)*, pages 20–29, New York, New York, March 1998. ACM Press.

[32] S.M. Berget, C. Moore, and P.A. Sharp. Spliced segments at the 5' terminus of adenovirus 2 late mRNA. *Proceedings of the National Academy of Sciences USA*, 74:3171–3175, 1977.

[33] P. Berman and S. Hannenhalli. Fast sorting by reversal. In *Seventh Annual Symposium on Combinatorial Pattern Matching*, volume 1075 of *Lecture Notes in Computer Science*, pages 168–185, Laguna Beach, California, June 1996. Springer-Verlag.

[34] P. Berman, Z. Zhang, Y.I. Wolf, E.V. Koonin, and W. Miller. Winnowing sequences from a database search. In S. Istrail, P.A. Pevzner, and M.S. Waterman, editors, *Proceedings of the Third Annual International Conference on Computational Molecular Biology (RECOMB-99)*, pages 50–58, New York, New York, March 1999. ACM Press.

[35] K. Biemann and H.A. Scoble. Characterization of tandem mass spectrometry of structural modifications in proteins. *Science*, 237:992–998, 1987.

[36] B.E. Blaisdell. A measure of the similarity of sets of sequences not requiring sequence alignment. *Proceedings of the National Academy of Sciences USA*, 16:5169–5174, 1988.

[37] A. Blum, T. Jiang, M. Li, J. Tromp, and M. Yannakakis. Linear approximation of shortest superstrings. *Journal of the ACM*, 41:630–647, 1994.

[38] H.L. Bodlaender, R.G. Downey, M.R. Fellows, and H.T. Wareham. The parameterized complexity of sequence alignment and consensus. *Theoretical Computer Science*, 147:31–54, 1995.

[39] M. Boehnke, K. Lange, and D.R. Cox. Statistical methods for multipoint radiation hybrid mapping. *American Journal of Human Genetics*, 49:1174–1188, 1991.

[40] K.S. Booth and G.S. Leuker. Testing for the consecutive ones property, interval graphs, and graph planarity using PQ-tree algorithms. *Journal of Computer and System Sciences*, 13:335–379, 1976.

[41] P. Bork and T.J. Gibson. Applying motif and profile searches. *Methods in Enzymology*, 266:162–184, 1996.

[42] M. Borodovsky and J. McIninch. Recognition of genes in DNA sequences with ambiguities. *BioSystems*, 30:161–171, 1993.

[43] M. Yu. Borodovsky, Yu.A. Sprizhitsky, E.I. Golovanov, and A.A. Alexandrov. Statistical features in the *E. coli* genome functional domains primary structure III. Computer recognition of protein coding regions. *Molecular Biology*, 20:1144–1150, 1986.

[44] D. Botstein, R.L. White, M. Skolnick, and R.W. Davis. Construction of a genetic linkage map in man using restriction fragment length polymorphisms. *American Journal of Human Genetics*, 32:314–331, 1980.

[45] R.S. Boyer and J.S. Moore. A fast string searching algorithm. *Communication of ACM*, 20:762–772, 1977.

[46] A. Brazma, I. Jonassen, I. Eidhammer, and D. Gilbert. Approaches to the automatic discovery of patterns in biosequences. *Journal of Computational Biology*, 5:279–305, 1998.

[47] V. Brendel, J.S. Beckman, and E.N. Trifonov. Linguistics of nucleotide sequences: morphology and comparison of vocabularies. *Journal of Biomolecular Structure and Dynamics*, 4:11–21, 1986.

[48] D. Breslauer, T. Jiang, and Z. Jiang. Rotations of periodic strings and short superstrings. *Journal of Algorithms*, 24:340–353, 1997.

[49] N. Broude, T. Sano, C. Smith, and C. Cantor. Enhanced DNA sequencing by hybridization. *Proceedings of the National Academy of Sciences USA*, 91:3072–3076, 1994.

[50] S. Brunak, J. Engelbrecht, and S. Knudsen. Prediction of human mRNA donor and acceptor sites from the DNA sequence. *Journal of Molecular Biology*, 220:49–65, 1991.

[51] W.J. Bruno, E. Knill, D.J. Balding, D.C. Bruce, N.A. Doggett, W.W. Sawhill, R.L. Stallings, C.C. Whittaker, and D.C. Torney. Efficient pooling designs for library screening. *Genomics*, 26:21–30, 1995.

[52] R. Bundschuh and T. Hwa. An analytic study of the phase transition line in local sequence alignment with gaps. In S. Istrail, P.A. Pevzner, and M.S. Waterman, editors, *Proceedings of the Third Annual International Conference on Computational Molecular Biology (RECOMB-99)*, pages 70–76, Lyon, France, April 1999. ACM Press.

[53] C. Burge, A.M. Campbell, and S. Karlin. Over- and under-representation of short oligonucleotides in DNA sequences. *Proceedings of the National Academy of Sciences USA*, 89:1358–1362, 1992.

[54] C. Burge and S. Karlin. Prediction of complete gene structures in human genomic DNA. *Journal of Molecular Biology*, 268:78–94, 1997.

[55] S. Burkhardt, A. Crauser, P. Ferragina, H.-P. Lenhof, E. Rivals, and M. Vingron. q-gram based database searching using a suffix array. In S. Istrail, P.A. Pevzner, and M.S. Waterman, editors, *Proceedings of the Third Annual International Conference on Computational Molecular Biology (RECOMB-99)*, pages 77–83, Lyon, France, April 1999. ACM Press.

[56] A. Caprara. Formulations and complexity of multiple sorting by reversals. In S. Istrail, P.A. Pevzner, and M.S. Waterman, editors, *Proceedings of the*

Third Annual International Conference on Computational Molecular Biology (RECOMB-99), pages 84–93, Lyon, France, April 1999. ACM Press.

[57] A. Caprara. Sorting by reversals is difficult. In S. Istrail, P.A. Pevzner, and M.S. Waterman, editors, *Proceedings of the First Annual International Conference on Computational Molecular Biology (RECOMB-97)*, pages 75–83, Santa Fe, New Mexico, January 1997. ACM Press.

[58] H. Carrillo and D. Lipman. The multiple sequence alignment problem in biology. *SIAM Journal on Applied Mathematics*, 48:1073–1082, 1988.

[59] R.P. Carstens, J.V. Eaton, H.R. Krigman, P.J. Walther, and M.A. Garcia-Blanco. Alternative splicing of fibroblast growth factor receptor 2 (FGF-R2) in human prostate cancer. *Oncogene*, 15:3059–3065, 1997.

[60] W.K. Cavenee, M.F. Hansen, M. Nordenskjold, E. Kock, I. Maumenee, J.A. Squire, R.A. Phillips, and B.L. Gallie. Genetic origin of mutations predisposing to retinoblastoma. *Science*, 228:501–503, 1985.

[61] W.I. Chang and J. Lampe. Theoretical and empirical comparisons of approximate string matching algorithms. In *Third Annual Symposium on Combinatorial Pattern Matching*, volume 644 of *Lecture Notes in Computer Science*, pages 175–184, Tucson, Arizona, April/May 1992. Springer-Verlag.

[62] W.I. Chang and E.L. Lawler. Sublinear approximate string matching and biological applications. *Algorithmica*, 12:327–344, 1994.

[63] K.M. Chao. Computing all suboptimal alignments in linear space. In *Fifth Annual Symposium on Combinatorial Pattern Matching*, volume 807 of *Lecture Notes in Computer Science*, pages 31–42, Asilomar, California, 1994. Springer-Verlag.

[64] K.M. Chao, W.R. Pearson, and W. Miller. Aligning two sequences within a specified diagonal band. *Computer Applications in Biosciences*, 8:481–487, 1992.

[65] M. Chee, R. Yang, E. Hubbel, A. Berno, X.C. Huang, D. Stern, J. Winkler, D.J. Lockhart, M.S. Morris, and S.P.A. Fodor. Accessing genetic information with high density DNA arrays. *Science*, 274:610–614, 1996.

[66] A. Chetverin and F. Kramer. Sequencing of pools of nucleic acids on oligonucleotide arrays. *BioSystems*, 30:215–232, 1993.

[67] L.T. Chow, R.E. Gelinas, T.R. Broker, and R.J. Roberts. An amazing sequence arrangement at the 5' ends of adenovirus 2 messenger RNA. *Cell*, 12:1–8, 1977.

[68] I. Chumakov, P. Rigault, S. Guillou, P. Ougen A. Billaut, G. Guasconi, P. Gervy, I. LeGall, P. Soularue, and L. Grinas et al. Continuum of overlapping clones spanning the entire human chromosome 21q. *Nature*, 359:380–387, 1992.

[69] G. Churchill. Stochastic models for heterogeneous DNA sequences. *Bulletin of Mathematical Biology*, 51:79–94, 1989.

[70] V. Chvátal and D. Sankoff. Longest common subsequences of two random sequences. *Journal of Applied Probability*, 12:306–315, 1975.

[71] V. Chvatal and D. Sankoff. An upper-bound techniques for lengths of common subsequences. In D. Sankoff and J.B. Kruskal, editors, *Time Warps, String Edits, and Macromolecules: The Theory and Practice of Sequence Comparison*, pages 353–357. Addison-Wesley, 1983.

[72] K.R. Clauser, P.R. Baker, and A.L. Burlingame. The role of accurate mass measurement (+/– 10ppm) in protein identification strategies employing MS or MS/MS and database searching. *Analytical Chemistry*, 71:2871–2882, 1999.

[73] F.S. Collins, M.L. Drumm, J.L. Cole, W.K. Lockwood, G.F. Vande Woude, and M.C. Iannuzzi. Construction of a general human chromosome jumping library, with application to cystic fibrosis. *Science*, 235:1046–1049, 1987.

[74] N.G. Copeland, N.A. Jenkins, D.J. Gilbert, J.T. Eppig, L.J. Maltals, J.C. Miller, W.F. Dietrich, A. Weaver, S.E. Lincoln, R.G. Steen, L.D. Steen, J.H. Nadeau, and E.S. Lander. A genetic linkage map of the mouse: Current applications and future prospects. *Science*, 262:57–65, 1993.

[75] T.H. Cormen, C.E. Leiserson, and R.L. Rivest. *Introduction to Algorithms*. The MIT Press, 1989.

[76] A. Coulson, J. Sulston, S. Brenner, and J. Karn. Toward a physical map of the genome of the nematode, *Caenorhabditis elegans*. *Proceedings of the National Academy of Sciences USA*, 83:7821–7825, 1986.

[77] D.R. Cox, M. Burmeister, E.R. Price, S. Kim, and R.M. Myers. Radiation hybrid mapping: a somatic cell genetic method for constructing high-resolution maps of mammalian chromosomes. *Science*, 250:245–250, 1990.

[78] E. Czabarka, G. Konjevod, M. Marathe, A. Percus, and D.C. Torney. Algorithms for optimizing production DNA sequencing. In *Proceedings of the Eleventh Annual ACM-SIAM Symposium on Discrete Algorithms (SODA 2000)*, pages 399–408, San Francisco, California, 2000. SIAM Press.

[79] V. Dancik, T. Addona, K. Clauser, J. Vath, and P.A. Pevzner. De novo peptide sequencing via tandem mass spectrometry. *Journal of Computational Biology*, 6:327–342, 1999.

[80] K.J. Danna, G.H. Sack, and D. Nathans. Studies of simian virus 40 DNA. VII. a cleavage map of the SV40 genome. *Journal of Molecular Biology*, 78:263–276, 1973.

[81] K.E. Davies, P.L. Pearson, P.S. Harper, J.M. Murray, T. O'Brien, M. Sarfarazi, and R. Williamson. Linkage analysis of two cloned DNA sequences flanking the Duchenne muscular dystrophy locus on the short arm of the human X chromosome. *Nucleic Acids Research*, 11:2303–2312, 1983.

[82] M.A. Dayhoff, R.M. Schwartz, and B.C. Orcutt. A model of evolutionary change in proteins. In *Atlas of Protein Sequence and Structure*, chapter 5, pages 345–352. 1978.

[83] J. Deken. Some limit results for longest common subsequences. *Discrete Mathematics*, 26:17–31, 1979.

[84] J. Deken. Probabilistic behavior of longest common subsequence length. In D. Sankoff and J.B. Kruskal, editors, *Time Warps, String Edits, and Macromolecules: The Theory and Practice of Sequence Comparison*, pages 359–362. Addison-Wesley, 1983.

[85] A. Dembo and S. Karlin. Strong limit theorem of empirical functions for large exceedances of partial sums of i.i.d. variables. *Annals of Probability*, 19:1737–1755, 1991.

[86] R.P. Dilworth. A decomposition theorem for partially ordered sets. *Annals of Mathematics*, 51:161–165, 1950.

[87] T. Dobzhansky and A.H. Sturtevant. Inversions in the chromosomes of *Drosophila pseudoobscura*. *Genetics*, 23:28–64, 1938.

[88] H. Donis-Keller, P. Green, C. Helms, S. Cartinhour, B. Weiffenbach, K. Stephens, T.P. Keith, D.W. Bowden, D.R. Smith, and E.S. Lander. A genetic linkage map of the human genome. *Cell*, 51:319–337, 1987.

[89] R.F. Doolittle, M.W. Hunkapiller, L.E. Hood, S.G. Devare, K.C. Robbins, S.A. Aaronson, and H.N. Antoniades. Simian sarcoma virus onc gene, v-sis, is derived from the gene (or genes) encoding a platelet-derived growth factor. *Science*, 221:275–277, 1983.

[90] R. Drmanac, S. Drmanac, Z. Strezoska, T. Paunesku, I. Labat, M. Zeremski, J. Snoddy, W.K. Funkhouser, B. Koop, and L. Hood. DNA sequence determination by hybridization: a strategy for efficient large-scale sequencing. *Science*, 260:1649–1652, 1993.

[91] R. Drmanac, I. Labat, I. Brukner, and R. Crkvenjakov. Sequencing of megabase plus DNA by hybridization: theory of the method. *Genomics*, 4:114–128, 1989.

[92] J. Dumas and J. Ninio. Efficient algorithms for folding and comparing nucleic acid sequences. *Nucleic Acids Research*, 10:197–206, 1982.

[93] R. Durbin, S. Eddy, A. Krogh, and G. Mitchinson. *Biological Sequence Analysis*. Cambridge University Press, 1998.

[94] M. Dyer, A. Frieze, and S. Suen. The probability of unique solutions of sequencing by hybridization. *Journal of Computational Biology*, 1:105–110, 1994.

[95] S.R. Eddy and R. Durbin. RNA sequence analysis using covariance models. *Nucleic Acids Research*, 22:2079–2088, 1994.

[96] N. El-Mabrouk, D. Bryant, and D. Sankoff. Reconstructing the pre-doubling genome. In S. Istrail, P.A. Pevzner, and M.S. Waterman, editors, *Proceedings of the Third Annual International Conference on Computational Molecular Biology (RECOMB-99)*, pages 154–163, Lyon, France, April 1999. ACM Press.

[97] J. Eng, A. McCormack, and J. Yates. An approach to correlate tandem mass spectral data of peptides with amino acid sequences in a protein database. *Journal of American Society for Mass Spectometry*, 5:976–989, 1994.

[98] G.A. Evans and K.A. Lewis. Physical mapping of complex genomes by cosmid multiplex analysis. *Proceedings of the National Academy of Sciences USA*, 86:5030–5034, 1989.

[99] W. Feldman and P.A. Pevzner. Gray code masks for sequencing by hybridization. *Genomics*, 23:233–235, 1994.

[100] D. Feng and R. Doolittle. Progressive sequence alignment as a prerequisite to correct phylogenetic trees. *Journal of Molecular Evolution*, 25:351–360, 1987.

[101] D. Fenyo, J. Qin, and B.T. Chait. Protein identification using mass spectrometric information. *Electrophoresis*, 19:998–1005, 1998.

[102] D. Fernandez-Baca, T. Seppalainen, and G. Slutzki. Bounds for parametric sequence comparison. In *Sixth International Symposium on String Processing and Information Retrieval*, pages 55–62, Cancun, Mexico, September 1999. IEEE Computer Society.

[103] J. Fernández-de Cossío, J. Gonzales, and V. Besada. A computer program to aid the sequencing of peptides in collision-activated decomposition experiments. *Computer Applications in Biosciences*, 11:427–434, 1995.

[104] J.W. Fickett. Recognition of protein coding regions in DNA sequences. *Nucleic Acids Research*, 10:5303–5318, 1982.

[105] J.W. Fickett. Finding genes by computer: the state of the art. *Trends in Genetics*, 12:316–320, 1996.

[106] J.W. Fickett and C.S. Tung. Assessment of protein coding measures. *Nucleic Acids Research*, 20:6441–6450, 1992.

[107] W.M. Fitch and T.F. Smith. Optimal sequence alignments. *Proceedings of the National Academy of Sciences USA*, 80:1382–1386, 1983.

[108] H. Fleischner. *Eulerian Graphs and Related Topics*. Elsevier Science Publishers, 1990.

[109] S.P.A. Fodor, R.P. Rava, X.C. Huang, A.C. Pease, C.P. Holmes, and C.L. Adams. Multiplex biochemical assays with biological chips. *Nature*, 364:555–556, 1993.

[110] S.P.A. Fodor, J.L. Read, M.S. Pirrung, L. Stryer, A.T. Lu, and D. Solas. Light-directed spatially addressable parallel chemical synthesis. *Science*, 251:767–773, 1991.

[111] S. Foote, D. Vollrath, A. Hilton, and D.C. Page. The human Y chromosome: overlapping DNA clones spanning the euchromatic region. *Science*, 258:60–66, 1992.

[112] D. Foulser and S. Karlin. Maximum success duration for a semi-markov process. *Stochastic Processes and their Applications*, 24:203–210, 1987.

[113] D. Frishman, A. Mironov, H.W. Mewes, and M.S. Gelfand. Combining diverse evidence for gene recognition in completely sequenced bacterial genomes. *Nucleic Acids Research*, 26:2941–2947, 1998.

[114] D.R. Fulkerson and O.A. Gross. Incidence matrices and interval graphs. *Pacific Journal of Mathematics*, 15:835–856, 1965.

[115] D.J. Galas, M. Eggert, and M.S. Waterman. Rigorous pattern-recognition methods for DNA sequences. Analysis of promoter sequences from Escherichia coli. *Journal of Molecular Biology*, 186:117–128, 1985.

[116] Z. Galil and R. Giancarlo. Speeding up dynamic programming with applications to molecular biology. *Theoretical Computer Science*, 64:107–118, 1989.

[117] J. Gallant, D. Maier, and J. Storer. On finding minimal length superstrings. *Journal of Computer and System Science*, 20:50–58, 1980.

[118] M. Gardner. On the paradoxial situations that arise from nontransitive relationships. *Scientific American*, pages 120–125, October 1974.

[119] M.R. Garey and D.S. Johnson. *Computers and Intractability: A Guide to the Theory of NP-Completeness*. W.H. Freeman and Co., 1979.

[120] W.H. Gates and C.H. Papadimitriou. Bounds for sorting by prefix reversals. *Discrete Mathematics*, 27:47–57, 1979.

[121] M.S. Gelfand. Computer prediction of exon-intron structure of mammalian pre-mRNAs. *Nucleic Acids Research*, 18:5865–5869, 1990.

[122] M.S. Gelfand. Statistical analysis and prediction of the exonic structure of human genes. *Journal of Molecular Evolution*, 35:239–252, 1992.

[123] M.S. Gelfand. Prediction of function in DNA sequence analysis. *Journal of Computational Biology*, 2:87–115, 1995.

[124] M.S. Gelfand and E.V. Koonin. Avoidance of palindromic words in bacterial and archaeal genomes: a close connection with restriction enzymes. *Nucleic Acids Research*, 25:2430–2439, 1997.

[125] M.S. Gelfand, A.A. Mironov, and P.A. Pevzner. Gene recognition via spliced sequence alignment. *Proceedings of the National Academy of Sciences USA*, 93:9061–9066, 1996.

[126] J.F. Gentleman and R.C. Mullin. The distribution of the frequency of occurrence of nucleotide subsequences, based on their overlap capability. *Biometrics*, 45:35–52, 1989.

[127] W. Gillett, J. Daues, L. Hanks, and R. Capra. Fragment collapsing and splitting while assembling high-resolution restriction maps. *Journal of Computational Biology*, 2:185–205, 1995.

[128] P.C. Gilmore and A.J. Hoffman. A characterization of comparability graphs and of interval graphs. *Canadian Journal of Mathematics*, 16:539–548, 1964.

[129] W. Gish and D.J. States. Identification of protein coding regions by database similarity search. *Nature Genetics*, 3:266–272, 1993.

[130] L. Goldstein and M.S. Waterman. Mapping DNA by stochastic relaxation. *Advances in Applied Mathematics*, 8:194–207, 1987.

[131] T.R. Golub, D.K. Slonim, P. Tamayo, C. Huard, M. Gaasenbeek, J.P. Mesirov, H. Coller, M.L. Loh, J.R. Downing, M.A. Caligiuri, C.D. Bloomfield, and E.S. Lander. Molecular classification of cancer: class discovery and class prediction by gene expression monitoring. *Science*, 286:531–537, 1999.

[132] M. Golumbic. *Algorithmic Graph Theory and Perfect Graphs*. Academic Press, 1980.

[133] G.H. Gonnet, M.A. Cohen, and S.A. Benner. Exhaustive matching of the entire protein sequence database. *Science*, 256:1443–1445, 1992.

[134] A. Gooley and N. Packer. The importance of co- and post-translational modifications in proteome projects. In W. Wilkins, K. Williams, R. Appel, and D. Hochstrasser, editors, *Proteome Research: New Frontiers in Functional Genomics*, pages 65–91. Springer-Verlag, 1997.

[135] O. Gotoh. Consistency of optimal sequence alignments. *Bulletin of Mathematical Biology*, 52:509–525, 1990.

[136] P. Green. Documentation for phrap. http://bozeman.mbt.washington.edu/phrap.docs/phrap.html.

[137] D.S. Greenberg and S. Istrail. Physical mapping by STS hybridization: algorithmic strategies and the challenge of software evaluation. *Journal of Computational Biology*, 2:219–273, 1995.

[138] M. Gribskov, J. Devereux, and R.R. Burgess. The codon preference plot: graphic analysis of protein coding sequences and prediction of gene expression. *Nucleic Acids Research*, 12:539–549, 1984.

[139] M. Gribskov, M. McLachlan, and D. Eisenberg. Profile analysis: detection of distantly related proteins. *Proceedings of the National Academy of Sciences USA*, 84:4355–4358, 1987.

[140] R. Grossi and F. Luccio. Simple and efficient string matching with k mismatches. *Information Processing Letters*, 33:113–120, 1989.

[141] L.J. Guibas and A.M. Odlyzko. String overlaps, pattern matching and non-transitive games. *Journal of Combinatorial Theory, Series A*, 30:183–208, 1981.

[142] R. Guigo, S. Knudsen, N. Drake, and T.F. Smith. Prediction of gene structure. *Journal of Molecular Biology*, 226:141–157, 1992.

[143] J.F. Gusella, N.S. Wexler, P.M. Conneally, S.L. Naylor, M.A. Anderson, R.E. Tanzi, P.C. Watkins, K. Ottina, M.R. Wallace, A.Y. Sakaguchi, A.B. Young, I. Shoulson, E. Bonilla, and J.B. Martin. A polymorphic DNA marker genetically linked to Huntington's disease. *Nature*, 306:234–238, 1983.

[144] D. Gusfield. Efficient methods for multiple sequence alignment with guaranteed error bounds. *Bulletin of Mathematical Biology*, 55:141–154, 1993.

[145] D. Gusfield. *Algorithms on Strings, Trees, and Sequences. Computer Science and Computational Biology*. Cambridge University Press, 1997.

[146] D. Gusfield, K. Balasubramanian, and D. Naor. Parametric optimization of sequence alignment. *Algorithmica*, 12:312–326, 1994.

[147] D. Gusfield, R. Karp, L. Wang, and P. Stelling. Graph traversals, genes and matroids: An efficient case of the travelling salesman problem. *Discrete Applied Mathematics*, 88:167–180, 1998.

[148] J.G. Hacia, J.B. Fan, O. Ryder, L. Jin, K. Edgemon, G. Ghandour, R.A. Mayer, B. Sun, L. Hsie, C.M. Robbins, L.C. Brody, D. Wang, E.S. Lander, R. Lipshutz, S.P. Fodor, and F.S. Collins. Determination of ancestral alleles for human single-nucleotide polymorphisms using high-density oligonucleotide arrays. *Nature Genetics*, 22:164–167, 1999.

[149] C.W. Hamm, W.E. Wilson, and D.J. Harvan. Peptide sequencing program. *Computer Applications in Biosciences*, 2:115–118, 1986.

[150] J.M. Hammersley. A few seedlings of research. In *Proceedings of the Sixth Berkeley Symposium on Mathematical Statististics and Probabilities*, pages 345–394, Berkeley, California, 1972.

[151] S. Hannenhalli. Polynomial algorithm for computing translocation distance between genomes. In *Sixth Annual Symposium on Combinatorial Pattern Matching*, volume 937 of *Lecture Notes in Computer Science*, pages 162–176, Helsinki, Finland, June 1995. Springer-Verlag.

[152] S. Hannenhalli, C. Chappey, E. Koonin, and P.A. Pevzner. Genome sequence comparison and scenarios for gene rearrangements: A test case. *Genomics*, 30:299–311, 1995.

[153] S. Hannenhalli and P.A. Pevzner. Transforming men into mice (polynomial algorithm for genomic distance problem). In *Proceedings of the 36th Annual IEEE Symposium on Foundations of Computer Science*, pages 581–592, Milwaukee, Wisconsin, 1995.

[154] S. Hannenhalli and P.A. Pevzner. Transforming cabbage into turnip (polynomial algorithm for sorting signed permutations by reversals). In *Proceedings of the 27th Annual ACM Symposium on the Theory of Computing*, pages 178–189, 1995 (full version appeared in Journal of ACM, 46: 1–27, 1999).

[155] S. Hannenhalli and P.A. Pevzner. To cut ... or not to cut (applications of comparative physical maps in molecular evolution). In *Seventh Anuual ACM-SIAM Symposium on Discrete Algorithms*, pages 304–313, Atlanta, Georgia, 1996.

[156] S. Hannenhalli, P.A. Pevzner, H. Lewis, S. Skeina, and W. Feldman. Positional sequencing by hybridization. *Computer Applications in Biosciences*, 12:19–24, 1996.

[157] W.S. Hayes and M. Borodovsky. How to interpret an anonymous bacterial genome: machine learning approach to gene identification. *Genome Research*, 8:1154–1171, 1998.

[158] S. Henikoff and J.G. Henikoff. Amino acid substitution matrices from protein blocks. *Proceedings of the National Academy of Sciences USA*, 89:10915–10919, 1992.

[159] G.Z. Hertz and G.D. Stormo. Identifying DNA and protein patterns with statistically significant alignments of multiple sequences. *Bioinformatics*, 15:563–577, 1999.

[160] N. Heuze, S. Olayat, N. Gutman, M.L. Zani, and Y. Courty. Molecular cloning and expression of an alternative hKLK3 transcript coding for a variant protein of prostate-specific antigen. *Cancer Research*, 59:2820–2824, 1999.

[161] M. H. Heydari and I. H. Sudborough. On the diameter of the pancake network. *Journal of Algorithms*, 25:67–94, 1997.

[162] D.G. Higgins, J.D. Thompson, and T.J. Gibson. Using CLUSTAL for multiple sequence alignments. *Methods in Enzymology*, 266:383–402, 1996.

[163] D.S. Hirschberg. A linear space algorithm for computing maximal common subsequences. *Communication of ACM*, 18:341–343, 1975.

[164] D.S. Hirschberg. Algorithms for the longest common subsequence problem. *Journal of ACM*, 24:664–675, 1977.

[165] J.D. Hoheisel, E. Maier, R. Mott, L. McCarthy, A.V. Grigoriev, L.C. Schalkwyk, D. Nizetic, F. Francis, and H. Lehrach. High resolution cosmid and P1 maps spanning the 14 Mb genome of the fission yeast *S. pombe*. *Cell*, 73:109–120, 1993.

[166] S. Hopper, R.S. Johnson, J.E. Vath, and K. Biemann. Glutaredoxin from rabbit bone marrow. *Journal of Biological Chemistry*, 264:20438–20447, 1989.

[167] Y. Hu, L.R. Tanzer, J. Cao, C.D. Geringer, and R.E. Moore. Use of long RT-PCR to characterize splice variant mRNAs. *Biotechniques*, 25:224–229, 1998.

[168] X. Huang, R.C. Hardison, and W. Miller. A space-efficient algorithm for local similarities. *Computer Applications in Biosciences*, 6:373–381, 1990.

[169] X. Huang and W. Miller. A time-efficient, linear-space local similarity algorithm. *Advances in Applied Mathematics*, 12:337–357, 1991.

[170] T.J. Hubbard, A.M. Lesk, and A. Tramontano. Gathering them into the fold. *Nature Structural Biology*, 4:313, 1996.

[171] E. Hubbell. Multiplex sequencing by hybridization. *Journal of Computational Biology*, 8, 2000.

[172] E. Hubbell and P.A. Pevzner. Fidelity probes for DNA arrays. In *Proceedings of the Seventh International Conference on Intelligent Systems for Molecular Biology*, pages 113–117, Heidelberg, Germany, August 1999. AAAI Press.

[173] T.J. Hudson, L.D. Stein, S.S. Gerety, J. Ma, A.B. Castle, J. Silva, D.K. Slonim, R. Baptista, L. Kruglyak, S.H. Xu, X. Hu, A.M.E. Colbert, C. Rosenberg, M.P. Reeve-Daly, S. Rozen, L. Hui, X. Wu, C. Vestergaard, K.M. Wilson, and J.S. Bae et al. An STS-based map of the human genome. *Science*, 270:1945–1954, 1995.

[174] J.W. Hunt and T.G. Szymanski. A fast algorithm for computing longest common subsequences. *Communication of ACM*, 20:350–353, 1977.

[175] R.M. Idury and M.S. Waterman. A new algorithm for DNA sequence assembly. *Journal of Computational Biology*, 2:291–306, 1995.

[176] C. Iseli, C.V. Jongeneel, and P. Bucher. ESTScan: a program for detecting, evaluating and reconstructing potential coding regions in EST sequences. In *Proceedings of the Seventh International Conference on Intelligent Systems for Molecular Biology*, pages 138–148, Heidelberg, Germany, August 6-10 1999. AAAI Press.

[177] A.G. Ivanov. Distinguishing an approximate word's inclusion on Turing machine in real time. *Izvestiia Academii Nauk SSSR, Series Math.*, 48:520–568, 1984.

[178] A. Jauch, J. Wienberg, R. Stanyon, N. Arnold, S. Tofanelli, T. Ishida, and T. Cremer. Reconstruction of genomic rearrangements in great apes gibbons by chromosome painting. *Proceedings of the National Academy of Sciences USA*, 89:8611–8615, 1992.

[179] T. Jiang and R.M. Karp. Mapping clones with a given ordering or interleaving. *Algorithmica*, 21:262–284, 1998.

[180] T. Jiang and M. Li. Approximating shortest superstrings with constraints. *Theoretical Computer Science*, 134:473–491, 1994.

[181] T. Jiang and M. Li. DNA sequencing and string learning. *Mathematical Systems Theory*, 29:387–405, 1996.

[182] R.J. Johnson and K. Biemann. Computer program (SEQPEP) to aid in the interpretation of high-energy collision tandem mass spectra of peptides. *Biomedical and Environmental Mass Spectrometry*, 18:945–957, 1989.

[183] Y.W. Kan and A.M. Dozy. Polymorphism of DNA sequence adjacent to human beta-globin structural gene: relationship to sickle mutation. *Proceedings of the National Academy of Sciences USA*, 75:5631–5635, 1978.

[184] S.K. Kannan and E.W. Myers. An algorithm for locating nonoverlapping regions of maximum alignment score. *SIAM Journal on Computing*, 25:648–662, 1996.

[185] H. Kaplan, R. Shamir, and R.E. Tarjan. Faster and simpler algorithm for sorting signed permutations by reversals. In *Proceedings of the Eighth Annual ACM-SIAM Symposium on Discrete Algorithms*, pages 344–351, New Orleans, Louisiana, January 1997.

[186] S. Karlin and S.F. Altschul. Methods for assessing the statistical significance of molecular sequence features by using general scoring schemes. *Proceedings of the National Academy of Sciences USA*, 87:2264–2268, 1990.

[187] S. Karlin and G. Ghandour. Multiple-alphabet amino acid sequence comparisons of the immunoglobulin kappa-chain constant domain. *Proceedings of the National Academy of Sciences USA*, 82:8597–8601, 1985.

[188] R.M. Karp, R.E. Miller, and A.L. Rosenberg. Rapid identification of repeated patterns in strings, trees and arrays. In *Proceedings of the Fourth Annual ACM Symposium on Theory of Computing*, pages 125–136, Denver, Colorado, May 1972.

[189] R.M. Karp and M.O. Rabin. Efficient randomized pattern-matching algorithms. *IBM Journal of Research and Development*, 31:249–260, 1987.

[190] R.M. Karp and R. Shamir. Algorithms for optical mapping. In S. Istrail, P.A. Pevzner, and M.S. Waterman, editors, *Proceedings of the Second Annual International Conference on Computational Molecular Biology (RECOMB-98)*, pages 117–124, New York, New York, March 1998. ACM Press.

[191] J. Kececioglu and R. Ravi. Of mice and men: Evolutionary distances between genomes under translocation. In *Proceedings of the 6th Annual ACM-SIAM Symposium on Discrete Algorithms*, pages 604–613, New York, New York, 1995.

[192] J. Kececioglu and D. Sankoff. Exact and approximation algorithms for the reversal distance between two permutations. In *Fourth Annual Symposium on Combinatorial Pattern Matching*, volume 684 of *Lecture Notes in Computer Science*, pages 87–105, Padova, Italy, 1993. Springer-Verlag.

[193] J. Kececioglu and D. Sankoff. Efficient bounds for oriented chromosome inversion distance. In *Fifth Annual Symposium on Combinatorial Pattern Matching*, volume 807 of *Lecture Notes in Computer Science*, pages 307–325, Asilomar, California, 1994. Springer-Verlag.

[194] J. Kececioglu and D. Sankoff. Exact and approximation algorithms for the inversion distance between two permutations. *Algorithmica*, 13:180–210, 1995.

[195] J.D. Kececioglu and E.W. Myers. Combinatorial algorithms for DNA sequence assembly. *Algorithmica*, 13:7–51, 1995.

[196] T.J. Kelly and H.O. Smith. A restriction enzyme from *Hemophilus influenzae*. II. Base sequence of the recognition site. *Journal of Molecular Biology*, 51:393–409, 1970.

[197] K. Khrapko, Y. Lysov, A. Khorlin, V. Shik, V. Florent'ev, and A. Mirzabekov. An oligonucleotide hybridization approach to DNA sequencing. *FEBS Letters*, 256:118–122, 1989.

[198] J.F.C. Kingman. Subadditive ergodic theory. *Annals of Probability*, 6:883–909, 1973.

[199] J. Kleffe and M. Borodovsky. First and second moment of counts of words in random texts generated by Markov chains. *Computer Applications in Biosciences*, 8:433–441, 1992.

[200] M. Knill, W.J. Bruno, and D.C. Torney. Non-adaptive group testing in the presence of errors. *Discrete Applied Mathematics*, 88:261–290, 1998.

[201] D.E. Knuth. Permutations, matrices and generalized Young tableaux. *Pacific Journal of Mathematics*, 34:709–727, 1970.

[202] D.E. Knuth. *The Art of Computer Programming*, chapter 2. Addison-Wesley, second edition, 1973.

[203] D.E. Knuth, J.H. Morris, and V.R. Pratt. Fast pattern matching in strings. *SIAM Journal on Computing*, 6:323–350, 1977.

[204] Y. Kohara, K. Akiyama, and K. Isono. The physical map of the whole E. coli chromosome: application of a new strategy for rapid analysis and sorting of a large genomic library. *Cell*, 50:495–508, 1987.

[205] G.A. Komatsoulis and M.S. Waterman. Chimeric alignment by dynamic programming: Algorithm and biological uses. In *Proceedings of the First Annual International Conference on Computational Molecular Biology (RECOMB-97)*, pages 174–180, Santa Fe, New Mexico, January 1997. ACM Press.

[206] A. Kotzig. Moves without forbidden transitions in a graph. *Matematicky Casopis*, 18:76–80, 1968.

[207] R.G. Krishna and F. Wold. Posttranslational modifications. In R.H. Angeletti, editor, *Proteins - Analysis and Design*, pages 121–206. Academic Press, 1998.

[208] A. Krogh, M. Brown, I.S. Mian, K. Sjölander, and D. Haussler. Hidden Markov models in computational biology: Applications to protein modeling. *Journal of Molecular Biology*, 235:1501–1531, 1994.

[209] A. Krogh, I.S. Mian, and D. Haussler. A Hidden Markov Model that finds genes in E. coli DNA. *Nucleic Acids Research*, 22:4768–4778, 1994.

[210] S. Kruglyak. Multistage sequencing by hybridization. *Journal of Computational Biology*, 5:165–171, 1998.

[211] J.B. Kruskal and D. Sankoff. An anthology of algorithms and concepts for sequence comparison. In D. Sankoff and J.B. Kruskal, editors, *Time Warps, String Edits, and Macromolecules: The Theory and Practice of Sequence Comparison*, pages 265–310. Addison-Wesley, 1983.

[212] G.M. Landau and J.P. Schmidt. An algorithm for approximate tandem repeats. In *Fourth Annual Symposium on Combinatorial Pattern Matching*, volume 684 of *Lecture Notes in Computer Science*, pages 120–133, Padova, Italy, 2-4 June 1993. Springer-Verlag.

[213] G.M. Landau and U. Vishkin. Efficient string matching in the presence of errors. In *26th Annual Symposium on Foundations of Computer Science*, pages 126–136, Los Angeles, California, October 1985.

[214] E.S. Lander and M.S. Waterman. Genomic mapping by fingerprinting random clones: a mathematical analysis. *Genomics*, 2:231–239, 1988.

[215] K. Lange, M. Boehnke, D.R. Cox, and K.L. Lunetta. Statistical methods for polyploid radiation hybrid mapping. *Genome Research*, 5:136–150, 1995.

[216] E. Lawler and S. Sarkissian. Adaptive error correcting codes based on cooperative play of the game of "Twenty Questions Game with a Liar". In *Proceedings of Data Compression Conference DCC '95*, page 464, Los Alamitos, California, 1995. IEEE Computer Society Press.

[217] C.E. Lawrence, S.F. Altschul, M.S. Boguski, J.S. Liu, A.F. Neuwald, and J.C. Wootton. Detecting subtle sequence signals: a Gibbs sampling strategy for multiple alignment. *Science*, 262:208–214, October 1993.

[218] J.K. Lee, V. Dancik, and M.S. Waterman. Estimation for restriction sites observed by optical mapping using reversible-jump Markov chain Monte Carlo. In S. Istrail, P.A. Pevzner, and M.S. Waterman, editors, *Proceedings of the Second Annual International Conference on Computational Molecular Biology (RECOMB-98)*, pages 147–152, New York, New York, March 1998. ACM Press.

[219] V.I. Levenshtein. Binary codes capable of correcting deletions, insertions and reversals. *Soviet Physics Doklady*, 6:707–710, 1966.

[220] B. Lewin. *Genes VII*. Oxford University Press, 1999.

[221] M. Li, B. Ma, and L. Wang. Finding similar regions in many strings. In *Proceedings of the 31st ACM Annual Symposium on Theory of Computing*, pages 473–482, Atlanta, Georgia, May 1999.

[222] S.Y.R. Li. A martingale approach to the study of ocurrence of sequence patterns in repeated experiments. *Annals of Probability*, 8:1171–1176, 1980.

[223] J. Lingner, T.R. Hughes, A. Shevchenko, M. Mann, V. Lundblad, and T.R. Cech. Reverse transcriptase motifs in the catalytic subunit of telomerase. *Science*, 276:561–567, 1997.

[224] D.J. Lipman, S. F Altschul, and J.D. Kececioglu. A tool for multiple sequence alignment. *Proceedings of the National Academy of Sciences USA*, 86:4412–4415, 1989.

[225] D.J. Lipman and W.R. Pearson. Rapid and sensitive protein similarity searches. *Science*, 227:1435–1441, 1985.

[226] R.J. Lipshutz, D. Morris, M. Chee, E. Hubbell, M.J. Kozal, N. Shah, N. Shen, R. Yang, and S.P.A. Fodor. Using oligonucleotide probe arrays to access genetic diversity. *Biotechniques*, 19:442–447, 1995.

[227] B.F. Logan and L.A. Shepp. A variational problem for random Young tableaux. *Advances in Mathematics*, 26:206–222, 1977.

[228] Y. Lysov, V. Florent'ev, A. Khorlin, K. Khrapko, V. Shik, and A. Mirzabekov. DNA sequencing by hybridization with oligonucleotides. *Doklady Academy Nauk USSR*, 303:1508–1511, 1988.

[229] C.A. Makaroff and J.D. Palmer. Mitochondrial DNA rearrangements and transcriptional alterations in the male sterile cytoplasm of Ogura radish. *Molecular Cellular Biology*, 8:1474–1480, 1988.

[230] M. Mann and M. Wilm. Error-tolerant identification of peptides in sequence databases by peptide sequence tags. *Analytical Chemistry*, 66:4390–4399, 1994.

[231] M. Mann and M. Wilm. Electrospray mass-spectrometry for protein characterization. *Trends in Biochemical Sciences*, 20:219–224, 1995.

[232] D. Margaritis and S.S. Skiena. Reconstructing strings from substrings in rounds. In *Proceedings of the 36th Annual Symposium on Foundations of Computer Science*, pages 613–620, Los Alamitos, California, October 1995.

[233] A.M. Maxam and W. Gilbert. A new method for sequencing DNA. *Proceedings of the National Academy of Sciences USA*, 74:560–564, 1977.

[234] G. Mayraz and R. Shamir. Construction of physical maps from oligonucleotide fingerprints data. *Journal of Computational Biology*, 6:237–252, 1999.

[235] F.R. McMorris, C. Wang, and P. Zhang. On probe interval graphs. *Discrete Applied Mathematics*, 88:315–324, 1998.

[236] W. Miller and E.W. Myers. Sequence comparison with concave weighting functions. *Bulletin of Mathematical Biology*, 50:97–120, 1988.

[237] A. Milosavljevic and J. Jurka. Discovering simple DNA sequences by the algorithmic significance method. *Computer Applications in Biosciences*, 9:407–411, 1993.

[238] B. Mirkin and F.S. Roberts. Consensus functions and patterns in molecular sequences. *Bulletin of Mathematical Biology*, 55:695–713, 1993.

[239] A.A. Mironov and N.N. Alexandrov. Statistical method for rapid homology search. *Nucleic Acids Research*, 16:5169–5174, 1988.

[240] A.A. Mironov, J.W. Fickett, and M.S. Gelfand. Frequent alternative splicing of human genes. *Genome Research*, 9:1288–1293, 1999.

[241] A.A. Mironov and P.A. Pevzner. SST versus EST in gene recognition. *Microbial and Comparative Genomics*, 4:167–172, 1999.

[242] A.A. Mironov, M.A. Roytberg, P.A. Pevzner, and M.S. Gelfand. Performance guarantee gene predictions via spliced alignment. *Genomics*, 51:332–339, 1998.

[243] S. Muthukrishnan and L. Parida. Towards constructing physical maps by optical mapping: An effective, simple, combinatorial approach. In S. Istrail, P.A. Pevzner, and M.S. Waterman, editors, *Proceedings of the First Annual International Conference on Computational Molecular Biology (RECOMB-97)*, pages 209–219, Santa Fe, New Mexico, January 1997. ACM Press.

[244] M. Muzio, A.M. Chinnaiyan, F.C. Kischkel, K. O'Rourke, A. Shevchenko, J. Ni, C. Scaffidi, J.D. Bretz, M. Zhang, R. Gentz, M. Mann, P.H. Krammer, M.E. Peter, and V.M. Dixit. FLICE, a novel FADD-homologous ICE/CED-3-like protease, is recruited to the CD95 (Fas/APO-1) death-inducing signaling complex. *Cell*, 85:817–827, 1996.

[245] E.W. Myers. A sublinear algorithm for approximate keyword searching. *Algorithmica*, 12:345–374, 1994.

[246] E.W. Myers and W. Miller. Optimal alignments in linear space. *Computer Applications in Biosciences*, 4:11–17, 1988.

[247] G. Myers. Whole genome shotgun sequencing. *IEEE Computing in Science and Engineering*, 1:33–43, 1999.

[248] J.H. Nadeau and B.A. Taylor. Lengths of chromosomal segments conserved since divergence of man and mouse. *Proceedings of the National Academy of Sciences USA*, 81:814–818, 1984.

[249] K. Nakata, M. Kanehisa, and C. DeLisi. Prediction of splice junctions in mRNA sequences. *Nucleic Acids Research*, 13:5327–5340, 1985.

[250] D. Naor and D. Brutlag. On near-optimal alignments of biological sequences. *Journal of Computational Biology*, 1:349–366, 1994.

[251] S.B. Needleman and C.D. Wunsch. A general method applicable to the search for similarities in the amino acid sequence of two proteins. *Journal of Molecular Biology*, 48:443–453, 1970.

[252] L. Newberg and D. Naor. A lower bound on the number of solutions to the probed partial digest problem. *Advances in Applied Mathematics*, 14:172–183, 1993.

[253] R. Nussinov, G. Pieczenik, J.R. Griggs, and D.J. Kleitman. Algorithms for loop matchings. *SIAM Journal on Applied Mathematics*, 35:68–82, 1978.

[254] S. O'Brien and J. Graves. Report of the committee on comparative gene mapping in mammals. *Cytogenetics and Cell Genetics*, 58:1124–1151, 1991.

[255] S. Ohno. *Sex chromosomes and sex-linked genes*. Springer-Verlag, 1967.

[256] S. Ohno, U. Wolf, and N.B. Atkin. Evolution from fish to mammals by gene duplication. *Hereditas*, 59:708–713, 1968.

[257] M.V. Olson, J.E. Dutchik, M.Y. Graham, G.M. Brodeur, C. Helms, M. Frank, M. MacCollin, R. Scheinman, and T. Frank. Random-clone strategy for genomic restriction mapping in yeast. *Proceedings of the National Academy of Sciences USA*, 83:7826–7830, 1986.

[258] O. Owolabi and D.R. McGregor. Fast approximate string matching. *Software Practice and Experience*, 18:387–393, 1988.

[259] J.D. Palmer and L.A. Herbon. Plant mitochondrial DNA evolves rapidly in structure, but slowly in sequence. *Journal of Molecular Evolution*, 27:87–97, 1988.

[260] A.H. Paterson, T.H. Lan, K.P. Reischmann, C. Chang, Y.R. Lin, S.C. Liu, M.D. Burow, S.P. Kowalski, C.S. Katsar, T.A. DelMonte, K.A. Feldmann, K.F. Schertz, and J.F. Wendel. Toward a unified genetic map of higher plants, transcending the monocot-dicot divergence. *Nature Genetics*, 15:380–382, 1996.

[261] S.D. Patterson and R. Aebersold. Mass spectrometric approaches for the identification of gel-separated proteins. *Electrophoresis*, 16:1791–1814, 1995.

[262] H. Peltola, H. Soderlund, and E. Ukkonen. SEQAID: a DNA sequence assembling program based on a mathematical model. *Nucleic Acids Research*, 12:307–321, 1984.

[263] M. Perlin and A. Chakravarti. Efficient construction of high-resolution physical maps from yeast artificial chromosomes using radiation hybrids: inner product mapping. *Genomics*, 18:283–289, 1993.

[264] P.A. Pevzner. l-tuple DNA sequencing: computer analysis. *Journal of Biomolecular Structure and Dynamics*, 7:63–73, 1989.

[265] P.A. Pevzner. Multiple alignment, communication cost, and graph matching. *SIAM Journal on Applied Mathematics*, 52:1763–1779, 1992.

[266] P.A. Pevzner. Statistical distance between texts and filtration methods in rapid similarity search algorithm. *Computer Applications in Biosciences*, 8:121–27, 1992.

[267] P.A. Pevzner. DNA physical mapping and alternating Eulerian cycles in colored graphs. *Algorithmica*, 13:77–105, 1995.

[268] P.A. Pevzner. DNA statistics, overlapping word paradox and Conway equation. In H.A. Lim, J.W. Fickett, C.R. Cantor, and R.J. Robbins, editors, *Proceedings of the Second International Conference on Bioinformatics, Supercomputing, and Complex Genome Analysis*, pages 61–68, St. Petersburg Beach, Florida, June 1993. World Scientific.

[269] P.A. Pevzner, M.Y. Borodovsky, and A.A. Mironov. Linguistics of nucleotide sequences. I: The significance of deviations from mean statistical characteristics and prediction of the frequencies of occurrence of words. *Journal of Biomolecular Structure and Dynamics*, 6:1013–1026, 1989.

[270] P.A. Pevzner, V. Dancik, and C.L. Tang. Mutation-tolerant protein identification by mass-spectrometry. In R. Shamir, S. Miyano, S. Istrail, P.A. Pevzner, and M.S. Waterman, editors, *Proceedings of the Fourth Annual International Conference on Computational Molecular Biology (RECOMB-00)*, pages 231–236, Tokyo, Japan, April 2000. ACM Press.

[271] P.A. Pevzner and R. Lipshutz. Towards DNA sequencing chips. In *Proceedings of the 19th International Conference on Mathematical Foundations of Computer Science*, volume 841 of *Lecture Notes in Computer Science*, pages 143–158, Kosice, Slovakia, 1994.

[272] P.A. Pevzner, Y. Lysov, K. Khrapko, A. Belyavski, V. Florentiev, and A. Mirzabekov. Improved chips for sequencing by hybridization. *Journal of Biomolecular Structure and Dynamics*, 9:399–410, 1991.

[273] P.A. Pevzner and M.S. Waterman. Generalized sequence alignment and duality. *Advances in Applied Mathematics*, 14(2):139–171, 1993.

[274] P.A. Pevzner and M.S. Waterman. Multiple filtration and approximate pattern matching. *Algorithmica*, 13:135–154, 1995.

[275] P.A. Pevzner and M.S. Waterman. Open combinatorial problems in computational molecular biology. In *Third Israeli Symposium on the Theory of Computing and Systems*, Tel-Aviv, Israel, January 1995.

[276] S. Pilpel. Descending subsequences of random permutations. *Journal of Combinatorial Theory, Series A*, 53:96–116, 1990.

[277] A. Pnueli, A. Lempel, and S. Even. Transitive orientation of graphs and identification of permutation graphs. *Canadian Journal of Mathematics*, 23:160–175, 1971.

[278] J.H. Postlethwait, Y.L. Yan, M.A. Gates, S. Horne, A. Amores, A. Brownlie, A. Donovan, E.S. Egan, A. Force, Z. Gong, C. Goutel, A. Fritz, R. Kelsh, E. Knapik, E. Liao, B. Paw, D. Ransom, A. Singer, M. Thomson, T.S. Abduljabbar, P. Yelick, D. Beier, J.S. Joly, D. Larhammar, and F. Rosa et al. Vertebrate genome evolution and the zebrafish gene map. *Nature Genetics*, 345-349:18, 1998.

[279] A. Poustka, T. Pohl, D.P. Barlow, G. Zehetner, A. Craig, F. Michiels, E. Ehrich, A.M. Frischauf, and H. Lehrach. Molecular approaches to mammalian genetics. *Cold Spring Harbor Symposium on Quantitative Biology*, 51:131–139, 1986.

[280] F.P. Preparata, A.M. Frieze, and E. Upfal. On the power of universal bases in sequencing by hybridization. In S. Istrail, P.A. Pevzner, and M.S. Waterman, editors, *Proceedings of the Third Annual International Conference on Computational Molecular Biology (RECOMB-99)*, pages 295–301, Lyon, France, April 1999. ACM Press.

[281] B. Prum, F. Rudolphe, and E. De Turckheim. Finding words with unexpected frequences in DNA sequences. *Journal of Royal Statistical Society, Series B*, 57:205–220, 1995.

[282] M. Regnier and W. Szpankowski. On the approximate pattern occurrences in a text. In *Compression and Complexity of Sequences 1997*, pages 253–264, 1998.

[283] K. Reinert, H.-P. Lenhof, P. Mutzel, K. Mehlhorn, and J.D. Kececioglu. A branch-and-cut algorithm for multiple sequence alignment. In S. Istrail, P.A. Pevzner, and M.S. Waterman, editors, *Proceedings of the First Annual International Conference on Computational Molecular Biology (RECOMB-97)*, pages 241–250, Santa Fe, New Mexico, January 1997. ACM Press.

[284] G. Rettenberger, C. Klett, U. Zechner, J. Kunz, W. Vogel, and H. Hameister. Visualization of the conservation of synteny between humans and pigs by hetereologous chromosomal painting. *Genomics*, 26:372–378, 1995.

[285] I. Rigoutsos and A. Floratos. Combinatorial pattern discovery in biological sequences. *Bioinformatics*, 14:55–67, 1998.

[286] J.C. Roach, C. Boysen, K. Wang, and L. Hood. Pairwise end sequencing: a unified approach to genomic mapping and sequencing. *Genomics*, 26:345–353, 1995.

[287] G.de E. Robinson. On representations of the symmetric group. *American Journal of Mathematics*, 60:745–760, 1938.

[288] E. Rocke and M. Tompa. An algorithm for finding novel gapped motifs in DNA sequences. In S. Istrail, P.A. Pevzner, and M.S. Waterman, editors, *Proceedings of the Second Annual International Conference on Computational Molecular Biology (RECOMB-98)*, pages 228–233, New York, New York, March 1998. ACM Press.

[289] J. Rosenblatt and P.D. Seymour. The structure of homometric sets. *SIAM Journal on Alg. Discrete Methods*, 3:343–350, 1982.

[290] M.A. Roytberg. A search for common pattern in many sequences. *Computer Applications in Biosciences*, 8:57–64, 1992.

[291] A.R. Rubinov and M.S. Gelfand. Reconstruction of a string from substring precedence data. *Journal of Computational Biology*, 2:371–382, 1995.

[292] B.E. Sagan. *The Symmetric Group: Representations, Combinatorial Algorithms, and Symmetric Functions.* Wadsworth Brooks Cole Mathematics Series, 1991.

[293] M.F. Sagot, A. Viari, and H. Soldano. Multiple sequence comparison—a peptide matching approach. *Theoretical Computer Science*, 180:115–137, 1997.

[294] T. Sakurai, T. Matsuo, H. Matsuda, and I. Katakuse. PAAS 3: A computer program to determine probable sequence of peptides from mass spectrometric data. *Biomedical Mass Spectrometry*, 11:396–399, 1984.

[295] S.L. Salzberg, A.L. Delcher, S. Kasif, and O. White. Microbial gene identification using interpolated Markov models. *Nucleic Acids Research*, 26:544–548, 1998.

[296] S.L. Salzberg, D.B. Searls, and S. Kasif. *Computational Methods in Molecular Biology*. Elsevier, 1998.

[297] F. Sanger, S. Nilken, and A.R. Coulson. DNA sequencing with chain terminating inhibitors. *Proceedings of the National Academy of Sciences USA*, 74:5463–5468, 1977.

[298] D. Sankoff. Minimum mutation tree of sequences. *SIAM Journal on Applied Mathematics*, 28:35–42, 1975.

[299] D. Sankoff. Simultaneous solution of the RNA folding, alignment and protosequence problems. *SIAM Journal on Applied Mathematics*, 45:810–825, 1985.

[300] D. Sankoff. Edit distance for genome comparison based on non-local operations. In *Third Annual Symposium on Combinatorial Pattern Matching*, volume 644 of *Lecture Notes in Computer Science*, pages 121–135, Tucson, Arizona, 1992. Springer-Verlag.

[301] D. Sankoff and M. Blanchette. Multiple genome rearrangements. In S. Istrail, P.A. Pevzner, and M.S. Waterman, editors, *Proceedings of the Second Annual International Conference on Computational Molecular Biology (RECOMB-98)*, pages 243–247, New York, New York, March 1998. ACM Press.

[302] D. Sankoff, R. Cedergren, and Y. Abel. Genomic divergence through gene rearrangement. In *Molecular Evolution: Computer Analysis of Protein and Nucleic Acid Sequences*, chapter 26, pages 428–438. Academic Press, 1990.

[303] D. Sankoff and M. Goldstein. Probabilistic models of genome shuffling. *Bulletin of Mathematical Biology*, 51:117–124, 1989.

[304] D. Sankoff, G. Leduc, N. Antoine, B. Paquin, B. Lang, and R. Cedergren. Gene order comparisons for phylogenetic inference: Evolution of the mitochondrial genome. *Proceedings of the National Academy of Sciences USA*, 89:6575–6579, 1992.

[305] D. Sankoff and S. Mainville. Common subsequences and monotone subsequences. In D. Sankoff and J.B. Kruskal, editors, *Time Warps, String Edits, and Macromolecules: The Theory and Practice of Sequence Comparison*, pages 363–365. Addison-Wesley, 1983.

[306] C. Schensted. Longest increasing and decreasing subsequences. *Canadian Journal of Mathematics*, 13:179–191, 1961.

[307] H. Scherthan, T. Cremer, U. Arnason, H. Weier, A. Lima de Faria, and L. Fronicke. Comparative chromosomal painting discloses homologous segments in distantly related mammals. *Nature Genetics*, 6:342–347, 1994.

[308] J.P. Schmidt. All highest scoring paths in weighted grid graphs and their application to finding all approximate repeats in strings. *SIAM Journal on Computing*, 27:972–992, 1998.

[309] W. Schmitt and M.S. Waterman. Multiple solutions of DNA restriction mapping problem. *Advances in Applid Mathematics*, 12:412–427, 1991.

[310] M. Schoniger and M.S. Waterman. A local algorithm for DNA sequence alignment with inversions. *Bulletin of Mathematical Biology*, 54:521–536, 1992.

[311] D.C. Schwartz, X. Li, L.I. Hernandez, S.P. Ramnarain, E.J. Huff, and Y.K. Wang. Ordered restriction maps of Saccharomyces cerevisiae chromosomes constructed by optical mapping. *Science*, 262:110–114, 1993.

[312] D. Searls and S. Dong. A syntactic pattern recognition system for DNA sequences. In H.A. Lim, J.W. Fickett, C.R. Cantor, and R.J. Robbins, editors, *Proceedings of the Second International Conference on Bioinformatics, Supercomputing, and Complex Genome Analysis*, pages 89–102, St. Petersburg Beach, Florida, June 1993. World Scientific.

[313] D. Searls and K. Murphy. Automata-theoretic models of mutation and alignment. In *Proceedings of the Third International Conference on Intelligent Systems for Molecular Biology*, pages 341–349, Cambridge, England, 1995.

[314] S.S. Skiena, W.D. Smith, and P. Lemke. Reconstructing sets from interpoint distances. In *Proceedings of Sixth Annual Symposium on Computational Geometry*, pages 332–339, Berkeley, California, June, 1990.

[315] S.S. Skiena and G. Sundaram. A partial digest approach to restriction site mapping. *Bulletin of Mathematical Biology*, 56:275–294, 1994.

[316] S.S. Skiena and G. Sundram. Reconstructing strings from substrings. *Journal of Computational Biology*, 2:333–354, 1995.

[317] D. Slonim, L. Kruglyak, L. Stein, and E. Lander. Building human genome maps with radiation hybrids. In S. Istrail, P.A. Pevzner, and M.S. Waterman, editors, *Proceedings of the First Annual International Conference on Computational Molecular Biology (RECOMB-97)*, pages 277–286, Santa Fe, New Mexico, January 1997. ACM Press.

[318] H.O. Smith, T.M. Annau, and S. Chandrasegaran. Finding sequence motifs in groups of functionally related proteins. *Proceedings of the National Academy of Sciences USA*, 87:826–830, 1990.

[319] H.O. Smith and K.W. Wilcox. A restriction enzyme from Hemophilus influenzae. I. Purification and general properties. *Journal of Molecular Biology*, 51:379–391, 1970.

[320] T.F. Smith and M.S. Waterman. Identification of common molecular subsequences. *Journal of Molecular Biology*, 147:195–197, 1981.

[321] E.E. Snyder and G.D. Stormo. Identification of coding regions in genomic DNA sequences: an application of dynamic programming and neural networks. *Nucleic Acids Research*, 21:607–613, 1993.

[322] E.E. Snyder and G.D. Stormo. Identification of protein coding regions in genomic DNA. *Journal of Molecular Biology*, 248:1–18, 1995.

[323] V.V. Solovyev, A.A. Salamov, and C.B. Lawrence. Predicting internal exons by oligonucleotide composition and discriminant analysis of spliceable open reading frames. *Nucleic Acids Research*, 22:5156–63, 1994.

[324] E.L. Sonnhammer, S.R. Eddy, and R. Durbin. Pfam: a comprehensive database of protein domain families based on seed alignments. *Proteins*, 28:405–420, 1997.

[325] E. Southern. United Kingdom patent application GB8810400. 1988.

[326] R. Staden. Methods for discovering novel motifs in nucleic acid seqences. *Computer Applications in Biosciences*, 5:293–298, 1989.

[327] R. Staden and A.D. McLachlan. Codon preference and its use in identifying protein coding regions in long DNA sequences. *Nucleic Acids Research*, 10:141–156, 1982.

[328] J.M. Steele. An Efron-Stein inequality for nonsymmetric statistics. *Annals of Statistics*, 14:753–758, 1986.

[329] M. Stefik. Inferring DNA structure from segmentation data. *Artificial Intelligence*, 11:85–144, 1978.

[330] E.E. Stuckle, C. Emmrich, U. Grob, and P.J. Nielsen. Statistical analysis of nucleotide sequences. *Nucleic Acids Research*, 18:6641–6647, 1990.

[331] A.H. Sturtevant and T. Dobzhansky. Inversions in the third chromosome of wild races of *Drosophila pseudoobscura*, and their use in the study of the history of the species. *Proceedings of the National Academy of Sciences USA*, 22:448–450, 1936.

[332] S.H. Sze and P.A. Pevzner. Las Vegas algorithms for gene recognition: subotimal and error tolerant spliced alignment. *Journal of Computational Biology*, 4:297–310, 1997.

[333] J. Tarhio and E. Ukkonen. A greedy approximation algorithm for constructing shortest common superstrings. *Theoretical Computer Science*, 57:131–145, 1988.

[334] J. Tarhio and E. Ukkonen. Boyer-Moore approach to approximate string matching. In J.R. Gilbert and R. Karlsson, editors, *Proceedings of the*

Second Scandinavian Workshop on Algorithm Theory, number 447 in Lecture Notes in Computer Science, pages 348–359, Bergen, Norway, 1990. Springer-Verlag.

[335] J.A. Taylor and R.S. Johnson. Sequence database searches via *de novo* peptide sequencing by tandem mass spectrometry. *Rapid Communications in Mass Spectrometry*, 11:1067–1075, 1997.

[336] W.R. Taylor. Multiple sequence alignment by a pairwise algorithm. *Computer Applications in Biosciences*, 3:81–87, 1987.

[337] S.M. Tilghman, D.C. Tiemeier, J.G. Seidman, B.M. Peterlin, M. Sullivan, J.V. Maizel, and P. Leder. Intervening sequence of DNA identified in the structural portion of a mouse beta-globin gene. *Proceedings of the National Academy of Sciences USA*, 75:725–729, 1978.

[338] M. Tompa. An exact method for finding short motifs in sequences with application to the Ribosome Binding Site problem. In *Proceedings of the Seventh International Conference on Intelligent Systems for Molecular Biology*, pages 262–271, Heidelberg, Germany, August 1999. AAAI Press.

[339] E. Uberbacher and R. Mural. Locating protein coding regions in human DNA sequences by a multiple sensor - neural network approach. *Proceedings of the National Academy of Sciences USA*, 88:11261–11265, 1991.

[340] E. Ukkonen. Approximate string matching with q-grams and maximal matches. *Theoretical Computer Science*, 92:191–211, 1992.

[341] S. Ulam. Monte-Carlo calculations in problems of mathematical physics. In *Modern mathematics for the engineer*, pages 261–281. McGraw-Hill, 1961.

[342] A.M. Vershik and S.V. Kerov. Asymptotics of the Plancherel measure of the symmetric group and the limiting form of Young tableaux. *Soviet Mathematical Doklady*, 18:527–531, 1977.

[343] M. Vihinen. An algorithm for simultaneous comparison of several sequences. *Computer Applications in Biosciences*, 4:89–92, 1988.

[344] M. Vingron and P. Argos. Motif recognition and alignment for many sequences by comparison of dot-matrices. *Journal of Molecular Biology*, 218:33–43, 1991.

[345] M. Vingron and P.A. Pevzner. Multiple sequence comparison and consistency on multipartite graphs. *Advances in Applied Mathematics*, 16:1–22, 1995.

[346] M. Vingron and M.S. Waterman. Sequence alignment and penalty choice. Review of concepts, studies and implications. *Journal of Molecular Biology*, 235:1–12, 1994.

[347] T.K. Vintsyuk. Speech discrimination by dynamic programming. *Comput.*, 4:52–57, 1968.

[348] A. Viterbi. Error bounds for convolutional codes and an asymptotically optimal decoding algorithm. *IEEE Transactions on Information Theory*, 13:260–269, 1967.

[349] D.G. Wang, J.B. Fan, C.J. Siao, A. Berno, P. Young, R. Sapolsky, G. Ghandour, N. Perkins, E. Winchester, J. Spencer, L. Kruglyak, L. Stein, L. Hsie, T. Topaloglou, E. Hubbell, E. Robinson, M. Mittmann, M.S. Morris, N. Shen, D. Kilburn, J. Rioux, C. Nusbaum, S. Rozen, T.J. Hudson, and E.S. Lander et al. Large-scale identification, mapping, and genotyping of single-nucleotide polymorphisms in the human genome. *Science*, 280:1074–1082, 1998.

[350] L. Wang and D. Gusfield. Improved approximation algorithms for tree alignment. In *Seventh Annual Symposium on Combinatorial Pattern Matching*, volume 1075 of *Lecture Notes in Computer Science*, pages 220–233, Laguna Beach, California, 10-12 June 1996. Springer-Verlag.

[351] L. Wang and T. Jiang. On the complexity of multiple sequence alignment. *Journal of Computational Biology*, 1:337–348, 1994.

[352] L. Wang, T. Jiang, and E.L. Lawler. Approximation algorithms for tree alignment with a given phylogeny. *Algorithmica*, 16:302–315, 1996.

[353] M.D. Waterfield, G.T. Scrace, N. Whittle, P. Stroobant, A. Johnsson, A. Wasteson, B. Westermark, C.H. Heldin, J.S. Huang, and T.F. Deuel. Platelet-derived growth factor is structurally related to the putative transforming protein p28sis of simian sarcoma virus. *Nature*, 304:35–39, 1983.

[354] M.S. Waterman. Secondary structure of single-stranded nucleic acids. *Studies in Foundations and Combinatorics, Advances in Mathematics Supplementary Studies*, 1:167–212, 1978.

[355] M.S. Waterman. Sequence alignments in the neighborhood of the optimum with general application to dynamic programming. *Proceedings of the National Academy of Sciences USA*, 80:3123–3124, 1983.

[356] M.S. Waterman. Efficient sequence alignment algorithms. *Journal of Theoretical Biology*, 108:333–337, 1984.

[357] M.S. Waterman. *Introduction to Computational Biology*. Chapman Hall, 1995.

[358] M.S. Waterman, R. Arratia, and D.J. Galas. Pattern recognition in several sequences: consensus and alignment. *Bulletin of Mathematical Biology*, 46:515–527, 1984.

[359] M.S. Waterman and M. Eggert. A new algorithm for best subsequence alignments with application to tRNA–rRNA comparisons. *Journal of Molecular Biology*, 197:723–728, 1987.

[360] M.S. Waterman, M. Eggert, and E. Lander. Parametric sequence comparisons. *Proceedings of the National Academy of Sciences USA*, 89:6090–6093, 1992.

[361] M.S. Waterman and J.R. Griggs. Interval graphs and maps of DNA. *Bulletin of Mathematical Biology*, 48:189–195, 1986.

[362] M.S. Waterman and M.D. Perlwitz. Line geometries for sequence comparisons. *Bulletin of Mathematical Biology*, 46:567–577, 1984.

[363] M.S. Waterman and T.F. Smith. Rapid dynamic programming algorithms for RNA secondary structure. *Advances in Applied Mathematics*, 7:455–464, 1986.

[364] M.S. Waterman, T.F. Smith, and W.A. Beyer. Some biological sequence metrics. *Advances in Mathematics*, 20:367–387, 1976.

[365] M.S. Waterman and M. Vingron. Rapid and accurate estimates of statistical significance for sequence data base searches. *Proceedings of the National Academy of Sciences USA*, 91:4625–4628, 1994.

[366] G.A. Watterson, W.J. Ewens, T.E. Hall, and A. Morgan. The chromosome inversion problem. *Journal of Theoretical Biology*, 99:1–7, 1982.

[367] J. Weber and G. Myers. Whole genome shotgun sequencing. *Genome Research*, 7:401–409, 1997.

[368] W.J. Wilbur and D.J. Lipman. Rapid similarity searches of nucleic acid protein data banks. *Proceedings of the National Academy of Sciences USA*, 80:726–730, 1983.

[369] K.H. Wolfe and D.C. Shields. Molecular evidence for an ancient duplication of the entire yeast genome. *Nature*, 387:708–713, 1997.

[370] F. Wolfertstetter, K. Frech, G. Herrmann, and T. Werner. Identification of functional elements in unaligned nucleic acid sequences. Computer Applications in Biosciences, 12:71–80, 1996.

[371] S. Wu and U. Manber. Fast text searching allowing errors. Communication of ACM, 35:83–91, 1992.

[372] G. Xu, S.H. Sze, C.P. Liu, P.A. Pevzner, and N. Arnheim. Gene hunting without sequencing genomic clones: finding exon boundaries in cDNAs. Genomics, 47:171–179, 1998.

[373] J. Yates, J. Eng, and A. McCormack. Mining genomes: Correlating tandem mass-spectra of modified and unmodified peptides to sequences in nucleotide databases. Analytical Chemistry, 67:3202–3210, 1995.

[374] J. Yates, J. Eng, A. McCormack, and D. Schieltz. Method to correlate tandem mass spectra of modified peptides to amino acid sequences in the protein database. Analytical Chemistry, 67:1426–1436, 1995.

[375] J. Yates, P. Griffin, L. Hood, and J. Zhou. Computer aided interpretation of low energy MS/MS mass spectra of peptides. In J.J. Villafranca, editor, Techniques in Protein Chemistry II, pages 477–485. Academic Press, 1991.

[376] P. Zhang, E.A. Schon, S.G. Fischer, E. Cayanis, J. Weiss, S. Kistler, and P.E. Bourne. An algorithm based on graph theory for the assembly of contigs in physical mapping. Computer Applications in Biosciences, 10:309–317, 1994.

[377] Z. Zhang. An exponential example for a partial digest mapping algorithm. Journal of Computational Biology, 1:235–239, 1994.

[378] D. Zidarov, P. Thibault, M.J. Evans, and M.J. Bertrand. Determination of the primary structure of peptides using fast atom bombardment mass spectrometry. Biomedical and Environmental Mass Spectrometry, 19:13–16, 1990.

[379] R. Zimmer and T. Lengauer. Fast and numerically stable parametric alignment of biosequences. In S. Istrail, P.A. Pevzner, and M.S. Waterman, editors, Proceedings of the First Annual International Conference on Computational Molecular Biology (RECOMB-97), pages 344–353, Santa Fe, New Mexico, January 1997. ACM Press.

[380] M. Zuker. RNA folding. Methods in Enzymology, 180:262–288, 1989.

[381] M. Zuker and D. Sankoff. RNA secondary structures and their prediction. Bulletin of Mathematical Biology, 46:591–621, 1984.

Index